QTL Mapping of Seed Quality Traits in Crops

QTL Mapping of Seed Quality Traits in Crops

Guest Editor

Abdelmajid Kassem

MDPI

Basel • Beijing • Wuhan • Barcelona • Belgrade • Novi Sad • Cluj • Manchester

Guest Editor
Abdelmajid Kassem
Biological and Forensic Sciences
Fayetteville State University
Fayetteville
United States

Editorial Office
MDPI AG
Grosspeteranlage 5
4052 Basel, Switzerland

This is a reprint of the Special Issue, published open access by the journal *Plants* (ISSN 2223-7747), freely accessible at: www.mdpi.com/journal/plants/special_issues/QTL_Seed.

For citation purposes, cite each article independently as indicated on the article page online and using the guide below:

Lastname, A.A.; Lastname, B.B. Article Title. *Journal Name* **Year**, *Volume Number*, Page Range.

ISBN 978-3-7258-3350-4 (Hbk)
ISBN 978-3-7258-3349-8 (PDF)
https://doi.org/10.3390/books978-3-7258-3349-8

Contents

About the Editor

Abdelmajid Kassem

Dr. My Abdelmajid Kassem is a distinguished plant biologist, bioinformatics researcher, and data scientist with over 25 years of experience in academia. He is currently a Professor at the Department of Biological and Forensic Sciences, Fayetteville State University. His research focuses on quantitative trait loci (QTL) mapping, genomic selection, and the integration of bioinformatics and machine learning in plant genetics.

Dr. Kassem has published extensively in peer-reviewed journals, with over 70+ scientific articles, book chapters, and a book covering plant genomics, biotechnology, and computational biology. He has successfully supervised and mentored numerous undergraduate and graduate students, fostering their careers in plant science and bioinformatics. His expertise extends to high-throughput genomic data analysis, functional genomics, and metabolomics, particularly in crop improvement and breeding programs.

In addition to his academic and research contributions, Dr. Kassem is actively involved in organizing international scientific conferences, including the American Moroccan Agricultural, Health, and Life Sciences (AMAHLS) Conference, where he serves as a co-organizer. He is also an advocate for interdisciplinary education, having developed and taught bioinformatics, and he has introduced data science courses at his department and institution.

Dr. Kassem holds a Ph.D. in Plant Biology and an M.S. in Data Science, equipping him with a unique blend of expertise in both biological and computational sciences. His work aims to bridge the gap between traditional plant breeding and modern genomic technologies, contributing to sustainable agriculture and global food security.

Preface

Quantitative trait locus (QTL) mapping has transformed plant genetics, providing crucial insights into the genetic architecture of complex agronomic traits. This Special Issue, "QTL Mapping of Seed Quality Traits in Crops", brings together 11 groundbreaking research articles that explore the identification and characterization of QTLs associated with key seed quality traits, including protein content, sugar accumulation, frost tolerance, seed longevity, and biochemical composition across various crop species.

The contributions in this issue highlight the latest advancements in QTL mapping methodologies, integrating high-throughput genomics, phenotypic analysis, and bioinformatics to unravel the genetic determinants of seed quality. These studies span multiple crops such as soybean, wheat, common bean, mustard, and Capsicum annuum, reflecting the broad applicability of QTL mapping in improving seed traits critical for food security, nutritional enhancement, and environmental adaptation.

A key focus of this issue is the translational potential of QTL discoveries. The identification of candidate genes linked to desirable seed traits opens new possibilities for molecular breeding and genetic improvement. By bridging fundamental research and practical applications, the findings presented here serve as a valuable resource for researchers, breeders, and agricultural scientists dedicated to enhancing seed quality through targeted breeding strategies.

As we face increasing global challenges in agriculture, including climate change and the demand for high-yield and nutrient-rich crops, QTL mapping remains a vital tool in plant science. This Special Issue not only advances our understanding of seed quality genetics but also provides a foundation for future innovations in crop improvement. I extend my sincere gratitude to all contributing authors, reviewers, and the editorial team for their efforts in making this issue a significant addition to the field of plant genetics.

Abdelmajid Kassem
Guest Editor

plants

Editorial

QTL Mapping of Seed Quality Traits in Crops

Moulay Abdelmajid Kassem

Plant Genomics and Biotechnology Laboratory, Department of Biological and Forensic Sciences, Fayetteville State University, Fayetteville, NC 28301, USA; mkassem@uncfsu.edu

The ability to map quantitative trait loci (QTLs) has revolutionized plant genetics, providing an essential toolkit for dissecting the genetic basis of agronomic traits. This Special Issue, "QTL Mapping of Seed Quality Traits in Crops", published in the journal Plants, features 11 cutting-edge research articles that exemplify the current advances in QTL mapping and its application to seed quality traits, including protein content, sugar accumulation, frost tolerance, seed longevity, and more. These contributions highlight the importance of QTL mapping as a powerful approach to improve seed quality, which is crucial for food security, nutritional improvement, and environmental adaptability in crops.

1. Advances in Understanding Seed Quality Traits Across Crop Species

Seed quality traits are often complex, influenced by multiple genetic loci and environmental factors. This complexity necessitates robust tools and methodologies, such as QTL mapping and candidate gene identification, to unravel the genetic architecture underlying these traits. The articles in this Special Issue reflect a diverse set of approaches and crop species, from soybean and wheat to common bean and mustard. Together, they provide an invaluable resource for researchers and breeders striving to improve seed quality.

2. Highlights of the Contributions

2.1. Protein and Nutritional Composition in Soybean

Soybean, a major global source of plant-based protein, has been the focus of multiple studies in this issue. Two recombinant inbred line (RIL) populations were studied in relation to seed protein content [1], revealing QTLs and candidate genes associated with this critical trait. Additionally, QTL mapping for seed sugar contents in the well-studied soybean population 'Forrest' × 'Williams 82' identified genomic regions influencing seed carbohydrates [2].

2.2. Frost Tolerance and Seed Longevity in Wheat and Capsicum

Understanding environmental stress responses is critical for sustainable agriculture. The identification of novel QTLs associated with frost tolerance in winter wheat [3] and seed longevity in *Capsicum annuum* [4] provides valuable insights into breeding crops resilient to cold climates and storage conditions.

2.3. Seed Morphology and Biochemical Composition

Digital image processing techniques were applied to assess seed size, shape, and color in bread wheat, enabling precise phenotypic measurements and QTL mapping [5]. Similarly, tocopherol content in soybean [6] and glucosinolate accumulation in mustard [7] were mapped to candidate loci, advancing our understanding of seed biochemical traits.

check for updates

Received: 9 January 2025
Revised: 28 January 2025
Accepted: 29 January 2025
Published: 6 February 2025

Citation: Kassem, M.A. QTL Mapping of Seed Quality Traits in Crops. *Plants* **2025**, *14*, 482. https://doi.org/10.3390/plants14030482

2.4. *Diversity in Common Bean and Wheat*

Natural variation in Portuguese common bean populations provided insights into the genetic architecture of nutritional traits [8], while high-density SNP-based genetic linkage maps were utilized to identify QTLs for grain traits in wheat [9].

2.5. *Integration of Databases and Bioinformatics*

Modern plant breeding relies heavily on data integration and bioinformatics. An article focused on leveraging crop databases for candidate gene identification [10] underscores the increasing role of computational tools in QTL analysis.

2.6. *Isoflavone and Nutraceutical Properties in Soybean*

Isoflavones are bioactive compounds with significant health benefits. Mapping QTLs for seed isoflavone content in soybean [11] highlights the potential for enhancing nutraceutical properties through targeted breeding.

3. Looking Ahead: Bridging the Gap Between Research and Application

The findings presented in this Special Issue demonstrate how QTL mapping has matured as a discipline, integrating high-throughput phenotyping, genomics, and bioinformatics. While significant progress has been made, the transition from mapping to application in breeding programs remains a challenge. Bridging this gap will require greater emphasis on candidate gene validation, functional genomics, and the incorporation of genomic selection in breeding pipelines.

4. Concluding Remarks

As the global population grows and climate change exerts increasing pressure on agricultural systems, the need for high-quality seeds is becoming ever more critical. The research presented in this Special Issue represents a step forward in addressing this challenge, offering novel genetic insights and practical tools for breeding programs. I hope these studies will inspire further research and innovation in QTL mapping and beyond.

Acknowledgments: I would like to extend my heartfelt gratitude to the authors, reviewers, and editorial staff whose dedication made this Special Issue possible. Their contributions have not only advanced the field of QTL mapping but have also provided practical insights for improving seed quality traits in crops. Special thanks go to the editorial team at Plants for their unwavering support and professionalism.

Conflicts of Interest: The authors declare no conflict of interest.

References

1. Park, H.R.; Seo, J.H.; Kang, B.K.; Kim, J.H.; Heo, S.V.; Choi, M.S.; Ko, J.Y.; Kim, C.S. QTLs and Candidate Genes for Seed Protein Content in Two Recombinant Inbred Line Populations of Soybean. *Plants* **2023**, *12*, 3589. [CrossRef] [PubMed]
2. Knizia, D.; Bellaloui, N.; Yuan, J.; Lakhssasi, N.; Anil, E.; Vuong, T.; Embaby, M.; Nguyen, H.T.; Mengistu, A.; Meksem, K.; et al. Quantitative Trait Loci and Candidate Genes That Control Seed Sugars Contents in the Soybean 'Forrest' by 'Williams 82' Recombinant Inbred Line Population. *Plants* **2023**, *12*, 3498. [CrossRef] [PubMed]
3. Bolouri, P.; Haliloğlu, K.; Mohammadi, S.A.; Türkoğlu, A.; İlhan, E.; Niedbała, G.; Szulc, P.; Niazian, M. Identification of Novel QTLs Associated with Frost Tolerance in Winter Wheat (*Triticum aestivum* L.). *Plants* **2023**, *12*, 1641. [CrossRef]
4. Arif, M.A.R.; Tripodi, P.; Waheed, M.Q.; Afzal, I.; Pistrick, S.; Schütze, G.; Börner, A. Genetic Analyses of Seed Longevity in *Capsicum annuum* L. in Cold Storage Conditions. *Plants* **2023**, *12*, 1321. [CrossRef]
5. Arif, M.A.R.; Komyshev, E.G.; Genaev, M.A.; Koval, V.S.; Shmakov, N.A.; Börner, A.; Afonnikov, D.A. QTL Analysis for Bread Wheat Seed Size, Shape and Color Characteristics Estimated by Digital Image Processing. *Plants* **2022**, *11*, 2105. [CrossRef] [PubMed]

6. Knizia, D.; Yuan, J.; Lakhssassi, N.; El Baze, A.; Cullen, M.; Vuong, T.; Mazouz, H.; Nguyen, H.T.; Kassem, M.A.; Meksem, K. QTL and Candidate Genes for Seed Tocopherol Content in 'Forrest' by 'Williams 82' Recombinant Inbred Line (RIL) Population of Soybean. *Plants* **2022**, *11*, 1258. [CrossRef] [PubMed]
7. Tandayu, E.; Borpatragohain, P.; Mauleon, R.; Kretzschmar, T. Genome-Wide Association Reveals Trait Loci for Seed Glucosinolate Accumulation in Indian Mustard (*Brassica juncea* L.). *Plants* **2022**, *11*, 364. [CrossRef] [PubMed]
8. Mendes, F.A.; Leitão, S.T.; Correia, V.; Mecha, E.; Rubiales, D.; Bronze, M.R.; Vaz Patto, M.C. Portuguese Common Bean Natural Variation Helps to Clarify the Genetic Architecture of the Legume's Nutritional Composition and Protein Quality. *Plants* **2022**, *11*, 26. [CrossRef]
9. Lv, D.; Zhang, C.; Yv, R.; Yao, J.; Wu, J.; Song, X.; Jian, J.; Song, P.; Zhang, Z.; Han, D.; et al. Utilization of a Wheat50K SNP Microarray-Derived High-Density Genetic Map for QTL Mapping of Plant Height and Grain Traits in Wheat. *Plants* **2021**, *10*, 1167. [CrossRef] [PubMed]
10. Brown, A.V.; Grant, D.; Nelson, R.T. Using Crop Databases to Explore Phenotypes: From QTL to Candidate Genes. *Plants* **2021**, *10*, 2494. [CrossRef] [PubMed]
11. Knizia, D.; Yuan, J.; Bellaloui, N.; Vuong, T.; Usovsky, M.; Song, Q.; Betts, F.; Register, T.; Williams, E.; Lakhssassi, N.; et al. The Soybean High Density 'Forrest' by 'Williams 82' SNP-Based Genetic Linkage Map Identifies QTL and Candidate Genes for Seed Isoflavone Content. *Plants* **2021**, *10*, 2029. [CrossRef]

plants

MDPI

Article

QTLs and Candidate Genes for Seed Protein Content in Two Recombinant Inbred Line Populations of Soybean

Hye Rang Park , Jeong Hyun Seo *, Beom Kyu Kang, Jun Hoi Kim, Su Vin Heo, Man Soo Choi, Jee Yeon Ko and Choon Song Kim

Department of Southern Area Crop Science, National Institute of Crop Science, Rural Development Administration, Miryang 50424, Republic of Korea; hrpark6@korea.kr (H.R.P.); hellobk01@korea.kr (B.K.K.); itomi123@korea.kr (J.H.K.); hsb3937@korea.kr (S.V.H.); mschoi73@korea.kr (M.S.C.); kjeeyeon@korea.kr (J.Y.K.); kcs3925@korea.kr (C.S.K.)

* Correspondence: next0501@korea.kr; Tel.: +82-55-350-1236

Abstract: This study aimed to discover the quantitative trait loci (QTL) associated with a high seed protein content in soybean and unravel the potential candidate genes. We developed two recombinant inbred line populations: YS and SI, by crossing Saedanbaek (high protein) with YS2035-B-91-1-B-1 (low protein) and Saedanbaek with Ilmi (low protein), respectively, and evaluated the protein content for three consecutive years. Using single-nucleotide polymorphism (SNP)-marker-based linkage maps, four QTLs were located on chromosomes 15, 18, and 20 with high logarithm of odds values (5.9–55.0), contributing 5.5–66.0% phenotypic variance. In all three experimental years, *qPSD20-1* and *qPSD20-2* were stable and identified in overlapping positions in the YS and SI populations, respectively. Additionally, novel QTLs were identified on chromosomes 15 and 18. Considering the allelic sequence variation between parental lines, 28 annotated genes related to soybean seed protein—including starch, lipid, and fatty acid biosynthesis-related genes—were identified within the QTL regions. These genes could potentially affect protein accumulation during seed development, as well as sucrose and oil metabolism. Overall, this study offers insights into the genetic mechanisms underlying a high soybean protein content. The identified potential candidate genes can aid marker-assisted selection for developing soybean lines with an increased protein content.

Keywords: quantitative trait loci; soybean protein; high protein; genetic map; single-nucleotide polymorphism

Citation: Park, H.R.; Seo, J.H.; Kang, B.K.; Kim, J.H.; Heo, S.V.; Choi, M.S.; Ko, J.Y.; Kim, C.S. QTLs and Candidate Genes for Seed Protein Content in Two Recombinant Inbred Line Populations of Soybean. *Plants* **2023**, *12*, 3589. https://doi.org/10.3390/plants12203589

Academic Editor: Abdelmajid Kassem

Received: 11 September 2023
Revised: 11 October 2023
Accepted: 12 October 2023
Published: 16 October 2023

1. Introduction

Soybean [*Glycine max* (L.) Merr.] is an important legume crop globally known for its high-quality protein and oil content [1,2]. Asian countries, such as Korea, Japan, China, and Indonesia, have a strong cultural tradition of consuming soy-based products. Recently, the consumption of traditional soy-based products has surged globally, dominating the global protein market [3–6]. This substantial growth is attributed to changing dietary preferences and the shifting behavior of consumers towards more sustainable and environmentally friendly food choices [7–10].

Soybean protein research has gained increasing interest because of its significance. Many researchers have aimed to explore the genetic aspects of the protein traits in soybeans through quantitative trait loci (QTL) and genome-wide association studies (GWAS) [2,4]. The seed protein traits in soybeans are linked with the seed oil content and weight. These quantitative traits are complex and influenced by multiple genes and environmental factors [4,11,12]. In particular, soybean seed storage proteins are influenced by multiple factors, including major transcription factors, phytohormones, protein accumulation, storage protein regulation and deposition, and environmental factors [4]. Since the publication of the soybean reference genome, research on the genetics of these factors has been actively conducted [13]. Researchers have actively reported the genetic regions associated with the protein content using traditional linkage analyses to identify QTLs [11], high-density single-nucleotide polymorphism (SNP) linkage maps for precise QTL detection [4], and GWAS to unravel the genetic basis of soybean protein traits [12,14]. These analyses have been conducted on a wide range of soybean resource populations, genetic resources with a high protein content in backcross, and recombinant inbred line (RIL) populations [12,14,15].

Numerous studies have mapped seed protein and oil content to specific genomic regions, primarily located on chromosomes 15 and 20 [14,16–20]. For example, cqSeed protein-03 has been identified as a major QTL for seed protein on chromosome 20. This QTL has been extensively represented in several large populations using various mapping methods since the publication of the reference genome. However, an accurate identification of the precise location and reliable candidate genes is challenging [11,19–23]. Only recently have some studies successfully fine-mapped cqSeed protein-003 across several mapping populations and narrowed its interval to 77.8 kb [24]. These studies identified an insertion/deletion within the CCT domain of *Glyma.20g085100* and showed a strong correlation with the seed protein content. The function of *Glyma.20g085100* has been confirmed using RNA interference (RNAi) in transgenic soybean plants [24,25].

Furthermore, through the fine mapping of the QTL detected on chromosome 15 [26], a specific allele derived from wild soybean was found to confer simultaneous effects on the 100-seed weight, protein content, and oil content traits that are negatively correlated [19,25,27,28]. This allele was localized to a 329 kb region on chromosome 15 [26]. In addition, *GmSWEET10a*, *GmSWEET10b*, and *GmSWEET39*—other representative genes on chromosome 15—are sugar transporters that affect seed protein and oil content [29–31]. *GmST05* (*Glyma.05g244100*) affects both the seed protein and oil content, in addition to its role in controlling seed size. This effect is likely achieved by regulating *GmSWEET39* transcription [32]. However, despite these findings, a comprehensive understanding of the genetic factors influencing soybean protein traits remains elusive.

GWAS signals for protein content were identified on chromosomes 15 and 20, exhibiting a greater prominence in Korean accessions. The frequency of alleles linked to a high protein content was lower in Chinese and US accessions. In Korea, soybean breeding and pedigree programs have focused on breeding for traits specifically related to soy-based food, with a particular emphasis on protein content [33]. Furthermore, the lack of thorough validation for diverse genetic backgrounds and limited utilization in practical breeding programs have been challenging. The main reason for this limitation is the modest effect of these QTLs on phenotypic variation [34]. To overcome this, it is crucial to further validate and evaluate these QTLs for their effective incorporation into breeding programs.

'Danbaekkong', previously utilized in several studies, has proven valuable for detecting the QTLs associated with protein content [20,33,35,36]. Several studies have extensively employed 'Danbaekkong' in QTL identification and breeding programs utilizing Danbaekkong-derived RIL populations [20,35]. However, Saedanbaek (SD) possesses a genetically distinct background from that of Danbaekkong. The high protein traits found in SD can be traced back to BARC-10 (MD87L, PI 572270), a breeding material recognized for its high protein content that is officially registered in the US National Plant Germplasm System [37]. Therefore, because SD genetically differs from the widely used high-protein cultivar 'Danbaekkong', it could unravel new allelic sources to increase the protein content.

The primary focus of this study was to identify the QTLs specifically linked to the seed protein content using two populations of RILs derived from SD, an elite high-protein cultivar as one of the parental lines, over three years. This finding will enhance our understanding of the genetic factors influencing seed protein content and provide valuable insights for future breeding efforts to improve soybean protein traits.

2. Results

2.1. Phenotypic Variation in the Seed Protein Content

The seed protein contents (%) in the parental lines [YS2035-B-91-1-B-1 (YS2035), Saedanbaek (SD), and Ilmi (IM)] and the two RIL mapping populations [YS2035 × SD (YS) and SD × IM (SI)] were assessed in 2020, 2021, and 2022. The protein contents of YS2035, SD, and IM in the parental lines were 47.3, 54.2, and 42.4%; 46.8, 54.3, and 44.4%; and 43.3, 50.6, and 41.2% in 2020, 2021, and 2022, respectively. The average seed protein content of SD (53.1%) was significantly higher than that of YS2035 (45.6%) and IM (42.7%; Figure 1 and Supplementary Table S1). The average protein content in the YS and SI populations ranged from 39.4 to 52.3% and 39.6 to 52.1%, respectively. Substantial variations in the protein content were observed between the parental lines, and the H^2 values in the YS and SI populations were 0.84 and 0.86, respectively (Supplementary Table S1). The protein content (%) in both populations showed a normal distribution and slightly transgressive inheritance. However, this transgressive inheritance was specifically prominent in the SI population, especially in 2021 and 2022 (Figure 1). An analysis of variance (ANOVA) revealed that the year differences and genotype × year interaction effects were highly significant in both the YS and SI populations (Supplementary Table S2).

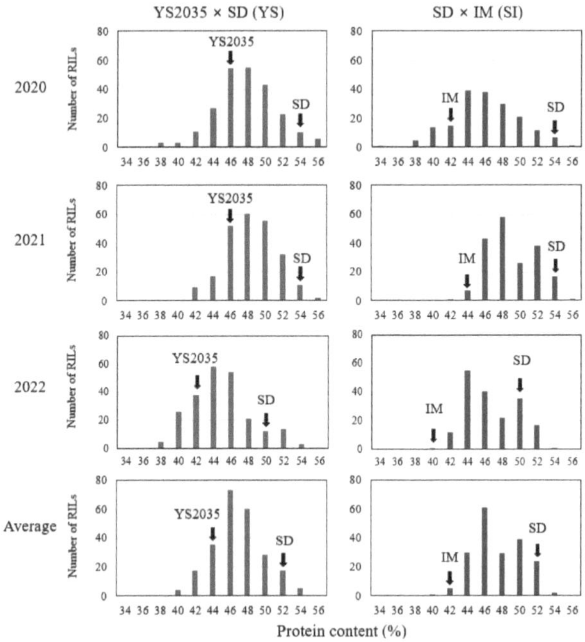

Figure 1. Frequency distribution of RIL protein content in the two mapping populations evaluated in 2020, 2021, and 2022. The parental values are shown using arrows. YS2035; YS2035-B-91-1-B-1, SD; Saedanbaek, IM; Ilmi, YS; YS2035 × SD, SI; and SD × IM.

2.2. Linkage Map Construction

A total of 180,375 high-quality SNPs markers were genotyped, of which 27,724 in the YS populations and 27,896 in the SI populations were polymorphic between the respective parental lines. After deleting redundant markers with >5% missing values, 2254 and 3544 SNPs were selected and used to construct linkage maps for the YS and SI populations, respectively. The polymorphic SNP markers were distributed across all 20 chromosomes with an average of 113 and 177 markers per chromosome and covered a total of 5339 and 3248 cM genetic distances in the YS and SI linkage maps, respectively. The average distances between the adjacent SNPs in the YS and SI populations were 2.5 and 0.9 cM, respectively. The average lengths (cM) in the YS and SI populations were 267 cM and 162 cM, respectively (Supplementary Tables S3 and S4). The YS population exhibited the lowest number of SNP markers on chromosome 18 (62), while the highest number was found on chromosome 16 (206). In the SI population, the lowest number of SNP markers were observed on chromosome 17 (88), while the highest number were on chromosome 5 (235) (Supplementary Tables S3 and S4). Despite using SD as the common parental line, the differences in the genomic length coverage between the two linkage maps could be attributed to the genetic differences between the other two parental lines (YS2035 and IM). Based on the above results, it was used more accurately for QTL mapping (Supplementary Table S5).

2.3. QTL Analysis

Across the three years of the experiment, specific QTLs with marker intervals (left–right) for seed protein were detected on chromosomes 15, 16, 17, 18, and 20 in the YS population and on chromosomes 9, 15, and 20 in the SI population (Figure 2 and Supplementary Tables S5 and S6). The differences in detecting different QTLs in the two mapping populations could be attributed to the genetic differences between YS2035 and IM. We selected four QTLs considering the logarithm of odds (LOD) and phenotypic variance explained (PVE) value of five or more years and environment. Subsequently, three out of four QTLs were detected on chromosomes 15, 18, and 20 in the YS population, and one QTL was detected on chromosome 20 in the SI population (Table 1). The LOD of the identified QTLs ranged from 5.9 to 55.0, and the PVE varied from 5.5 to 66.0%. The major QTLs—*qPSD20-1* (LOD, 20.9–30.6; PVE, 22.5–35.4%) in the YS population and *qPSD20-2* (LOD, 23.0–55.0; PVE, 34.1–66.0%) in the SI population on chromosome 20—were consistently detected across all three years. Most of the QTLs identified in relation to the IciM-ADD values were designated as *qPYS16*, as they originated from YS as the parent chromosome 16 in the YS population; however, all of the QTLs were named *qPSD*, because the QTLs appeared in the parent SD regardless of the populations (Supplementary Table S6). The major *qPSD20-1* spanned from 31,781,045 to 31,961,695 bp in the YS population. In addition, another major QTL on chromosome 20, *qPSD20-2*, spanning from 30,395,400 to 31,781,045 bp on the physical map, was stably detected for three consecutive years in the SI population. In addition, the QTLs on chromosomes 15 and 18 were detected in more than one year. The physical positions of the markers flanking *qPSD15-1* on chromosome 15 in the YS population detected in 2020 and 2021 were from 7,930,801 to 8,678,412 bp. The physical positions of the QTL *qPSD18-1* detected in 2020 were from 46,911,930 to 47,526,734 bp. A total of 181 genes were identified in the four QTL regions (Table 1 and Supplementary Table S6).

Table 1. Quantitative trait loci (QTL) associated with high protein identified in the two recombinant inbred line (RIL) mapping populations derived from 'YS2035' × 'Saedanbaek' and 'Saedanbaek' × 'Ilmi'.

Population [1]	Marker [2]	Chr [3]	Genetic Position (cM)	Physical Position of Markers (bp) [4]	Year	Gene Name	Gene No.	LOD [5]	PVE [6] (%)	Add [7]	Reference
Y × S	*qPSD15-1*	15	305	7,930,801– 8,678,412	2020 2021 Average [8]	*Glyma.15g101800–* *Glyma.15g110600*	89	14.0 14.6 12.3	17.5 17.1 13.8	−2.7 −2.1 −1.7	
Y × S	*qPSD18-1*	18	75	46,911,930– 47,526,734	2022 Average	*Glyma.18g193300–* *Glyma.18g197100*	39	6.7 5.9	7.0 5.5	−0.9 −0.6	[38–40]
Y × S	*qPSD20-1*	20	96	31,781,045– 31,961,695	2020 2021 2022 Average	*Glyma.20g085100–* *Glyma.20g085700*	7	21.1 24.7 20.9 30.6	22.5 29.1 24.7 35.4	−1.8 −1.6 −1.8 −1.6	[14,24,25]
S × I	*qPSD20-2*	20	68	30,395,400– 31,781,045	2020 2021 2022 Average	*Glyma.20g081000–* *Glyma.20g085450*	46	23.0 48.2 55.0 52.7	34.1 59.7 66.0 61.5	2.4 2.1 2.4 2.3	[14,15,19, 24,25,41]

[1] Y × S, YS2035 × Saedanbaek; S × I, Saedanbaek × Ilmi. [2] *qPSD*, 'Saedanbaek' contributed to the allele. [3] Chr, Chromosome. [4] Physical position of the markers, the soybean reference genome (*Glycine max* Wm82.a2.v1) was used to determine the physical positions of the markers. [5] Logarithm of odds value at the peak likelihood of QTL. [6] Phenotypic variation explained (PVE) by QTL. [7] Additive effect. [8] Average values for three years: 2020, 2021, and 2022.

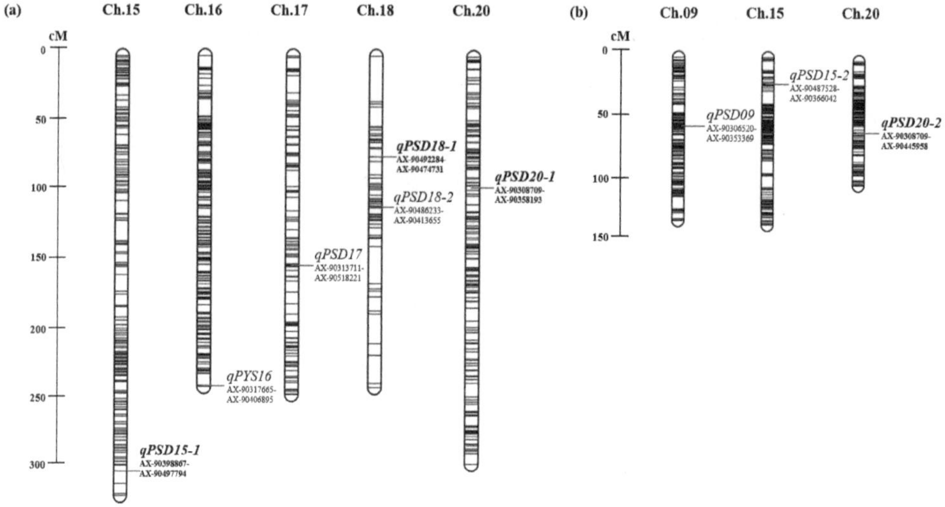

Figure 2. Quantitative trait loci (QTL) associated with seed protein content in (**a**) YS2035 × Saedanbaek (YS) and (**b**) Saedanbaek × Ilmi (SI) mapping populations. The bars inside each chromosome represent the position of markers used to construct the linkage map. The QTLs and marker positions are shown using red bars. The genetic distance (cM) of chromosomes was displayed as the rulers on the left side in the YS and SI populations. The main selected QTLs are highlighted by the bold font. Genetic map details are provided in Supplementary Table S5.

2.4. Phenotypic Variation According to the Allele Patterns

The top 20 and bottom 20 RILs with high and low seed protein content were selected from the YS and SI populations (Table 2). These genotypes were assessed using representative markers linked to *qPSD15-1*, *qPSD18-1*, and *qPSD20-1*, located on chromosomes 15, 18, and 20, respectively, which are associated with genes that promote a high protein content.

The average protein content in the RILs with SD genotypes (high protein) was 53.1%, while that in the RILs with the YS and IM genotypes (low protein) was 45.6% and 42.7%, respectively, over the three years (Supplementary Table S1). The protein content in the top 20 RILs ranged from 51.1 to 52.3%, whereas that in the bottom 20 RILs ranged from 39.4 to 41.6% in both the YS and SI populations (Supplementary Table S1). Through the genome sequencing of the three parents—SD, YS, and IM—both populations included SD, and the QTL regions were all derived from SD, so the RILs in the SI population were included in the top 20 proteins. An analysis of the allelic patterns in the top 20 RILs revealed that the SD was predominantly present in these recombinants at the three loci. In contrast, the alleles of the bottom 20 recombinants at the representative markers were mostly derived from IM or YS (Table 2). In both populations, the RILs with the SD allele in the *qPSD15-1* marker exhibited an average protein content of 46.6%, whereas those with the YS or IM allele showed a protein content of 45.2% (Figure 3a). Similarly, the RILs with the SD allele for the *qPSD18-1* marker had an average protein content of 47.0%, whereas those with the YS or IM allele displayed 45.5% (Figure 3b). The RILs with the SD allele at the *qPSD20-1* marker had an average protein content of 49.3%, whereas those with the YS or IM allele had a protein content of 44.9% (Figure 3c). According to the combination of the allele patterns of *qPSD15-1*, *qPSD18-1*, and *qPSD20-1*, the protein content of the RILs with all low-protein alleles (AAA) was 43.3%, whereas that of the RILs with all high-protein parental alleles (BBB) was 49.6%. The RILs with an SD allele at *qPSD15-1* (BAA) exhibited an average protein content of 44.6%, while those with an SD allele at *qPSD18-1* (ABA) had a protein content of 44.0%. The RILs with an SD allele at *qPSD20-1* (AAB) showed a protein content of 48.1%. Furthermore, recombinants harboring SD alleles at *qPSD15-1* and *qPSD18-1* (BBA) had an average protein content of 45.7%. Similarly, the protein content in the RIL with SD alleles at both *qPSD18-1* and *qPSD20-1* (ABB) was 48.9%, while that in the RILs with SD alleles at both *qPSD15-1* and *qPSD20-1* (BAB) was 48.6% (Figure 3d).

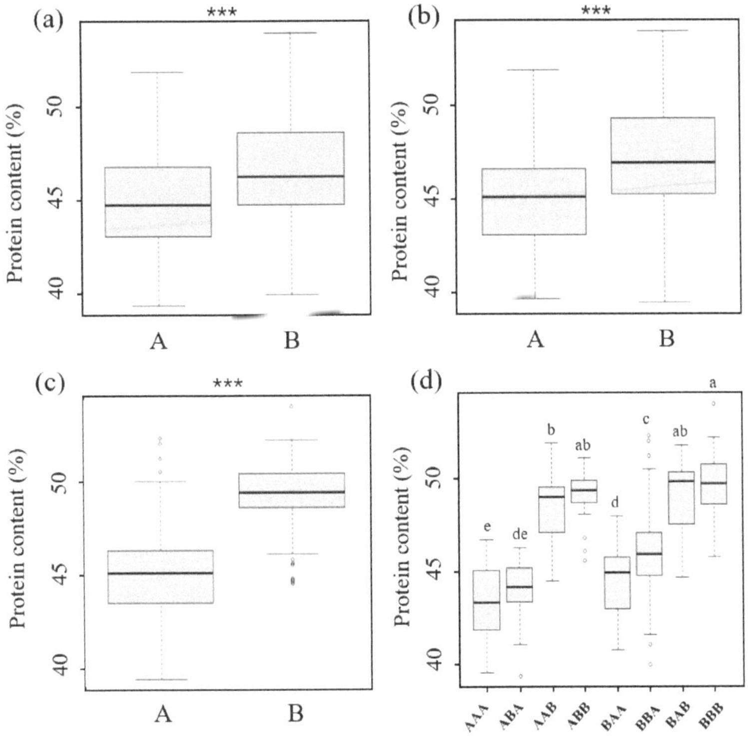

Figure 3. Boxplots of protein content and the allele effect of the major QTLs on chromosomes 15, 18, and 20. Trait values of the recombinants with high (SD; B) or low parent (YS/IM; A) alleles (**a**) *qPSD15-1*, (**b**) *qPSD18-1*, (**c**) *qPSD20-1* markers and (**d**) with all three markers. Asterisks indicate significant differences between parental lines in the RILs of 'YS2035' and 'Saedanbaek' (YS) or 'Saedanbaek' and 'Ilmi' (SI) at $p < 0.001$. The center bold line represents the median. Different lowercase letters indicate significant differences between genotypes; $p < 0.05$; Duncan's multiple range test (DMRT).

Table 2. Genotypes of the top and bottom 20 RILs in YS2035 × Saedanbaek (YS) and Saedanbaek × Ilmi (SI) populations based on high and low protein content at markers linked to *qPSD15-1*, *qPSD18-1*, and *qPSD20-1* markers. 'A' genotype shows that the selected RILs derived line was homogeneous for the allele from YS2035-B-91-1-B-1 (YS: low protein) and Ilmi (IM: low protein), 'B' genotype shows that the line was homogeneous for the allele from Saedanbaek (SD: high protein).

Top 20 RILs with High Protein Content	Genotype of the Marker Linked to the QTLs			Protein Content (%)	Bottom 20 RILs with Low Protein Content	Genotype of the Marker Linked to the QTLs			Protein Content (%)
	qPSD15-1	qPSD18-1	qPSD20-1			qPSD15-1	qPSD18-1	qPSD20-1	
YS-196	B	B	A	52.3	YS-229	A	B	A	39.4
SI-400	B	B	B	52.1	SI-465	A	A	A	39.6
YS-068	B	B	B	52.1	YS-111	A	A	A	39.8
YS-080	B	A	B	52.0	YS-209	A	A	A	39.9
YS-005	B	A	B	52.0	YS-036	B	B	A	40.0
SI-317	A	A	B	51.9	YS-037	A	A	A	40.3
YS-190	B	A	B	51.8	YS-117	A	A	A	40.4
YS-109	B	B	B	51.7	SI-337	A	A	A	40.6
YS-199	B	B	B	51.6	YS-043	B	A	A	40.8
SI-428	B	B	B	51.5	YS-118	B	A	A	40.9

Table 2. *Cont.*

Top 20 RILs with High Protein Content	Genotype of the Marker Linked to the QTLs			Protein Content (%)	Bottom 20 RILs with Low Protein Content	Genotype of the Marker Linked to the QTLs			Protein Content (%)
	qPSD15-1	qPSD18-1	qPSD20-1			qPSD15-1	qPSD18-1	qPSD20-1	
YS-173	B	A	B	51.4	SI-361	B	B	A	41.1
SI-423	B	A	B	51.4	YS-063	A	B	A	41.1
YS-205	B	A	B	51.3	YS-227	B	A	A	41.2
SI-445	B	B	B	51.3	YS-048	B	A	A	41.2
YS-090	B	B	B	51.3	SI-473	B	A	A	41.2
YS-008	B	A	B	51.2	YS-015	B	A	A	41.4
SI-348	B	B	B	51.2	YS-235	A	A	A	41.6
YS-202	B	B	B	51.2	YS-100	B	B	A	41.6
SI-326	B	B	B	51.1	SI-502	A	A	A	41.6
SI-345	A	B	B	51.1	YS-045	B	A	A	41.6

2.5. SNP Variation Analysis and Variant Annotation

Next, we annotated the polymorphic SNPs in the genes mapped to the *qPSD15-1*, *qPSD18-1*, and *qPSD20-1* QTLs using the whole-genome sequencing of SD, YS2035, and IM. After verifying the tri-parent SNP selection based on the soybean reference genome, only genes that exhibited differences from SD were screened for SNPs in YS2035 and IM. In total, 28 genes were selected among the 181 genes mapped to the intervals of the four QTLs (Table 3). The variant annotations identified 9 frameshift, 96 missense, 4 stop-gains, and other variants in 89 genes mapped to *qPSD15-1*. The genes mapped to *qPSD18-1* (39) comprised 6 frameshift, 91 missense, 3 stop-gains, and others. In all three experimental years, *qPSD20-1* and *qPSD20-2* were consistently identified in the overlapping genomic regions in the YS and SI populations. *Glyma.20g085100* was selected in a common SNP from two populations as the missense variant. In the *qPSD20-1* and *qPSD20-2* regions, we identified 7 and 46 genes, respectively. In these genes, we detected 1 frameshift, 55 missense, 1 stop-gain, and more (Table 1 and Supplementary Table S7). Of the 28 annotated genes in the QTL regions, 6 harbored stop-gain, 13 harbored frameshift, and 9 harbored missense variants (Table 3). The annotation of these 28 genes identified their association with several biological processes, including starch biosynthetic, carbohydrate metabolic process, sucrose metabolic process, fatty acid biosynthesis, lipid metabolic process, and protein polymerization (Table 3).

Table 3. Candidate genes for seed protein content identified on the reference genome based on the QTL-linked SNPs in the YS and SI mapping populations.

Population	Marker	Gene ID	Annotation Description	Biological Process	Reference	SNP Type
Y × S	*qPSD15-1*	*Glyma.15g102100*	Alpha/Beta hydrolase domain-containing protein	NA		Stop gain
		Glyma.15g102202	Elongation factor Tu GTP binding domain	Translational elongation		Frameshift variant
		Glyma.15g102252	Elongation factor Tu C-terminal domain	Translational elongation		Frameshift variant
		Glyma.15g102800	Mediator of RNA polymerase II transcription subunit 33a	Phenylpropanoid metabolic process		Stop gain
		Glyma.15g103100	Mitochondrial editing factor 18	RNA modification		Frameshift variant
		Glyma.15g107200	GPI-anchored protein	Biological process		Stop gain
		Glyma.15g108000	Starch/carbohydrate-binding module (family 53)	Starch biosynthetic process		Frameshift variant

Table 3. *Cont.*

Population	Marker	Gene ID	Annotation Description	Biological Process	Reference	SNP Type
Y × S	*qPSD15-1*	*Glyma.15g108900*	Glycosyl hydrolases family 17	Carbohydrate metabolic process		Frameshift variant
		Glyma.15g109800	Peroxisomal membrane protein 2	Biological process		Frameshift variant
		Glyma.15g109900	F-BOX protein with a domain protein	NA		Frameshift variant
	qPSD18-1	*Glyma.18g193300*	Laccase	Iron ion transport		Frameshift variant
		Glyma.18g193600	Fructose-1,6-bisphosphatase, N-terminal domain	Sucrose metabolic process	[38]	Frameshift variant
		Glyma.18g194700	NA	NA		Stop gain
		Glyma.18g194900	NA	NA		Frameshift variant
		Glyma.18g195000	NA	Biological process		Frameshift variant
		Glyma.18g195700	Alpha-carboxyltransferase aCT-1 precursor	Fatty acid biosynthesis	[39,40]	Missense variant
		Glyma.18g195900	Carboxyl transferase domain	Fatty acid biosynthesis	[39,40]	Missense variant
		Glyma.18g196000	Carboxyl transferase domain	Fatty acid biosynthesis	[39,40]	Missense variant
		Glyma.18g196600	NA	NA		Stop gain
		Glyma.18g197100	NA	NA		Frameshift variant
	qPSD20-1	*Glyma.20g085100*	*POWR1* CCT motif family protein	Biological process	[14,24,25]	Missense variant
		Glyma.20g085700	Unknown protein	NA	[15]	Stop gain
S × I	*qPSD20-2*	*Glyma.20g081500*	Lipase containing protein	Lipid catabolic process		Missense variant
		Glyma.20g082450	Ammonium transporter 1	Ammonium transport	[15]	Missense variant
		Glyma.20g082700	Sugar efflux transporter SWEET52	Carbohydrate transport	[42,43]	Missense variant
		Glyma.20g084000	Small nuclear ribonucleoprotein F	Spliceosomal snRNP assembly	[15]	Missense variant
		Glyma.20g084051	Far1-relate	Regulation of transcription	[15]	Missense variant
		Glyma.20G084500	WD40 repeat protein	Innate immune response	[15]	Missense variant
		Glyma.20g085100	*POWR1* CCT motif family protein	Biological process	[14,24,25]	Missense variant

The soybean reference genome (*Glycine max* Wm82.a4.v1) was used to annotate genes.

3. Discussion

This study aimed to discover new genes associated with the protein content in soybeans using two RIL populations derived from soybean cultivars with contrasting protein contents. Among the three parental lines, SD—developed in 2010 as a high-protein cultivar (48.2%) in South Korea—is widely recommended for soybean foods, such as tofu and soybean paste [44]. Here, the average protein content measured in SD over three years was 53.1% (Supplementary Table S1), whereas the average protein content of cultivated soybeans is approximately 40% [45]. A high broad-sense heritability for protein content was observed in average years 0.84 and 0.86 in the YS and SI populations, respectively (Supplementary Table S1), suggesting a highly significant ($p < 0.001$) influence of genotype × year interaction on the traits. These results suggest that SD is a suitable candidate for conducting QTL analyses to map the genetic intervals associated with a high protein content in soybeans.

Since the first study on the QTLs associated with protein content—which identified cqProt-001 and cpProt-003 on chromosomes 15 and 20, respectively [16]—several studies have confirmed the involvement of these QTLs in regulating the protein content in soybeans [33]. Additionally, other QTLs have also been identified in soybeans. A QTL related to a high protein and low oil content contributed by PI407788A, a high protein cultivar, was identified on chromosome 15 [17]. The QTL, cqSeed protein-003, located on chromosome 20, is associated with protein and amino acid content and derived from another high-protein cultivar, Danbaekkong [20,35]. Bandillo et al. [46] used SoySNP50K data to explore the connection between genetic variations and protein content across more than 12,000 *G. max* accessions [47].

This study detected a high LOD value, PVE, and stability of the major QTLs *qPSD20-1* in the YS population and *qPSD20-2* in the SI population. The major seed protein content QTLs on chromosome 20, commonly referred to as the repeat overlapping interval, have been identified in numerous studies [14,15,19,24,25,41]. In other RIL populations derived using SD as a parental line, *qHPO20*—associated with seed protein and oil content, and mapped to a wide region (4.8–34.3 Mbp) on chromosome 20—was stably detected in three years [19]. Our study located *qPSD20-1* and *qPSD20-2* to narrower intervals (31.7–31.9 and 30.3–31.7 Mbp, respectively; Table 1) than those in previously reported studies. Concordant with our study, previous studies have identified major protein- and oil content-related QTLs and confirmed the association of genes with the traits on chromosome 20 [11,12,14,15,17,19,20,24,41]. Our stable and major QTLs on chromosome 20 identified here harbored eight genes in the YS and SI populations. In particular, *Glyma.20g085100* is an SNP found commonly in both populations. Another study identified *Glyma.20g085100*, underlying the major QTL located on chromosome 20, related to soybean seed protein and oil, harboring tandem repeats. This gene encodes the CCT domain [14,24,25]. The CCT-domain gene, *POWR1*, likely related to lipid metabolism and nutrient transport, plays a pleiotropic role in regulating soybean seed quality and yield [25]. The insertion of a transposable element into the CCT domain of *POWR1* led to an increased seed weight and oil content but decreased protein content. Conversely, the overexpression of *POWR1* in transgenic plants improved protein content but reduced seed weight and oil content [25]. Among these, one gene exhibited a stop-gain, and another showed a missense variant in the YS population, whereas seven genes displayed a missense variant in the SI population. (Table 3). *Glyma.20g081500* (lipase-containing protein) and *Glyma. 20g082700* (sugar efflux transporter SWEET52) are presumed to be involved in protein, carbohydrate, and lipid metabolism during soybean seed development. These studies have shown that these genes would affect protein content after seed maturity [38,42,43,48,49]. However, there are few specifically studied and identified genes within this interval. These genes have not been characterized in previous studies; therefore, understanding their role in regulating soybean protein content warrants further research.

The QTL *qPSD18-1* on chromosome 18 in the YS population was detected at 46.9–47.5 Mbp intervals in 2022 (Table 1 and Supplement Table S6). Among the genes underlying these QTLs, two displayed stop-gains, five showed frameshift variants, and three exhibited missense variants. *Glyma.18g193600* (fructose-1,6-bishosphatase) is thought to be related to seed sucrose development (Table 3). A recent GWAS study reported that

Glyma.18g193600 is likely to play a role in the interconnected process of sucrose biosynthesis in edamame beans [38]. Among the potential candidate genes identified here, *Glyma.18g195700, Glyma.18g195900,* and *Glyma.18g196000* (fatty acid biosynthesis) might be related to soybean storage proteins [39,40]. In soybean seeds, storage proteins— essential nutritional components—are initially synthesized as precursors in sucrose and oil [38,39,49].

The QTL *qPSD15-1* on chromosome 15 in the YS population, as a novel QTL, was detected at 7.9–8.6 Mbp intervals in 2020 and 2021 (Table 1 and Supplement Table S6). Among this QTL, three displayed stop-gains and seven exhibited frameshift variants. *Glyma.15g108000* (the starch/carbohydrate-binding module) and *Glyma.15g108900* (carbohydrate metabolic process) are involved in carbohydrate biosynthesis, and related genes are being published in chromosome 15 (Table 3). Recently, *Glyma.15g049200* was identified as one of the candidate genes through fine mapping within the QTL regions simultaneously associated with soybean seed weight, protein content, and oil content [26,31]. Moreover, the QTLs on chromosome 15 exhibit pleiotropic effects on soybean seed protein and oil content. Certain sugar transporters, such as *GmSWEET10a, GmSWEET39* (*Glyma.15g049200*), and *GmSWEET10b* (*Glyma.8g183500*) have been identified in these regions [31,35]. During soybean domestication, the SWEET paralogs *GmSWEET10a* and *GmSWEET10b* went through stepwise selection, influencing seed size, oil, and protein levels by regulating the sugar distribution from the seed coat to the embryo [31,33]. In addition to the major QTL, the minor QTLs on chromosome 15 with overlapping positions, as detected in previous studies [50], may also contribute to seed protein and oil content.

During soybean seed development, storage proteins are transported for carbohydrate and lipid synthesis [4,32]. The genes *Glyma.15g108000, Glyma.15g108900, Gylma.18g193600,* and *Glyma.20g082700* (related to starch and carbohydrates synthesis during seed development), *Glyma.18g195700, Glyma.18g195900, Glyma.18g196000* (associated with fatty acid biosynthesis), and *Glyma.20g081500* (related to lipid catabolic processes during seed development) are likely to regulate protein accumulation. These candidate genes may regulate protein accumulation by influencing the sugar delivery from the seed coat integument to the embryo [38,40,43]. Several studies on soybean RIL populations have reported that seed protein, sucrose, and oil content show negative correlations [4,19,25,27,28,32,51]. Most candidate genes identified in this study have not been previously reported to be associated with soy protein. Therefore, further studies are required to gain valuable insights for soybean protein research.

QTLs related to seed protein content have been extensively studied using GWAS, QTL analyses, fine mapping, and haplotype mapping [11,12,14,15,19,20,24,25,46]. Here, we identified novel regions on chromosomes 15, 18, and 20, which showed consistent associations with soybean protein contents. These findings suggest that the newly identified QTLs, along with previously recognized ones, are likely to further elucidate the genetic factors associated with protein-related traits. However, it remains difficult to identify genes that are directly involved in regulating protein traits, warranting further studies using genetic resources with a high protein content.

4. Materials and Methods

4.1. Plant Materials

Two RIL populations involving three parental lines: SD (high-seed protein cultivar) [44], YS2035 (low-seed protein line), and IM (low-seed protein cultivar) [52], were used here. The YS (YS2035 × SD) and SI (SD × IM) mapping populations were developed using the single-seed descent method from the F2 to the F5:10 and F5:7 generations, respectively. The YS and SD mapping populations, comprising 237 and 189 RILs, respectively, and the parental lines were cultivated under experimental field conditions at the Miryang farm in South Korea (35°29′46.5″ N 128°44′29.9″ E) in 2020, 2021, and 2022. The populations were planted in rows measuring 4 m in length, with spacings of 70 cm between each row and

15 cm between individual plants. Fertilizers and pesticides were administered following established cultivation methods in South Korea [53].

4.2. *Analysis of Crude Seed Protein Concentrations*

The protein content was measured using 15 mg of seed powder in both mapping populations, comprising 426 RILs, and assessed each year using the Dumas method [54] with a Rapid N Cube (Elementar Analysen System, Hanau, Germany), following the manufacturer's instructions [55]. The protein analysis of each individual RIL was performed three times per year.

4.3. *Genomic DNA Extraction and Genotyping*

Genomic DNA was extracted from dry seeds of each line of the mapping populations and the three parental lines using a Maxwell RSC 48 instrument (Promega Madison, WI, USA), following the manufacturer's instructions. The DNA quality was assessed using a NanoDrop ND-2000 (Thermo Fisher Scientific, Waltham, MA, USA), and each DNA sample was diluted to a concentration of 10 ng/µL for genotyping. The mapping populations and the parental lines were genotyped using the 180K Axiom SoyaSNP array [56].

4.4. *Genetic Linkage Map Construction and QTL Analysis*

The SNP markers showing polymorphism between the parental lines were identified from the Axiom 180 K SoyaSNP array genotyping data to construct the genetic linkage map. The genetic linkage maps of the two mapping populations were constructed using the QTL IciMapping software version 4.2 [57]. The grouping threshold was set at a 3.0 logarithm of odds (LOD), nnTwoOpt was used as the ordering algorithm, and the sum of the adjacent recombination fractions was used for rippling, following the methodology described in a previous study [37]. Missing data with >5% were used to remove the redundant markers. The mapping of each linkage group was performed using Kosambi's mapping function. The association between each trait and the SNP markers was assessed using the inclusive composite interval mapping (IciM) function of the IciMapping software, with a 1000 permutation test. The QTLs were named by combining abbreviated letters *q* for QTL, *P* for seed protein, and *SD* for the parent Saedanbaek (SD), followed by the chromosome name and nth QTL on the chromosome. For instance, *qPSD*15-2 represents the second QTL identified on chromosome 15.

4.5. *Prediction of Novel Candidate QTL and Genes*

Firstly, the QTLs detected for more than two years were selected. Statistically significant QTLs associated with soybean seed protein content were identified by examining the genotypes within the QTL regions using SNP markers. We performed the genome sequencing of SD, YS2035, and IM using the Illumina Hiseq X sequencing platform (Illumina, San Diego, CA, USA). Reads were mapped using Bowtie 2 (v2.2.4) and variants were called with Freebayes (v1.3.4). After verifying tri-parent SNP selection based on the soybean reference genome, only genes that exhibited differences from SD were screened for SNPs in YS2035 and IM. The QTL regions were further investigated using SoyBase (www.soybase.org (accessed on 5 September 2023)) to identify the candidate genes. Annotated information on the candidate genes was obtained from the soybean reference genome (Wm82. a4. v1). The candidate genes were presented based on their gene descriptions and SNP variations within the QTL regions.

4.6. *Statistical Analysis*

To assess the phenotypic variations in protein within the populations, various statistical tests were performed, including an analysis of variance (ANOVA), Student's *t*-test, and Duncan's multiple range test (DMRT). The statistical analyses were conducted using

R V3.6.3 software [58]. The broad-sense heritability (H^2) for the mean values in each environment was calculated using an equation with some modifications [59].

$$H^2 = \sigma^2{}_G/(\sigma^2{}_G + \sigma^2{}_{GY}/Y + \sigma^2{}_{e/rY}) \quad (1)$$

where $\sigma^2{}_{GY}$ and $\sigma^2{}_e$ are the components of genotype × year and error variances, respectively. The component of genotype × year variance ($\sigma^2{}_{GY}$) and the mean square of error ($\sigma^2{}_e$) was estimated with reference [60].

5. Conclusions

We conducted a three-year field study using two RIL populations derived from a cross between the elite cultivar SD and either YS2035 or IM and identified several QTLs on chromosomes 15, 18, and 20. In all three experimental years, *qPSD20-1* and *qPSD20-2* were consistently identified in the overlapping genomic region in the YS and SI populations. These QTLs have been previously reported in various studies related to soybean protein content, whereas the other identified QTLs are novel. This suggests that the regulation of protein content in soybean seed may be influenced by sucrose and oil biosynthesis. Therefore, the potential utility of the results from this study for protein in soybean seed is expected to increase.

Supplementary Materials: The following supporting information can be downloaded at: https://www.mdpi.com/article/10.3390/plants12203589/s1, Table S1: Mean and range of the seed protein content (%) of the parental and recombinant inbred lines (RILs) in two mapping populations over three years; Table S2: Analysis of variance (ANOVA) for the soybean seed protein content of the mapping populations and their parental lines, 'Saedanbaek', 'YS2035', and 'Ilmi' in 2020, 2021, and 2022; Table S3: Summary of the genetic linkage map of the RIL population derived from the cross between 'YS2035' and 'Saedanbaek'; Table S4: Summary of the genetic linkage map of the RIL population derived from the cross between 'Saedanbaek' and 'Ilmi'; Table S5: High-density genetic linkage maps in the cross between 'YS2035' and 'Saedanbaek' (Y × S) or 'Saedanbaek' and 'Ilmi' (S × I); Table S6: QTLs associated with seed protein content identified in the recombinant inbred line populations derived from the cross between 'YS2035' and 'Saedanbaek' (Y × S) or 'Saedanbaek' and 'Ilmi' (S × I); Table S7: Characteristics of singe nucleotide polymorphisms (SNP) between 'YS2035', 'Saedanbaek', and 'Saedanbaek' and 'Ilmi' in the QTL regions for seed protein content identified on chromosome 15, 18, and 20.

Author Contributions: Conceptualization, H.R.P., J.H.S., B.K.K., J.H.K., S.V.H., M.S.C., J.Y.K. and C.S.K.; methodology, H.R.P., J.H.S., B.K.K., J.H.K., S.V.H., M.S.C., J.Y.K. and C.S.K.; investigation, J.H.S.; resources J.H.S., B.K.K., J.H.K. and M.S.C.; data curation, H.R.P. and J.H.S.; writing—original draft, H.R.P.; writing—review and editing, H.R.P. and J.H.S.; visualization, J.H.S.; funding acquisition, J.H.S. All authors have read and agreed to the published version of the manuscript.

Funding: This research was supported by the National Institute of Crop Science (NICS), Rural Development Administration (RDA) of the Republic of Korea (No. PJ016054012023).

Data Availability Statement: The data sets generated in this study are included in this published article and its Supplementary Materials.

Conflicts of Interest: The authors declare no conflict of interest.

References

1. Henchion, M.; Hayes, M.; Mullen, A.M.; Fenelon, M.; Tiwari, B. Future Protein Supply and Demand: Strategies and Factors Influencing a Sustainable Equilibrium. *Foods* **2017**, *6*, 53. [CrossRef] [PubMed]
2. Gupta, S.K.; Manjaya, J.G. Advances in improvement of soybean seed composition traits using genetic, genomic and biotechnological approaches. *Euphytica* **2022**, *218*, 99. [CrossRef]
3. Qin, P.; Wang, T.; Luo, Y. A review on plant-based proteins from soybean: Health benefits and soy product development. *J. Agric. Food Res.* **2022**, *7*, 100265. [CrossRef]
4. Liu, S.; Liu, Z.; Hou, X.; Li, X. Genetic mapping and functional genomics of soybean seed protein. *Mol. Breed.* **2023**, *43*, 29. [CrossRef] [PubMed]

5. Thrane, M.; Paulsen, P.V.; Orcutt, M.W.; Krieger, T.M. Soy Protein: Impacts, Production, and Applications. In *Sustainable Protein Sources*; Elsevier: Amsterdam, The Netherlands, 2017; pp. 23–45. ISBN 9780128027769.
6. Soystats. *2023 SOYSTATS: A Reference Guide to Important Soybean Facts and Figures*; American Soybean Association: St. Louis, MO, USA, 2023.
7. Caputo, V.; Sogari, G.; Van Loo, E.J. Do plant-based and blend meat alternatives taste like meat? A combined sensory and choice experiment study. *Appl. Econ. Perspect. Policy* **2022**, *45*, 86–105. [CrossRef]
8. Zhang, T.; Dou, W.; Zhang, X.; Zhao, Y.; Zhang, Y.; Jiang, L.; Sui, X. The Development History and Recent Updates on Soy Protein-Based Meat Alternatives. *Trends Food Sci. Technol.* **2021**, *109*, 702–710. [CrossRef]
9. Mariotti, F.; Gardner, C.D. Dietary Protein and Amino Acids in Vegetarian Diets-A Review. *Nutrients* **2019**, *11*, 2661. [CrossRef]
10. Detzel, A.; Kruger, M.; Busch, M.; Blanco-Gutierrez, I.; Varela, C.; Manners, R.; Bez, J.; Zannini, E. Life cycle assessment of animal-based foods and plant-based protein-rich alternatives: An environmental perspective. *J. Sci. Food Agric.* **2022**, *102*, 5098–5110. [CrossRef]
11. Bolon, Y.T.; Joseph, B.; Cannon, S.B.; Graham, M.A.; Diers, B.W.; Farmer, A.D.; May, G.D.; Muehlbauer, G.J.; Specht, J.E.; Tu, Z.J.; et al. Complementary Genetic and Genomic Approaches Help Characterize the Linkage Group I Seed Protein QTL in Soybean. *BMC Plant Biol.* **2010**, *10*, 41. [CrossRef]
12. Hwang, E.-Y.; Song, Q.; Jia, G.; Specht, J.E.; Hyten, D.L.; Costa, J.; Cregan, P.B. A Genome-Wide Associa-tion Study of Seed Protein and Oil Content in Soybean. *BMC Genom.* **2014**, *15*, 1. [CrossRef]
13. Schmutz, J.; Cannon, S.B.; Schlueter, J.; Ma, J.; Mitros, T.; Nelson, W.; Hyten, D.L.; Song, Q.; Thelen, J.J.; Cheng, J.; et al. Genome sequence of the palaeopolyploid soybean. *Nature* **2010**, *463*, 178–183. [CrossRef] [PubMed]
14. Marsh, J.I.; Hu, H.; Petereit, J.; Bayer, P.E.; Valliyodan, B.; Batley, J.; Nguyen, H.T.; Edwards, D. Haplotype mapping uncovers unexplored variation in wild and domesticated soybean at the major protein locus cqProt-003. *Theor. Appl. Genet.* **2022**, *135*, 1443–1455. [CrossRef] [PubMed]
15. Kim, W.J.; Kang, B.H.; Moon, C.Y.; Kang, S.; Shin, S.; Chowdhury, S.; Choi, M.S.; Park, S.K.; Moon, J.K.; Ha, B.K. Quantitative Trait Loci (QTL) Analysis of Seed Protein and Oil Content in Wild Soybean (*Glycine soja*). *Int. J. Mol. Sci.* **2023**, *24*, 4077. [CrossRef]
16. Diers, B.W.; Keim, P.; Fehr, W.R.; Shoemaker, R.C. RFLP Analysis of Soybean Seed Protein and Oil Content. *Theor. Appl. Genet.* **1992**, *83*, 608–612. [CrossRef]
17. Kim, M.; Schultz, S.; Nelson, R.L.; Diers, B.W. Identification and Fine Mapping of a Soybean Seed Protein QTL from PI 407788A on Chromosome 15. *Crop Sci.* **2016**, *56*, 219–225. [CrossRef]
18. Brzostowski, L.F.; Diers, B.W. Agronomic Evaluation of a High Protein Allele from PI407788A on Chromosome 15 across Two Soybean Backgrounds. *Crop Sci.* **2017**, *57*, 2972–2978. [CrossRef]
19. Seo, J.-H.; Kim, K.-S.; Ko, J.-M.; Choi, M.-S.; Kang, B.-K.; Kwon, S.-W.; Jun, T.-H.; Singh, R. Quantitative trait locus analysis for soybean (*Glycine max*) seed protein and oil concentrations using selected breeding populations. *Plant Breed.* **2019**, *138*, 95–104. [CrossRef]
20. Warrington, C.V.; Abdel-Haleem, H.; Hyten, D.L.; Cregan, P.B.; Orf, J.H.; Killam, A.S.; Bajjalieh, N.; Li, Z.; Boerma, H.R. QTL for seed protein and amino acids in the Benning x Danbaekkong soybean population. *Theor. Appl. Genet.* **2015**, *128*, 839–850. [CrossRef]
21. Chung, J.; Babka, H.L.; Graef, G.L.; Staswick, P.E.; Lee, D.J.; Cregan, P.B.; Shoemaker, R.C.; Specht, J.E. The Seed Protein, Oil, and Yield QTL on Soybean Linkage Group I. *Crop Sci.* **2003**, *43*, 1053–1067. [CrossRef]
22. Pandurangan, S.; Pajak, A.; Molnar, S.J.; Cober, E.R.; Dhaubhadel, S.; Hernandez-Sebastia, C.; Kaiser, W.M.; Nelson, R.L.; Huber, S.C.; Marsolais, F. Relationship between asparagine metabolism and protein concentration in soybean seed. *J. Exp. Bot.* **2012**, *63*, 3173–3184. [CrossRef]
23. Mao, T.; Jiang, Z.; Han, Y.; Teng, W.; Zhao, X.; Li, W.; Morris, B. Identification of quantitative trait loci underlying seed protein and oil contents of soybean across multi-genetic backgrounds and environments. *Plant Breed.* **2013**, *132*, 630–641. [CrossRef]
24. Fliege, C.E.; Ward, R.A.; Vogel, P.; Nguyen, H.; Quach, T.; Guo, M.; Viana, J.P.G.; Dos Santos, L.B.; Specht, J.E.; Clemente, T.E.; et al. Fine mapping and cloning of the major seed protein quantitative trait loci on soybean chromosome 20. *Plant J.* **2022**, *110*, 114–128. [CrossRef] [PubMed]
25. Goettel, W.; Zhang, H.; Li, Y.; Qiao, Z.; Jiang, H.; Hou, D.; Song, Q.; Pantalone, V.R.; Song, B.H.; Yu, D.; et al. POWR1 is a domestication gene pleiotropically regulating seed quality and yield in soybean. *Nat. Commun.* **2022**, *13*, 3051. [CrossRef] [PubMed]
26. Yang, H.; Wang, W.; He, Q.; Xiang, S.; Tian, D.; Zhao, T.; Gai, J. Identifying a wild allele conferring small seed size, high protein content and low oil content using chromosome segment substitution lines in soybean. *Theor. Appl. Genet.* **2019**, *132*, 2793–2807. [CrossRef] [PubMed]
27. Wilcox, J.R.; Guodong, Z. Relationships between Seed Yield and Seed Protein in Determinate and Indeterminate Soybean Populations. *Crop Sci.* **1997**, *37*, 361–364. [CrossRef]
28. Cober, E.R.; Voldeng, H.D. Developing High-Protein, High-Yield Soybean Populations and Lines. *Crop Sci.* **2000**, *40*, 39–42. [CrossRef]
29. Miao, L.; Yang, S.; Zhang, K.; He, J.; Wu, C.; Ren, Y.; Gai, J.; Li, Y. Natural variation and selection in GmSWEET39 affect soybean seed oil content. *N. Phytol.* **2020**, *225*, 1651–1666. [CrossRef]

30. Wang, S.; Liu, S.; Wang, J.; Yokosho, K.; Zhou, B.; Yu, Y.C.; Liu, Z.; Frommer, W.B.; Ma, J.F.; Chen, L.Q.; et al. Simultaneous changes in seed size, oil content and protein content driven by selection of SWEET homologues during soybean domestication. *Natl. Sci. Rev.* **2020**, *7*, 1776–1786. [CrossRef]
31. Zhang, H.; Goettel, W.; Song, Q.; Jiang, H.; Hu, Z.; Wang, M.L.; An, Y.C. Selection of GmSWEET39 for oil and protein improvement in soybean. *PLoS Genet.* **2020**, *16*, e1009114. [CrossRef]
32. Duan, Z.; Li, Q.; Wang, H.; He, X.; Zhang, M. Genetic regulatory networks of soybean seed size, oil and protein contents. *Front. Plant Sci.* **2023**, *14*, 1160418. [CrossRef]
33. Patil, G.; Mian, R.; Vuong, T.; Pantalone, V.; Song, Q.; Chen, P.; Shannon, G.J.; Carter, T.C.; Nguyen, H.T. Molecular mapping and genomics of soybean seed protein: A review and perspective for the future. *Theor. Appl. Genet.* **2017**, *130*, 1975–1991. [CrossRef] [PubMed]
34. Van, K.; McHale, L.K. Meta-Analyses of QTLs Associated with Protein and Oil Contents and Compositions in Soybean [*Glycine max* (L.) Merr.] Seed. *Int. J. Mol. Sci.* **2017**, *18*, 1180. [CrossRef] [PubMed]
35. Brzostowski, L.F.; Pruski, T.I.; Specht, J.E.; Diers, B.W. Impact of seed protein alleles from three soybean sources on seed composition and agronomic traits. *Theor. Appl. Genet.* **2017**, *130*, 2315–2326. [CrossRef] [PubMed]
36. Lee, C.; Choi, M.-S.; Kim, H.-T.; Yun, H.-T.; Lee, B.; Chung, Y.-S.; Kim, R.W.; Choi, H.-K. Soybean [*Glycine max* (L.) Merrill]: Importance as A Crop and Pedigree Reconstruction of Korean Varieties. *Plant Breed. Biotechnol.* **2015**, *3*, 179–196. [CrossRef]
37. Lee, J.S.; Kim, S.-M.; Kang, S. Fine mapping of quantitative trait loci for sucrose and oligosaccharide contents in soybean [*Glycine max* (L.) Merr.] using 180 K Axiom® SoyaSNP genotyping platform. *Euphytica* **2015**, *208*, 195–203. [CrossRef]
38. Wang, Z.; Yu, D.; Morota, G.; Dhakal, K.; Singer, W.; Lord, N.; Huang, H.; Chen, P.; Mozzoni, L.; Li, S.; et al. Genome-wide association analysis of sucrose and alanine contents in edamame beans. *Front. Plant Sci.* **2022**, *13*, 1086007. [CrossRef]
39. Arias, C.L.; Quach, T.; Huynh, T.; Nguyen, H.; Moretti, A.; Shi, Y.; Guo, M.; Rasoul, A.; Van, K.; McHale, L.; et al. Expression of AtWRI1 and AtDGAT1 during soybean embryo development influences oil and carbohydrate metabolism. *Plant Biotechnol. J.* **2022**, *20*, 1327–1345. [CrossRef]
40. Xu, W.; Wang, Q.; Zhang, W.; Zhang, H.; Liu, X.; Song, Q.; Zhu, Y.; Cui, X.; Chen, X.; Chen, H. Using transcriptomic and metabolomic data to investigate the molecular mechanisms that determine protein and oil contents during seed development in soybean. *Front. Plant Sci.* **2022**, *13*, 1012394. [CrossRef]
41. Wang, J.; Mao, L.; Zeng, Z.; Yu, X.; Lian, J.; Feng, J.; Yang, W.; An, J.; Wu, H.; Zhang, M.; et al. Genetic mapping high protein content QTL from soybean 'Nanxiadou 25′ and candidate gene analysis. *BMC Plant Biol.* **2021**, *21*, 388. [CrossRef]
42. Du, J.; Wang, S.; He, C.; Zhou, B.; Ruan, Y.L.; Shou, H. Identification of regulatory networks and hub genes controlling soybean seed set and size using RNA sequencing analysis. *J. Exp. Bot.* **2017**, *68*, 1955–1972. [CrossRef]
43. Hooker, J.C.; Nissan, N.; Luckert, D.; Zapata, G.; Hou, A.; Mohr, R.M.; Glenn, A.J.; Barlow, B.; Daba, K.A.; Warkentin, T.D.; et al. GmSWEET29 and Paralog GmSWEET34 Are Differentially Expressed between Soybeans Grown in Eastern and Western Canada. *Plants* **2022**, *11*, 2337. [CrossRef] [PubMed]
44. Kim, H.T.; Ko, J.M.; Baek, I.Y.; Jeon, M.K.; Han, W.Y.; Park, K.Y.; Lee, B.W.; Lee, Y.H.; Jung, C.S.; Oh, K.W.; et al. Soybean Cultivar for Tofu, 'Saedanbaek' with Disease Resistance, and High Protein Content. *Korean J. Breed. Sci.* **2014**, *46*, 295–301. [CrossRef]
45. Natarajan, S.; Luthria, D.; Bae, H.; Lakshman, D.; Mitra, A. Transgenic soybeans and soybean protein analysis: An overview. *J. Agric. Food Chem.* **2013**, *61*, 11736–11743. [CrossRef] [PubMed]
46. Bandillo, N.; Jarquin, D.; Song, Q.; Nelson, R.; Cregan, P.; Specht, J.; Lorenz, A. A Population Structure and Genome-Wide Association Analysis on the USDA Soybean Germplasm Collection. *Plant Genome* **2015**, *8*, e0024. [CrossRef]
47. Song, Q.; Hyten, D.L.; Jia, G.; Quigley, C.V.; Fickus, E.W.; Nelson, R.L.; Cregan, P.B. Fingerprinting Soybean Germplasm and Its Utility in Genomic Research. *G3* **2015**, *5*, 1999–2006. [CrossRef]
48. Brummer, E.C.; Graef, G.L.; Orf, J.; Wilcox, J.R.; Shoemaker, R.C. Mapping QTL for Seed Protein and Oil Content in Eight Soybean Populations. *Crop Sci.* **1997**, *37*, 370–378. [CrossRef]
49. Li, L.; Zheng, W.; Zhu, Y.; Ye, H.; Tang, B.; Arendsee, Z.W.; Jones, D.; Li, R.; Ortiz, D.; Zhao, X.; et al. QQS orphan gene regulates carbon and nitrogen partitioning across species via NF-YC interactions. *Proc. Natl. Acad. Sci. USA* **2015**, *112*, 14734–14739. [CrossRef]
50. Zhang, Y.; Li, W.; Lin, Y.; Zhang, L.; Wang, C.; Xu, R. Construction of a high-density genetic map and mapping of QTLs for soybean (*Glycine max*) agronomic and seed quality traits by specific length amplified fragment sequencing. *BMC Genom.* **2018**, *19*, 641. [CrossRef]
51. Kambhampati, S.; Aznar-Moreno, J.A.; Bailey, S.R.; Arp, J.J.; Chu, K.L.; Bilyeu, K.D.; Durrett, T.P.; Allen, D.K. Temporal changes in metabolism late in seed development affect biomass composition. *Plant Physiol.* **2021**, *186*, 874–890. [CrossRef]
52. Shin, D.C.; Baek, I.Y.; Kang, S.T.; Song, S.B.; Hur, S.O.; Kwack, Y.H.; Lim, M.S. A New Disease and Lodging Resistance, High Yielding Soybean Variety "Ilmikong". *Korean J. Breed. Sci.* **1998**, *30*, 397.
53. RDA (Rural Development Administration). *Agricultural Science Technology Standards for Investigation of Research*; Rural Development Administration: Jeonju, Republic of Korea, 2012.
54. Saint-Denis, T.; Goupy, J. Optimization of a nitrogen analyser based on the Dumas method. *Anal. Chim. Acta* **2004**, *515*, 191–198. [CrossRef]

55. Dhungana, S.K.; Seo, J.-H.; Kang, B.-K.; Park, J.-H.; Kim, J.-H.; Sung, J.-S.; Back, I.-Y.; Shin, S.-O.; Jung, C.-S. Protein, Amino Acid, Oil, Fatty Acid, Sugar, Anthocyanin, Isoflavone, Lutein, and Antioxidant Variations in Colored Seed-Coated Soybeans. *Plants* **2021**, *10*, 1765. [CrossRef] [PubMed]
56. Lee, Y.G.; Jeong, N.; Kim, J.H.; Lee, K.; Kim, K.H.; Pirani, A.; Ha, B.K.; Kang, S.T.; Park, B.S.; Moon, J.K.; et al. Development, validation and genetic analysis of a large soybean SNP genotyping array. *Plant J.* **2015**, *81*, 625–636. [CrossRef] [PubMed]
57. Meng, L.; Li, H.; Zhang, L.; Wang, J. QTL IciMapping: Integrated software for genetic linkage map construction and quantitative trait locus mapping in biparental populations. *Crop J.* **2015**, *3*, 269–283. [CrossRef]
58. R Core Team. R: A Language and Environment for Statistical Computing. R Foundation for Statistical Computing. 2021. Available online: https://www.r-project.org (accessed on 20 May 2023).
59. Toker, C. Estimates of broad-sense heritability for seed yield and yield criteria in faba bean (*Vicia faba* L.). *Hereditas* **2004**, *140*, 222–225. [CrossRef]
60. Dhungana, S.K.; Park, J.-H.; Oh, J.-H.; Kang, B.-K.; Seo, J.-H.; Sung, J.-S.; Kim, H.-S.; Shin, S.-O.; Back, I.-Y.; Jung, C.-S. Quantitative Trait Locus Mapping for Drought Tolerance in Soybean Recombinant Inbred Line Population. *Plants* **2021**, *10*, 1816. [CrossRef]

plants

MDPI

Article

Quantitative Trait Loci and Candidate Genes That Control Seed Sugars Contents in the Soybean 'Forrest' by 'Williams 82' Recombinant Inbred Line Population

Dounya Knizia [1], Nacer Bellaloui [2], Jiazheng Yuan [3], Naoufal Lakhssasi [1], Erdem Anil [1], Tri Vuong [4], Mohamed Embaby [1], Henry T. Nguyen [4], Alemu Mengistu [5], Khalid Meksem [1] and My Abdelmajid Kassem [3,*]

1 School of Agricultural Sciences, Southern Illinois University, Carbondale, IL 62901, USA; dounya.knizia@siu.edu (D.K.); naoufal.lakhssassi@siu.edu (N.L.); erdem.anil@siu.edu (E.A.); mohamed.embaby@siu.edu (M.E.); meksem@siu.edu (K.M.)
2 USDA, Agriculture Research Service, Crop Genetics Research Unit, 141 Experiment Station Road, Stoneville, MS 38776, USA; nacer.bellaloui@usda.gov
3 Plant Genomics and Biotechnology Lab, Department of Biological and Forensic Sciences, Fayetteville State University, Fayetteville, NC 28301, USA; jyuan@uncfsu.edu
4 Division of Plant Science and Technology, University of Missouri, Columbia, MO 65211, USA; vuongt@missouri.edu (T.V.); nguyenhenry@missouri.edu (H.T.N.)
5 USDA, Agriculture Research Service, Crop Genetics Research Unit, 605 Airways Blvd, Jackson, TN 38301, USA; alemu.mengistu@usda.gov
* Correspondence: mkassem@uncfsu.edu

Abstract: Soybean seed sugars are among the most abundant beneficial compounds for human and animal consumption in soybean seeds. Higher seed sugars such as sucrose are desirable as they contribute to taste and flavor in soy-based food. Therefore, the objectives of this study were to use the 'Forrest' by 'Williams 82' (F × W82) recombinant inbred line (RIL) soybean population (n = 309) to identify quantitative trait loci (QTLs) and candidate genes that control seed sugar (sucrose, stachyose, and raffinose) contents in two environments (North Carolina and Illinois) over two years (2018 and 2020). A total of 26 QTLs that control seed sugar contents were identified and mapped on 16 soybean chromosomes (chrs.). Interestingly, five QTL regions were identified in both locations, Illinois and North Carolina, in this study on chrs. 2, 5, 13, 17, and 20. Amongst 57 candidate genes identified in this study, 16 were located within 10 Megabase (MB) of the identified QTLs. Amongst them, a cluster of four genes involved in the sugars' pathway was collocated within 6 MB of two QTLs that were detected in this study on chr. 17. Further functional validation of the identified genes could be beneficial in breeding programs to produce soybean lines with high beneficial sucrose and low raffinose family oligosaccharides.

Keywords: soybean; RIL; Forrest; Williams 82; linkage map; RFOs; sucrose; raffinose; stachyose; SNPs

Citation: Knizia, D.; Bellaloui, N.; Yuan, J.; Lakhssasi, N.; Anil, E.; Vuong, T.; Embaby, M.; Nguyen, H.T.; Mengistu, A.; Meksem, K.; et al. Quantitative Trait Loci and Candidate Genes That Control Seed Sugars Contents in the Soybean 'Forrest' by 'Williams 82' Recombinant Inbred Line Population. *Plants* **2023**, *12*, 3498. https://doi.org/10.3390/plants12193498

Academic Editor: Zhaoshi Xu

Received: 29 August 2023
Revised: 3 October 2023
Accepted: 6 October 2023
Published: 8 October 2023

1. Introduction

Sugars, including sucrose, stachyose, glucose, raffinose, galactose, fructose, rhamnose, and starch, play a major role in seed and fruit development and seed desiccation tolerance (DT) [1–4]. Sucrose and raffinosaccharides (raffinose and stachyose), also called raffinose family oligosaccharides (RFOs), make up 5–7%, 1%, and 3–4% of total carbohydrates, respectively, of soybean seed dry weights [5]. RFOs are synthesized from sucrose through a series of additions of galactinol units and are involved in DT, freezing, stress tolerance, and seed longevity [6–9]. Galactinol synthase (GolS) is the key enzyme in the RFO biosynthetic pathway converting galactinol and myo-inositol as the main precursors to form RFOs. Galactinol synthase (GolS) converts myo-inositol and UDP-galactose into galactinol, while sucrose and galactinol are converted into raffinose by raffinose synthase [9,10]. In addition to being involved in stress tolerance, RFOs are reported to play a role in several signal transduction pathways [11], exports of specific mRNAs [12], and trafficking of certain vesicle membranes [13].

Like most seed components, seed sugars [4] are influenced by many factors, including abiotic and biotic stresses, and environmental factors, such as temperature, soil moisture, freezing, seed maturity, and growth conditions [1,14–19]. It was shown that stachyose contents increased drastically in drying seeds but not in seeds kept at high humidity levels, which reveals the critical role of stachyose in DT [1]. The effect of water deficit (WD) on enzymes involved in sugar biosynthetic pathways in soybean nodules was investigated. Sucrose synthase activity declined drastically with increased WD while sucrose content increased [14]. Other studies showed that WD impacts negatively on sucrose biosynthesis and translocation from sources to sinks more than other sugars' (raffinose and stachyose) biosynthesis [16,19]. Investigating 'Clark' and 'Harosoy' near-isogenic lines (NILs) revealed that Clark's sugar contents decreased with increased days of maturity for both cultivars while both positive and negative effects were observed concerning the effects of temperature in two different years (2004 and 2005) [15]. In 2004, seed sugar contents increased with temperature increase, while the contents decreased with increased temperatures in 2005 [15]. The effect of WD on several seed components, including sugars, was investigated in several susceptible and resistant Phomopsis soybean cultivars. Sugar (sucrose, raffinose, and stachyose) contents were higher in seeds of resistant maturity group III cultivars than their susceptible counterparts [16]. A recent study investigated the effect of soil moisture on seed sugars (sucrose, raffinose, stachyose) and starch contents among other compounds in two soybean cultivars in maturity group V (Asgrow, AG6332, and Progeny 5333RY) and showed that sucrose, stachyose, and raffinose contents, in addition to the mineral nutrient (N, P, K, and Ca) contents, decreased with increased soil moisture in both cultivars [17].

During recent decades, more than 53 QTLs that control seed sucrose and RFOs, other sugars (glucose, galactose, fructose, fucose, rhamnose), and starch contents have been reported using different biparental and natural populations and mapping methods including single marker analysis, interval mapping (IM), composite interval mapping (CIM), and genome-wide association studies (GWASs) [18,20]. However, to our knowledge, only a few of these studies identified candidate genes within these QTL regions, as summarized in [18]. There is *Glyma.01g127600*, which encodes for a protein phosphatase on chr. 1; *Glyma.03g019300*, which encodes for a MADS-box protein; *Glyma.03g064700*, which encodes for a phosphatidylinositol monophosphate-5-kinase on chr. 3; and *Glyma.06g194200*, which encodes for a gibberellin-regulated protein on chr. 6 [18,21].

To improve seed quality, several attempts to manipulate seed sugars, phytic acid, and the contents of other beneficial compounds have been made in recent years [22–24]. Monogastric animals (such as poultry and pigs) and humans do not produce α-galactosidase and cannot digest RFOs, which reduces gastrointestinal performance, flatulence, and diarrhea. Therefore, reducing raffinose and stachyose and increasing sucrose in soybean seed contents are desirable and the main goals in breeding programs [22–27]. The objective of this study was to genetically map QTLs for seed sucrose, raffinose, and stachyose contents using the 'Forrest' by 'Williams 82' RIL population, in addition to identifying candidate genes involved in soybean seed sugar biosynthesis.

2. Materials and Methods

2.1. Plant Materials

The 'Forrest' × 'Williams 82' RIL population (F × W82, n = 309) was previously studied and described in detail in our previous research [28,29]. The parents and RILs were evaluated in two locations: Spring Lake, NC (35.17° N, 78.97° W, 2018) and Carbondale, IL (37° N, 89° W, 2020). Briefly, seed parents and RIL seeds were grown in a randomized block design with 25 cm row spaces and three replicates. More details about growth conditions, crop management, and seed harvesting were described earlier [28,29].

2.2. Seed Sugar Quantification

RILs, parents (Forrest and Williams 82), and soybean germplasm seeds were harvested at maturity, and sugar (sucrose, raffinose, and stachyose) contents (%) were quantified using near-infrared reflectance (NIR) with an AD 7200 array feed analyzer (Perten, Springfield, IL, USA) as described earlier [15,30].

2.3. DNA Isolation, SNP Genotyping, and Genetic Map Construction

Parents' and RILs' genomic DNA was extracted using the cetyltrimethylammonium bromide (CTAB) method as previously described [31]. A NanoDrop spectrophotometer (NanoDrop Technologies Inc., Centreville, DE, USA) was used to quantify DNA samples (50 ng/µL), and genotyping was performed using the Illumina Infinium SoySNP6K BeadChips (Illumina, Inc., San Diego, CA, USA) as described earlier [15] at the Soybean Genomics and Improvement Laboratory (USDA-ARS, Beltsville, MD, USA). The F × W82 genetic linkage map was constructed using JoinMap 4.0 [28,32] as previously described to detect QTLs for seed isoflavones [28] and seed tocopherol contents [29].

2.4. Sugar QTL Detection

WinQTL Cartographer [33] interval mapping (IM) and composite interval mapping (CIM) methods were used to identify QTLs that control seed sugar (sucrose, stachyose, and raffinose) contents in this RIL population. The following parameters were used: Model 6, 1 cM step size, 10 cM window size, 5 control markers, and 1000 permutations. Furthermore, chromosomes were drawn using MapChart 2.2 [34].

2.5. Sugars Biosynthesis Candidate Genes' Identification

The Glyma numbers of the sucrose and RFO biosynthesis genes were obtained via reverse BLAST of the genes underlying the RFO pathway in *Arabidopsis* using the available data in SoyBase. The sequences of the *Arabidopsis* genes were obtained from the Phytozome database (https://phytozome-next.jgi.doe.gov; accessed on 15 August 2023). These sequences were used for Blast in SoyBase. The obtained genes that control the RFO pathway were mapped to the identified sugars' QTLs.

2.6. Expression Analysis

The expression analysis of the identified candidate genes was performed using the publicly available data from SoyBase [20] to produce the expression profiles of these candidate genes in the soybean reference genome Williams 82 in the Glyma1.0 Gene Models version.

2.7. Comparison of the Williams 82 and Forrest Sequences

Sequences of Forrest and Williams 82 cv. were obtained from the variant calling and haplotyping analysis, which was performed using 108 soybean germplasm sequenced lines as described previously [35].

3. Results

3.1. Sugar Frequency Distribution

The frequency distributions among sucrose, raffinose, and stachyose contents were quite different in the F × W82 RIL population based on the Shapiro–Wilk method for the normality test. Raffinose (2018), stachyose (2018), and sucrose (2020) were normally distributed. Only positive or negative skewness were identified in the RIL population, and all kurtosis values of these variables were positive (Table 1; Figure 1). In terms of coefficient of variation (CV), the value of sucrose 2018 showed the highest percentage of variation (62.86%) compared to other assessed traits, and the rest of the CVs appeared to be less varied within these two years. The histogram of sucrose 2018 was extremely skewed, and the other traits evaluated were normally distributed.

Table 1. Seed sugar contents' means, ranges, CVs, skewness, and kurtosis in the F × W82 RIL population evaluated in Spring Lake, NC (2018) and Carbondale, IL (2020). Mean and range values are expressed in μg/g of seed weight. ** $p < 0.01$, *** $p < 0.001$.

Year	Sugar	Mean	Range	CV (%)	SE	Skewness	Kurtosis	W Value ($p < 0.05$)
2018	Sucrose	2.58	22.7	62.86	0.12	12.2	161.38	0.22 ***
	Raffinose	0.67	0.26	9.16	0.01	0.18	3.26	0.99
	Stachyose	2.23	2.55	21.74	0.03	−0.07	2.85	0.99
2020	Sucrose	4.92	4.98	17.2	0.05	−0.13	3.15	0.99
	Raffinose	0.83	0.41	7.28	0.01	0.65	4.83	0.97 ***
	Stachyose	3.61	2.15	9.06	0.02	−0.48	3.8	0.98 **

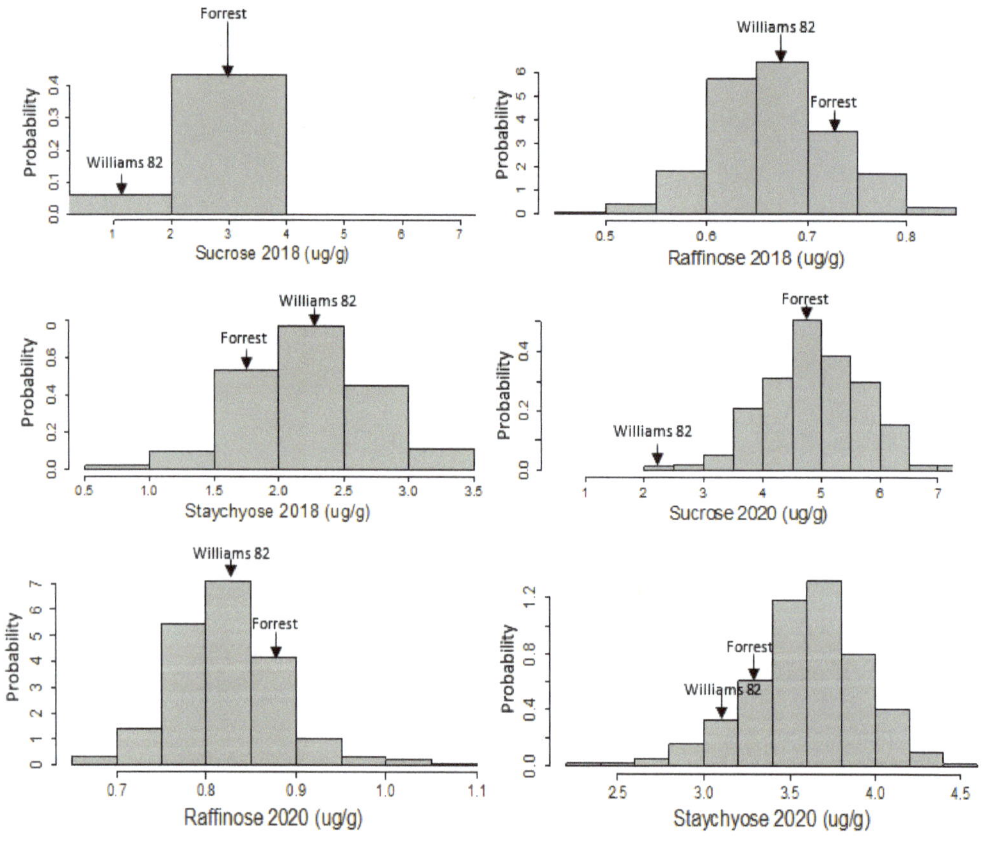

Figure 1. Frequency distribution of sugars (sucrose, raffinose, and stachyose) in the F × W82 RIL population grown in two environments over two years (Spring Lake, NC in 2018 and Carbondale, IL in 2020).

The broad-sense heritability ($h^2{}_b$) of seed sugar weight for sucrose, raffinose, and stachyose contents across two different environments appeared quite different. Stachyose had the highest heritability (92%), and the $h^2{}_b$ for sucrose was 36.8% (Table 2). However, no negative $h^2{}_b$ values for sugar contents were observed. The RIL–year interactions still played a significant role in the molecular formation among the three sugar contents in soybean seeds. The Sum Sq and Mean Sq to determine $\sigma_G{}^2$ and $\sigma_{GE}{}^2$ for each trait (Table 2) using the type I sum of squares (ANOVA (model)) function in the R program were implemented.

Table 2. Two-way ANOVA of seed sugar (sucrose, stachyose, and raffinose) contents in the F × W82 RIL population evaluated in Spring Lake, NC (2018) and Carbondale, IL (2020).

Response: Sucrose				
	Df	**Sum Sq**	**Mean Seq**	**H^2**
Line	369	1134.22	3.0738	0.378
Year	1	5.6	5.5975	
Line × Year	2	3.82	1.9108	
Residuals	0	0	NA	
Response: Raffinose				
	Df	Sum Sq	Mean Seq	H^2
Line	369	3.4552	0.0093891	0.739
Year	1	0.0253	0.0253139	
Line × Year	2	0.0048	0.0023972	
Residuals	0	0	NA	
Response: Stachyose				
	Df	Sum Sq	Mean Seq	H^2
Line	369	246.73	0.66865	0.92
Year	1	1.611	1.61115	
Line × Year	2	0.106	0.05307	
Residuals	0	0	NA	

3.2. Sugars Contents' QTLs

IM and CIM were used to identify QTLs for seed sugar contents in this F × W82 RIL population; however, only QTLs identified by CIM are presented here. The QTLs identified with the IM method are reported in Tables S1 and S2. A total of 26 QTLs that control seed sugar contents were identified in both NC-2018 (19 QTLs) and IL-2020 (7 QTLs) via CIM (Tables 3 and 4; Figure S1).

Table 3. Quantitative trait loci (QTLs) that control sugar (sucrose, stachyose, and raffinose) contents in F × W82 RIL population in Spring Lake, NC in 2018. These QTLs were identified via CIM method. * Indicates novel QTL.

Sugar	QTL	Chr.	Marker/Interval	Position (cM)	LOD	R^2	Add. Eff.
Sucrose	*qSUC-1*	1	Gm01_3504836-Gm01_3466825	0.01–12.1	39.19	20.46	−3.05
	qSUC-2	2	Gm02_5155733-Gm02_9925870	128.5–142.2	42.77	47.90	4.42
	qSUC-3	3	Gm03_4595422-Gm03_4113546	39.2–39.8	32.62	20.50	3.05
	qSUC-4 *	4	Gm04_7672403	6.5–16.5	54.35	37.50	4.62
	qSUC-5	5	Gm05_3867435-Gm05_3273418	31.5–37.01	20.65	17.51	2.60
	qSUC-6	6	Gm06_1737718-Gm06_5014399	48.5–52.4	5.36	10.50	−1.37
	qSUC-7	9	Gm09_1888876	173.9–178.1	32.62	20.50	3.05
	qSUC-8 *	10	Gm10_621706	214.01–216.01	34.25	19.10	−4.48
	qSUC-9	13	Gm13_3891723-Gm13_3524828	0.2–58.2	19.12	17.51	2.60
	qSUC-10	17	Gm17_4967175-Gm17_5294475	0.4–1.0	33.22	20.50	3.05
	qSUC-11 *	18	Gm18_1620585-Gm18_2020823	94.7–96.5	20.10	17.51	2.60
	qSUC-12	20	Gm19_2552468	172.11	6.98	9.10	1.41
Stachyose	*qSTA-1*	13	Gm13_3524828	96.2–98.2	2.52	14.8	0.19
	qSTA-2	13	Gm13_3884070-Gm13_3803273	121.8–123.2	2.60	5.2	0.11
	qSTA-3	19	Gm19_3789399-Gm19_4362616	98.01–124.1	4.21	8.5	−0.16
	qSTA-4	19	Gm19_4946208-Gm19_5032228	184.1–186.1	2.53	5.3	0.11
Raffinose	*qRAF-1*	9	Gm09_4024436-Gm09_4082234	108.01–110.9	2.26	4.6	−0.01
	qRAF-2	9	Gm09_1888876	173.9–178.1	2.47	7.6	0.08
	qRAF-3	12	Gm12_6023395-Gm12_2379195	114.6–118.6	2.15	4.7	−0.01

Table 4. Quantitative trait loci (QTLs) that control sugar (sucrose, stachyose, and raffinose) contents in F × W82 RIL population in Carbondale, IL in 2020. These QTLs were identified via CIM method.

Sugar	QTL	Chr.	Marker	Position (cM)	LOD	R^2	Add. Eff.
Sucrose	*qSUC-1*	2	Gm02_1199805-Gm02_1373746	196.4–205.6	2.63	3.60	−0.16
	qSUC-2	5	Gm05_3803682-Gm05_3748078	18.01–22.1	2.10	0.03	−0.14
	qSUC-3	8	Gm08_5960619-Gm08_8268861	47.1–55.9	2.37	0.04	0.16
Stachyose	*qSTA-1*	13	Gm13_2748576	0.5–4.5	2.03	0.09	0.21
	qSTA-2	16	Gm16_3183754-Gm16_3010888	81.6–94.7	2.85	3.92	0.10
	qSTA-3	17	Gm17_8449684-Gm17_8352493	136.5–136.7	2.37	3.00	−0.08
	qSTA-4	20	Gm20_294157-Gm20_1133712	145.4–148.5	3.59	4.50	−0.12

In Spring Lake, NC in 2018 (NC-2018), 12 QTLs that control seed sucrose content (qSUC-1–qSUC-12) were identified and mapped on Chrs. 1, 2, 3, 4, 5, 6, 9, 10, 13, 17, 18, and 19; 4 QTLs that control seed stachyose content (qSTA-1–qSTA-4) were identified and mapped on Chrs. 13 and 19; and 3 QTLs that control seed raffinose content (qRAF-1–qRAF-3) were identified and mapped on Chr. 9 and 12 (Tables 3 and 5; Figure S1). Likewise, in Carbondale, IL in 2020 (IL-2020), 3 QTLs that control seed sucrose content (qSUC-1–qSUC-3) were identified and mapped on Chrs. 2, 5, and 8; and 4 QTLs that control seed stachyose content (qSTA-1–qSTA-4) were identified and mapped on Chrs. 13, 16, 17, and 20 (Tables 4 and 6; Figure S1). No QTL that controls seed raffinose content was identified in this location.

Table 5. QTLs and candidate genes that control sugar (sucrose, stachyose, and raffinose) contents in F × W82 RIL population in Spring Lake, NC in 2018. These QTLs were identified via CIM method.

Sugar	QTL	Marker/Interval	LOD	R^2	Wm82.a2.v1	Start	End	Wm82.a1.v1.1	Start	End	Dis. (MB)
Sucrose	*qSUC-1*	Gm01_3504836-Gm01_3466825	39.19	20.46	.	.	.	.	.	.	.
	qSUC-2	Gm02_5155733-Gm02_9925870	42.77	47.9	*Glyma.02G016700*	1490049	1491170	*Glyma02g02030*	1475851	1476528	3.6
	qSUC-3	Gm03_4595422-Gm03_4113546	32.62	20.5	.	.	.	.	.	.	.
	qSUC-4	Gm04_7672403	54.35	37.5	.	.	.	.	.	.	.
	qSUC-5	Gm05_3867435-Gm05_3273418	20.65	17.51	*Glyma.05G040300*	3593378	3598821	*Glyma05g02510*	1870330	1875692	1.3
					Glyma.05G003900	307460	312091	*Glyma05g08950*	8806144	8810647	4.9
	qSUC-6	Gm06_1737718-Gm06_5014399	5.36	10.5	*Glyma.06G175500*	14845358	14849994	*Glyma06g18480*	14802178	14807061	9.7
					Glyma.06G179200	15217419	15223877	*Glyma06g18890*	15175181	15181763	10.16
	qSUC-7	Gm09_1888876	32.62	20.5	*Glyma.09G073600*	7809852	7816248	*Glyma09g08550*	7845409	7851685	5.9
					Glyma.09G016600	1285132	1290884	*Glyma09g01940*	1270010	1276140	0.6
	qSUC-8	Gm10_621706	34.25	19.1	*Glyma.10G017300*	1523661	1524691	*Glyma10g02170*	1519053	1519546	0.8
	qSUC-9	Gm13_3891723-Gm13_3524828	19.12	17.51	.	.	.	.	.	.	.
	qSUC-10	Gm17_4967175-Gm17_5294475	33.22	20.5	*Glyma.17G037400*	2732048	2737399	*Glyma17g04160*	2739794	2745132	2.2
					Glyma.17G045800	3404918	3410491	*Glyma17g05067*	3412682	3418160	1.5
					Glyma.17G035800	2629011	2639005	*Glyma17g03990*	2637080	2646732	2.3
					Glyma.17G111400	8744555	8747526	*Glyma17g11970*	9015075	9018145	3.7
	qSUC-11	Gm18_1620585-Gm18_2020823	20.1	17.51	.	.	.	.	.	.	.
	qSUC-12	Gm19_2552468	6.98	9.1	*Glyma.19G004400*	359933	363588	*Glyma19g00441*	238429	242106	2.3
Stachyose	*qSTA-1*	Gm13_3524828	2.52	14.8	.	.	.	.	.	.	.
	qSTA-2	Gm13_3884070-Gm13_3803273	2.6	5.2	.	.	.	.	.	.	.
	qSTA-3	Gm19_3789399-Gm19_4362616	4.21	8.5	*Glyma.19G004400*	359933	363588	*Glyma19g00440*	241366	241903	3.5
	qSTA-4	Gm19_4946208-Gm19_5032228	2.53	5.3	*Glyma.19G004400*	359933	363588	*Glyma19g00440*	241366	241903	4.7
Raffinose	*qRAF-1*	Gm09_4024436-Gm09_4082234	2.26	4.6	*Glyma.09G073600*	7809852	7816248	*Glyma09g08550*	7845409	7851685	3.7
					Glyma.09G016600	1285132	1290884	*Glyma09g01940*	1270010	1276140	2.7
					Glyma.09G167000	39103764	39109664	*Glyma09g29710*	36530532	36536435	2.5
	qRAF-2	Gm09_1888876	2.47	7.6	*Glyma.09G073600*	7809852	7816248	*Glyma09g08550*	7845409	7851685	5.9
					Glyma.09G016600	1285132	1290884	*Glyma09g01940*	1270010	1276140	0.6
	qRAF-3	Gm12_6023395-Gm12_2379195	2.15	4.7	.	.	.	.	.	.	.

Table 6. QTLs and candidate genes that control sugar (sucrose, stachyose, and raffinose) contents in F × W82 RIL population in Carbondale, IL in 2020. These QTLs were identified via CIM method.

Sugar	QTL	Marker	LOD	R^2	Wm82.a2.v1	Start	End	Wm82.a1.v1.1	Start	End	Dis. (MB)
Sucrose	*qSUC-1*	Gm02_1199805-Gm02_1373746	2.63	3.6	*Glyma.02G016700*	1490049	1491170	*Glyma02g02030*	1475851	1476528	0.2
	qSUC-2	Gm05_3803682-Gm05_3748078	2.1	0.03	*Glyma.05G040300*	3593378	3598821	*Glyma05g02510*	1870330	1875692	1.8
					Glyma.05G003900	307460	312091	*Glyma05g08950*	8806144	8810647	5.002
	qSUC-3	Gm08_5960619-Gm08_8268861	2.37	0.04	*Glyma.08G043800*	3450235	3451725	*Glyma08g04860*	3446035	3447462	2.5
					Glyma.08G143500	10949673	10956219	*Glyma08g15220*	11038816	11045375	2.7
					Glyma.08G011800	942037	944988	*Glyma08g01480*	939512	942346	5.01
					Glyma.08G023100	1852651	1856671	*Glyma08g02690*	1848105	1853380	4.1
Stachyose	*qSTA-1*	Gm13_2748576	2.03	0.09	.	.	.	.	.	.	.
	qSTA-2	Gm16_3183754-Gm16_3010888	2.85	3.92	.	.	.	.	.	.	.
	qSTA-3	Gm17_8449684-Gm17_8352493	2.37	3	*Glyma.17G037400*	2732048	2737399	*Glyma17g04160*	2739794	2745132	5.6
					Glyma.17G045800	3404918	3410491	*Glyma17g05067*	3412682	3418160	4.9
					Glyma.17G035800	2629011	2639005	*Glyma17g03990*	2637080	2646732	5.8
					Glyma.17G111400	8744555	8747526	*Glyma17g11970*	9015075	9018145	0.5
	qSTA-4	Gm20_294157-Gm20_1133712	3.59	4.5		.	.	.	.	.	.

No QTL for seed sugar contents was identified in other studies within the QTL regions on chr. 4 (qSUC-4-NC-2018, 6.5–16.5 cM), chr. 10 (qSUC-8-NC-2018, 214.1–216.1 cM), or chr. 18 (qSUC-11-NC-2018, 20.1–17.5 cM), which indicates they are novel QTL regions.

3.3. In Silico Sucrose, Raffinose, and Stachyose Biosynthetic Pathway Genes in Soybean

In the literature, the sugar (sucrose, raffinose, and stachyose) biosynthetic pathway was studied in many plants, including the plant model *Arabidopsis thaliana* [36,37] and the leguminous model *Medicago sativa* L. [38]. A reverse BLAST of the genes identified in *Arabidopsis thaliana* was conducted using SoyBase [20] to reconstruct the sugar (sucrose, raffinose, and stachyose) biosynthetic pathway in soybean (Figure 2).

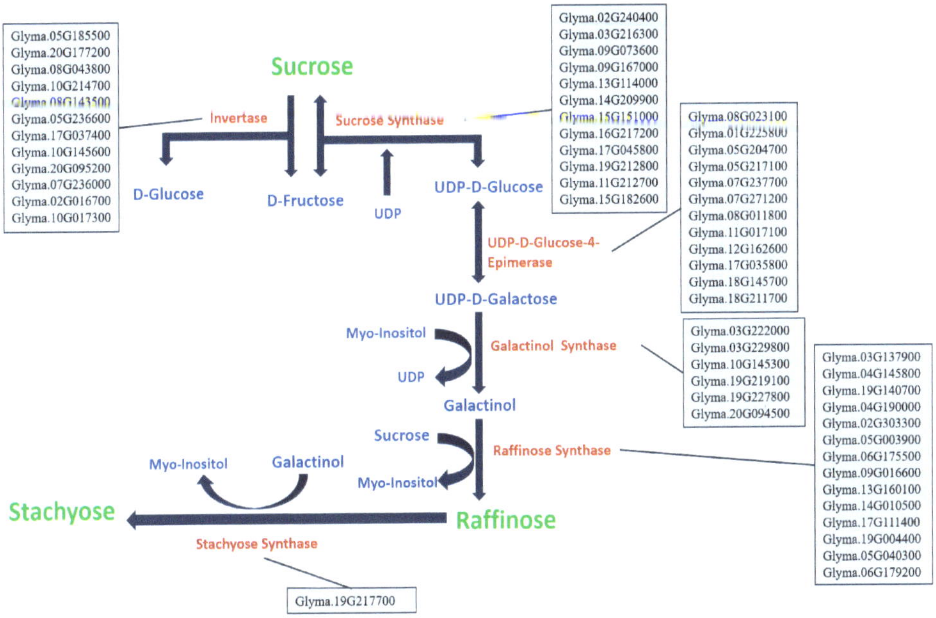

Figure 2. The sugar (sucrose, raffinose, and stachyose) biosynthetic pathway with the identified candidate genes in soybean. The genes are in Wm82.a2.v1 annotation.

A total of fifty-seven candidate genes were identified to underly the sugar (sucrose, raffinose, and stachyose) biosynthetic pathway (Figure 2). In this pathway, twelve candidate genes were identified for invertase: *Glyma.05G185500*, *Glyma.20G177200*, *Glyma.08G043800*, *Glyma.10G214700*, *Glyma.08G143500*, *Glyma.05G236600*, *Glyma.17G037400*, *Glyma.10G145600*, *Glyma.20G095200*, *Glyma.07G236000*, *Glyma.02G016700*, and *Glyma.10G017300*. Twelve candidate genes were identified for sucrose synthase: *Glyma.02G240400*, *Glyma.03G216300*, *Glyma.09G073600*, *Glyma.09G167000*, *Glyma.13G114000*, *Glyma.14G209900*, *Glyma.15G151000*, *Glyma.16G217200*, *Glyma.17G045800*, *Glyma.19G212800*, *Glyma.11G212700*, and *Glyma.15G18 2600*. Twelve candidate genes were identified for UDP-D-Glucose-4-Epimerase: *Glyma.08G0 23100*, *Glyma.01G225800*, *Glyma.05G204700*, *Glyma.05G217100*, *Glyma.07G237700*, *Glyma.07G 271200*, *Glyma.08G011800*, *Glyma.11G017100*, *Glyma.12G162600*, *Glyma.17G035800*, *Glyma.18 G145700*, and *Glyma.18G211700*. For galactinol synthase, six candidate genes were identified: *Glyma.03G222000*, *Glyma.03G229800*, *Glyma.10G145300*, *Glyma.19G219100*, *Glyma.19G2 27800*, and *Glyma.20G094500*. Fourteen candidate genes were identified for raffinose synthase: *Glyma.03G137900*, *Glyma.04G145800*, *Glyma.19G140700*, *Glyma.04G190000*, *Glyma.02G 303300*, *Glyma.05G003900*, *Glyma.06G175500*, *Glyma.09G016600*, *Glyma.13G160100*, *Glyma.14*

G010500, *Glyma.17G111400*, *Glyma.19G004400*, *Glyma.05G040300*, and *Glyma.06G179200*. For stachyose synthase, only one candidate gene was identified: *Glyma.19G217700* (Figure 2).

3.4. Association between the Identified Sugar (Sucrose, Raffinose, and Stachyose) Biosynthetic Pathway Candidate Genes and Reported QTLs

The identified genes for sugar (sucrose, raffinose, and stachyose) biosynthesis in soybean were mapped to the identified QTLs. Amongst fifty-seven candidate genes, sixteen were located less than 10 MB from the identified QTLs on chrs. 2, 5, 6, 8, 9, 10, 17, and 19 (Tables 3–6).

The sucrose synthase candidate gene *Glyma.09G073600* and the raffinose synthase candidate gene *Glyma.09G016600* are positioned close to *qSUC-7-IL-2018*, *qRAF-1-IL-2018*, and *qRAF-2-IL-2018* on Chr.9 (Tables 3–6). The invertase candidate gene *Glyma.02G016700* is located 3.6 and 0.2 MB away from *qSUC-1-IL-2018* and *qSUC-1-NC-2020*, respectively, on Chr. 2 (Tables 3–6). The raffinose synthase candidate genes *Glyma.05G003900* and *Glyma.05G040300* are located close to *qSUC-5-IL-2018* and *qSUC-2-NC-2020* on Chr. 5 (Tables 3–6). On chr. 6, the raffinose synthase candidate gene *Glyma.06G175500* is located close to *qSUC-6-IL-2018* (Tables 3–6). The invertase candidate genes *Glyma.08G043800* and *Glyma.08G143500*, and the UDP-D-Glucose-4-Epimerase candidate genes *Glyma.08G011800* and *Glyma.08G023100* on chr. 8 are located close to *qSUC-3-NC-2020* (Tables 3–6, S3 and S4). On chr. 10, the invertase candidate gene *Glyma.10G017300* is located close to *qSUC-8-IL-2018* (Tables 3–6). On Chr. 17, a cluster of four genes involved in the sugar pathway is collocated within 6 MB of two QTLs (qSUC-10-NC-2018 and qSTA-3-IL-2020) that were identified in this study. These genes are *Glyma.17G037400* encoding for an invertase, *Glyma.17G045800* encoding for sucrose synthase, *Glyma.17G111400* encoding for raffinose synthase, and *Glyma.17G035800* encoding for UDP-D-glucose-4-epimerase (Tables 3–6, Figure S3). The raffinose synthase candidate gene *Glyma.19G004400* is positioned close to *qSTA-3-IL-2018* and *qSTA-4-IL-2018* (Tables 3–6), as well as *qRAF-8-IL-2018* and *qRAF-9-IL-2018* identified using the IM method (Tables 3 and 4).

3.5. Association between the Identified Candidate Genes and the Previously Reported QTLs

Several studies have identified and mapped QTLs underlying the seed sugar content using different populations and mapping methods [39–42], as summarized in [18].

The identified genes have been mapped to the previously reported QTL regions associated with the seed sugar content using data from SoyBase [18,20,43]. In this study, 6 candidate genes were located within the identified seed sugar QTLs and 18 were located <9 MB away from these regions (Table 7). Among them is the invertase candidate gene *Glyma.08G143500*, which is located within the seed sucrose 1-2 QTL on Chr. 8 [20,39]. Also, the galactinol-sucrose galactosyl-transferase 6-related candidate gene *Glyma.13G160100* is situated within the seed sucrose 1-5 QTL [20,39] (Table 7). Likewise, the raffinose synthase candidate gene *Glyma.19G140700* is collocated within the seed sucrose 1-8 QTL [20,39], less than <0.5 MB away from seed sucrose 2-11 and seed sucrose 2-10 [20,41], and 1.9 MB from seed oligosaccharide 2-7 [20,40].

Table 7. Candidate genes controlling sugar (sucrose, stachyose, and raffinose) contents associated with previously reported QTLs.

Gene ID	Start	End	QTL	QTL Start	QTL End	Reference
Glyma.02G240400	42892680	42898279	Seed sucrose 2-2	39547350	41441274	[41]
			Seed oligosaccharide 1-1	39547350	41441274	[41]
Glyma.05G236600	41293446	41294570	Seed sucrose 1-1	3924139	4279362	[39]
Glyma.08G043800	3450235	3451725	Seed sucrose 1-3	7892162	8937354	[39]
Glyma.08G143500	10949673	10956219	Seed sucrose 1-2	10865328	13126779	[39]
Glyma.09G073600	7809852	7816248	Seed sucrose 4-2	2973041	5901485	[44]
Glyma.13G114000	22767704	22773231	Seed sucrose 1-5	26196486	28912864	[39]
Glyma.14G209900	47515899	47521687	Seed sucrose 3-1	38859467	40060720	[40]
			Seed oligosaccharide 2-1	38859467	40060720	[40]
Glyma.15G151000	12497113	12508050	Seed sucrose 3-3	13755345	17021739	[40]
			Seed oligosaccharide 2-3	13755345	17021739	[40]
Glyma.19G140700	40199041	40201038	Seed sucrose 1-8	40205349	40265091	[39]
			Seed oligosaccharide 2-7	42119600	43329204	[40]
Glyma.19G212800	46633685	46639818	Seed oligosaccharide 2-7	42119600	43329204	[40]
			qSU1901	45311975	45464136	[43]
Glyma.19G217700	47033812	47037286	Seed oligosaccharide 2-7	42119600	43329204	[40]
			qSU1901	45311975	45464136	[43]
Glyma.20G095200	33827363	33831352	Seed sucrose 1-4	2716974	25498552	[39]
Glyma.08G011800	942037	944988	Seed sucrose 1-3	7892162	8937354	[39]
			Seed sucrose 1-13	8283676	9192408	[39]
Glyma.08G023100	1852651	1856671	Seed sucrose 1-3	7892162	8937354	[39]
			Seed sucrose 1-13	8283676	9192408	[39]
Glyma.19G219100	47148224	47150373	Seed sucrose 1-8	40205349	40265091	[39]
			Seed sucrose 2-10	40637071	41616190	[41]
			Seed sucrose 2-11	40637071	41616190	[41]
			Seed oligosaccharide 2-7	42119600	43329204	[40]
Glyma.19G227800	47911129	47914214	Seed sucrose 1-8	40205349	40265091	[39]
			Seed sucrose 2-10	40637071	41616190	[41]
			Seed sucrose 2-11	40637071	41616190	[41]
			Seed oligosaccharide 2-7	42119600	43329204	[40]
Glyma.20G094500	33759416	33761555	Seed sucrose 1-4	2716974	25498552	[39]
Glyma.20G177200	41446962	41451980	qSU2002	40523599	41882459	[43]
Glyma.15G182600	17910130	17916426	Seed sucrose 3-3	13755345	17021739	[40]
			Seed oligosaccharide 2-3	13755345	17021739	[40]
Glyma.05G003900	307460	312091	Seed sucrose 1-1	3924139	4279362	[39]
Glyma.09G016600	1285132	1290884	Seed sucrose 4-2	2973041	5901485	[44]
Glyma.17G111400	8744555	8747526	qSS1701	7470395	10014816	[43]
			qSS1702	7969537	10599548	[43]
Glyma.13G160100	27576191	27579282	Seed sucrose 1-5	26196486	28912864	[39]
Glyma.19G004400	359933	363588	Seed sucrose 2-3	4244065	12744826	[41]
			Seed oligosaccharide 1-2	4244065	12744826	[41]
			Seed sucrose 2-6	9284015	34059981	[41]
			Seed oligosaccharide 1-5	9284015	34059981	[41]

The sucrose synthase candidate gene *Glyma.02G240400* was located close (<1.5 MB) to two QTLs controlling seed sugar contents, the seed sucrose 2-2 and seed oligosaccharide 1-1 [20,41]. Moreover, the raffinose synthase candidate gene *Glyma.05G003900* is located less than <4 MB away from the seed sucrose 1-1 [20,39]. The raffinose synthase candidate gene *Glyma.19G004400* is located less than 9 MB away from four QTLs controlling the sugar contents, namely seed sucrose 2-3, seed oligosaccharide 1-2, seed sucrose 2-6, and seed oligosaccharide 1-5 [20,41] (Table 7). On chr. 8, the seed sucrose 1-3 and seed sucrose 1-13 are located close to the invertase candidate genes *Glyma.08G043800*, and *Glyma.08G143500*, as well as UDP-D-glucose-4-epimerase candidate genes *Glyma.08G011800* and *Glyma.08G023100* [20,39] (Table 7). The sucrose synthase candidate gene *Glyma.09G073600* and the raffinose candidate gene *Glyma.09G016600* are positioned less than <2 MB away from the seed sucrose 4-2 [20,44] (Table 7). Interestingly, the sucrose synthase candidate genes *Glyma.15G182600* and *Glyma.15G151000* are located less than <1.25 MB from the seed sucrose 3-3 and seed oligosaccharide 2-3 [20,40].

3.6. Organ-Specific Expression of the Identified Candidate Genes

The expression pattern of the identified candidate genes was investigated in Williams 82 cv. using the RNA-seq data available in SoyBase [20]. The dataset includes several plant tissues, including leaves, nodules, roots, pods, and seeds (Figures 3A,B and S2). Four of the fifty-seven identified candidate genes have no available RNA-seq data, including the sucrose synthase candidate genes *Glyma.03G216300*, *Glyma.17G045800*, and *Glyma.19G212800*, as well as the UDP-D-glucose-4-epimerase candidate gene *Glyma.18G211700* (Figure S2). The raffinose synthase candidate gene *Glyma.04G145800* was not expressed in any of the analyzed tissues, whilst the rest of the identified genes showed different expression patterns.

The sucrose synthase candidate genes *Glyma.09G073600* and *Glyma.13G114000* presented a high expression profile in all the analyzed tissues except for the young leaves, while the raffinose synthase candidate gene *Glyma.17G111400* was abundantly expressed in all the analyzed tissues except for the seeds and nodules. Interestingly, the sucrose synthase candidate gene *Glyma.15G182600* was highly expressed in all the tissues excluding the young leaves and the nodules. The raffinose synthase candidate gene *Glyma.03G137900* was abundantly expressed in flowers, nodules, and roots. The raffinose synthase candidate gene *Glyma.14G010500* and the invertase candidate gene *Glyma.05G236600* were highly expressed in the flowers and pods. Also, the UDP-D-glucose-4-epimerase candidate gene *Glyma.05G204700* was abundantly expressed in the flowers and seeds. While the invertase candidate gene *Glyma.20G177200* was highly expressed in nodules and roots, the raffinose synthase candidate gene *Glyma.06G179200* was found to be highly expressed in seeds (Figures 3A and S2).

Seventeen of the identified candidate genes were situated less than 10 MB away from the identified QTL regions. *Glyma.09G073600* was highly expressed in seeds in Williams 82 cv., followed by *Glyma.17G111400*, *Glyma.17G035800*, and *Glyma.08G043800* with a moderated expression profile. The remaining genes had lower expression patterns, excluding the *Glyma.02G016700*, *Glyma.06G175500*, *Glyma.09G016600*, *Glyma.10G017300*, and *Glyma.19G004400* genes, which were not expressed in seeds in Williams 82 cv.

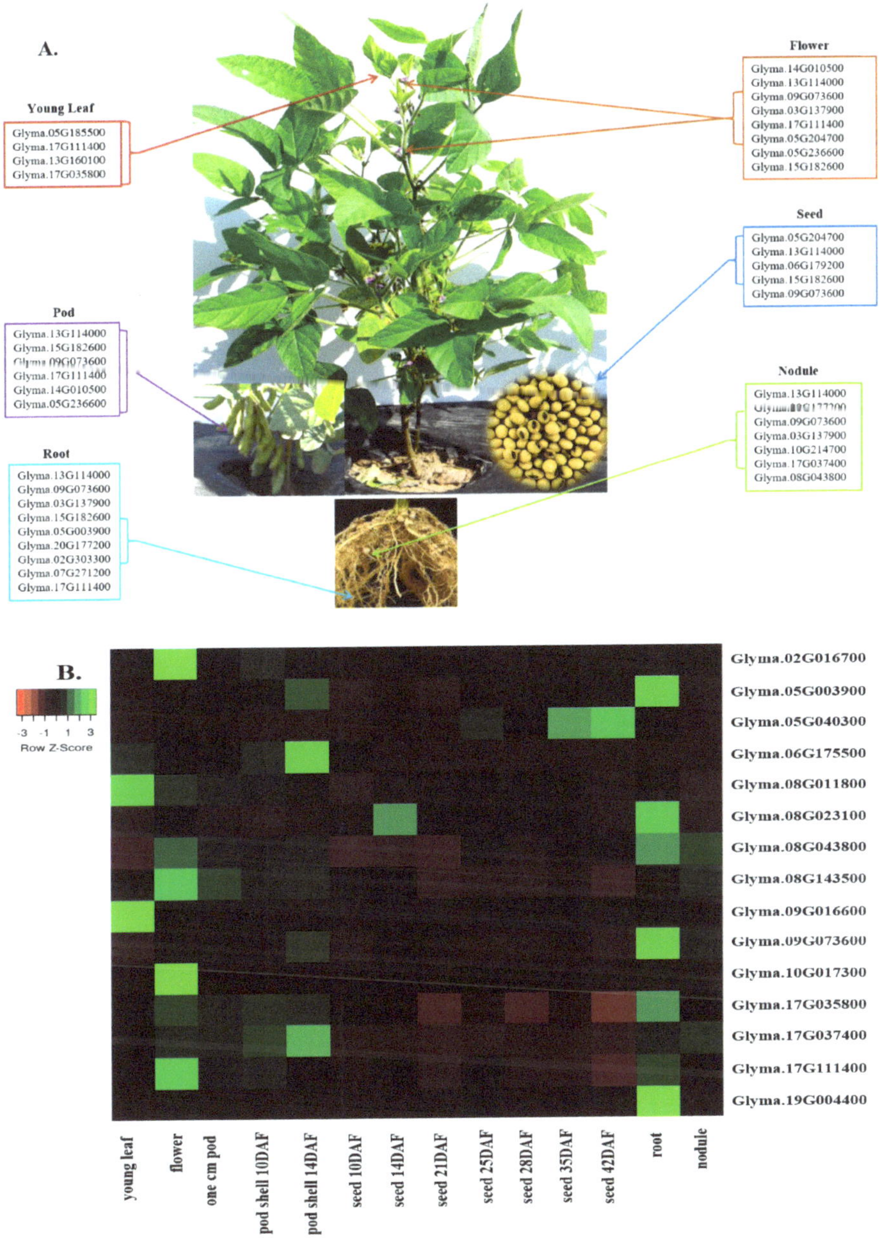

Figure 3. (**A**) Tissue-specific expression of the identified sugar candidate genes. (**B**) Expression HeatMap of the identified candidate genes located within 10 MB of the identified sugar QTL regions in Williams 82 (RPKM) were retrieved from publicly available RNA-seq data from the Soybase database [20]. RNA-seq data are not available in Soybase for the *Glyma.17G045800* candidate gene.

4. Comparison of the Williams 82 and Forrest Sequences

The sequences of the candidate genes located less than 10 MB from the identified QTLs were compared. The results showed that six of them had SNPs and InDels between the

Forrest and Williams 82 sequences: *Glyma.09G073600*, *Glyma.08G143500*, *Glyma.05G003900*, *Glyma.17G035800*, *Glyma.17G111400*, and *Glyma.09G016600* (Table S4, Figure 4).

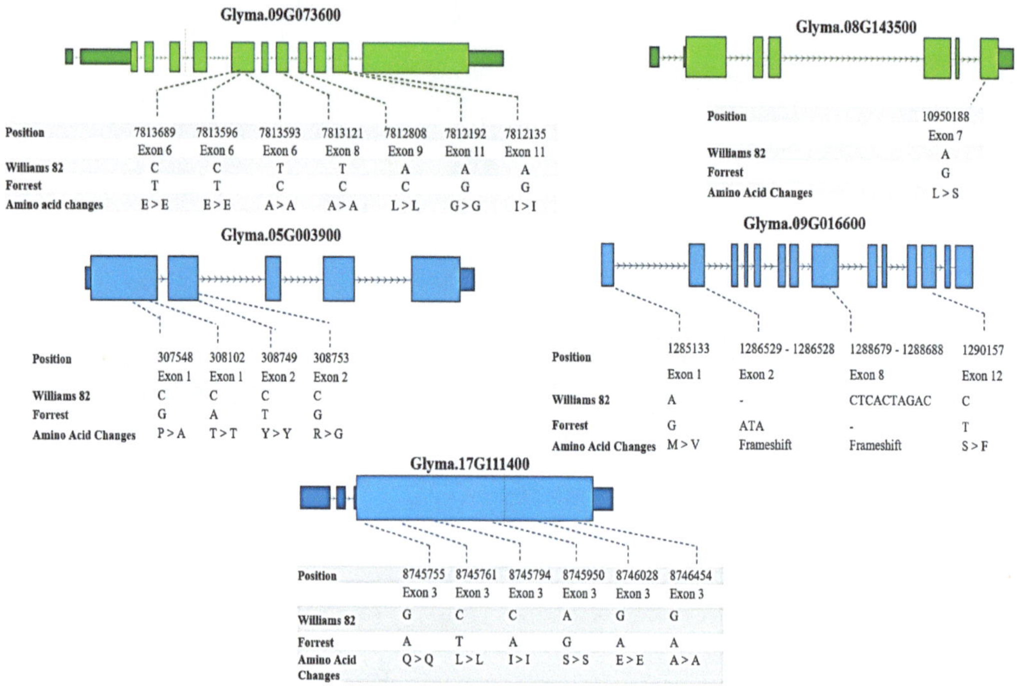

Figure 4. Positions of SNPs between Forrest and Williams 82 cultivars in *Glyma.09G073600*, *Glyma.08G143500*, *Glyma.05G003900*, *Glyma.17G111400*, and *Glyma.09G016600* coding sequences. In the gene model diagram, the light blue/light green boxes represent exons, blue/green bars represent introns, and dark blue/dark green boxes represent 3′UTR or 5′UTR. SNPs were positioned relative to the genomic position in the genome version W82.a2.

The sucrose synthase *Glyma.09G073600* had in total 28 SNPs and 7 InDels; three of these SNPs were located upstream of the 5′UTR, two are downstream of the 3′UTR, and seven were located in the exons (Table S4, Figure 4). For the invertase candidate gene *Glyma.08G143500*, there were 20 SNPs and 5 InDels. One of these SNPs was located in exon 7, causing a missense mutation, and two SNPs were located upstream of the 5′UTR (Table S4, Figure 4). The raffinose synthase candidate gene *Glyma.05G003900* had nine SNPs and one InDel; four of those SNPs were in the exons, from which two SNPs resulted in missense mutations (Table S4, Figure 4). Likewise, the raffinose synthase candidate gene *Glyma.09G016600* possessed 12 SNPs and 2 InDels. Amongst these SNPs, there were two located in exons, which resulted in missense mutations, in addition to the two InDels located in the exons (Table S4, Figure 4). For the raffinose candidate gene *Glyma.17G111400*, eight SNPs were found, of which one was located upstream of the 5′ UTR, another one was downstream of the 3′UTR, and the last six were in exons causing silent mutations (Table S4, Figure 4). Finally, the UDP-D-Glucose-4-Epimerase candidate gene *Glyma.17G035800* had two SNPs that were positioned in introns (Table S4).

5. Discussion

Soybean seed sugars play a major role in seed and fruit development. Recently, soy products became very popular as a result of a growing demand for vegan diets [45]. The quality and taste of these products are determined by the soybean seed sugar content [39].

These sugars include sucrose, raffinose, and stachyose which make up 5–7%, 1%, and 3–4% of total carbohydrates, respectively [5]. However, the raffinose and stachyose fermentation by human and monogastric animal intestine microbes leads to a reduced gastrointestinal performance, flatulence, and diarrhea. Thus, reducing raffinose and stachyose and increasing sucrose in soybean seed content are desirable [22,27].

Given the importance of the soybean seed sucrose content for the quality of soybean-based products for food and feed, breeding programs are focused on developing soybean seeds with a high sucrose content and low RFO content [43,46]. Thus, soybean varieties with a high sucrose content are valuable for soybean food and feed products [47].

The identification of QTLs associated with sugar components using different types of molecular markers is one of the breeding-process approaches that researchers use to breed for a high-sucrose soybean. In soybean and other crops, it is well established that seed sugar contents are complex polygenic traits, and many studies including this study have mapped QTLs for sugar contents using various mapping populations including biparental populations where parents may not necessarily have contrasting amounts of sugar contents, such as in the "MD96-5722" by "Spencer" RIL population [30].

In the current study, all seed sugar (sucrose, raffinose, and stachyose) phenotypic data, except one (sucrose, 2018), exhibited normal distributions in all environments studied (years and locations), showing that these traits are polygenic and complex, as shown previously [21,39–41,44,47–53].

The SNP-based genetic linkage map facilitated the identification of several QTLs including QTLs for seed isoflavone contents [28], seed tocopherol contents [29], and seed sugar (sucrose, stachyose, and raffinose) contents, as reported in the current study.

The heritability (H^2) of sucrose, stachyose, and raffinose was estimated to be 37.5%, 73.9%, and 92%, respectively. There is no doubt that the environment and the interactions of genotype and environment play a major role in the heritability of traits [28,29,43,54,55]. A trait biosynthesis that involves several genes is expected to have a lower heritability than a trait biosynthesis that involves fewer genes. Figure 2 shows the number of potential genes that are involved in sucrose biosynthesis versus those involved in raffinose and stachyose; it seems like there is a correlation between the heritability values and the number of genes involved in the biosynthesis pathway.

Among the identified sugar QTLs, there are novel QTL regions (qSUC-4, qSUC-8, and qSUC-11). All the other QTLs have been located within or very close to the previously reported sugar QTLs [30,39–41,44], as summarized in [18]. Five other genomic regions on chrs. 2, 6, 12, 16, and 19 harboring sugar QTLs either from this study or from other studies are of particular interest. On chr. 2, qSUC-2-NC-2018 may correspond to *suc 1-1* identified previously [39]. This QTL region contains the *Glyma.02G016700* candidate gene that encodes for invertase.

Interestingly, several QTLs have been identified previously, including a QTL that controls seed raffinose content within the qSUC-1-NC-2018 region (chr. 1) [30], two QTLs (suc 2-2 and suc 3-2) that control seed sucrose content within the qSUC-2-NC-2018 region (chr. 2) [20,40,41], a QTL that controls seed sucrose content (suc-001) within the qSUC-3-NC-2018 region (chr. 3), [30], 2 QTLs that control seed sucrose (suc 1-1 and suc 4-1) content within the qSUC-5-NC-2018 region (chr. 5) [39,44], a QTL that controls seed raffinose content (raf003 and raf004) within the qSUC-6-NC-2018 and qSUC-7-NC-2018 regions (chrs. 6 and 9) [30], a QTL that controls seed sucrose (suc 1-5) content within the qSUC-9-NC-2018 region (chr. 13) [39], and a QTL that controls seed sucrose (suc 1-4) content within the qSUC-12-NC-2018 region (chr. 20) [39].

Likewise, several other QTLs have been identified previously: a QTL that controls seed sucrose (suc 2-2, 3-2) content within the qSUC-1-IL-2020 region (chr. 2) [40,41], a QTL that control seed sucrose (suc 1-1, 4-1) content within the qSUC-2-IL-2020 (chr. 5) [39,44] and qSUC-3-IL-2020 (chr. 8) regions, and a QTL that control seed sucrose (suc 1-2, 1-3, 1-13) content within the qSUC-3-IL-2020 region (chr. 8) [39]. Within the QTL regions that were found to control seed stachyose contents (qSTA-1-IL-2020, qSTA-2-IL-2020, and qSTA-4-IL-

2020) reported in the current study on chrs. 13, 16, and 19, several QTLs that control seed sucrose (suc 1-4, 1-5, 3-5, 3-6) and seed raffinose (raff007) contents have been identified previously [39–41].

On chr. 6, qSUC-6-NC-2018 most likely corresponds to *suc 2-2* [41] and raffinose (*raf003*) QTL regions identified previously [30,39]. The QTL region contains the *Glyma.06G175500* candidate gene encoding for raffinose synthase. Interestingly, the genomic region on chr. 19 comprising a cluster of sucrose QTLs (suc 1-6 to 1-8, 2-3 to 2-11) [39,41] also contains two stachyose QTLs identified in this study (qSTA-3-NC-2018 and qSTA-4-NC-2018). The candidate gene *Glyma.19G004400*, which also encodes for raffinose synthase, was identified within this QTL region.

No candidate genes have been identified on chrs. 12 (qRAF-3-NC-2018), 16 (qSTA-2-NC-2018), or 20 (qSTA-4-NC-2018).

Remarkably, within the novel QTL regions reported here on chrs. 4, 10, and 18, seven candidate genes were identified, including the *Glyma.18G145700* encoding for UDP-D-glucose-4-epimerase on chr. 18 (Tables 5 and 6, and Figure 2).

Interestingly, five QTL regions were detected in both locations, IL and NC. The first QTL region contains qSUC-5-NC-2018 and qSUC-2-IL-2020, which were detected in the same location on chr. 5. Additionally, qSUC-9-NC-2018, qSTA-1-NC-2018, and qSTA-2-NC-2018 were located only 1 MB away from qSTA-1-IL-2020 on chr.13. Moreover, qSUC-12-NC-2018 was 1.3 MB away from qSTA-4-IL-2020 on chr. 20. Furthermore, qSUC-10-NC-2018 and qSTA-3-IL-2020 were positioned 3.1 MB away from each other on chr. 17. Additionally, qSUC-2-NC-2018 and qSUC-1-IL-2020 were located ~4 MB away on chr. 2. The QTL regions that were not detected in both locations may be affected by environmental conditions.

In a previous study [54], 31,245 SNPs and 323 soybean germplasm accessions grown in three different environments were used to identify 72 QTLs associated with individual sugars and 14 associated with total sugar [54]. In addition, ten candidate genes that are within the 100 Kb flanking regions of the lead SNPs in six chromosomes were significantly associated with sugar content in soybean, eight of which are involved in the sugar metabolism in soybean [54]. Amongst these candidate genes, the raffinose synthase gene *Glyma.05G003900* was also reported in this study.

A recent study used an RIL population from a cross of ZD27 by HF25 to identify 16 QTLs controlling seed sucrose and soluble sugar contents in soybean [43]. Amongst these QTLs, qSU1701 [43] with an LOD = 7.61 and phenotypic variation explained (PVE) = 16.8% was identified on chr. 17 to be associated with the seed sucrose content. This QTL region is collocated with qSUC-10-NC-2018 identified in this study for the same trait with an LOD = 33.2 and an R^2 = 20.5. On the same chr., qSS1701 [43] and qSS1702, identified to be associated with the seed soluble sugar content, are collocated with qSTA-3-IL-2020. These QTLs are positioned less than 8 MB away from a cluster of four genes involved in the sugars' pathway, including *Glyma.17G037400* encoding for invertase, *Glyma.17G045800* encoding for sucrose synthase, *Glyma.17G111400* encoding for raffinose synthase (showing 7 SNP variations in exons) (Figure 4), and *Glyma.17G035800* encoding for UDP-D-glucose-4-epimerase. Our results confirm that this region on chr. 17 is a major QTL associated with seed sugar contents in soybean. In the same study [43], qSU2001 identified on chr. 20 with LOD = 3.38 and PVE = 5.6% was collocated with qSUC-12-NC-2018, and it was 0.3 MB away from qSTA-4-IL-2020. The invertase candidate gene Glyma.20G177200 is positioned within qSU2002 [43] identified on chr. 20 with LOD = 7.9 and PVE = 14.4%. These results confirm that this region on chr. 20 is involved in soybean seed sugar contents. On chr. 3, qSS0301 was previously identified [43] to be associated with soluble sugar contents in soybean with an LOD = 5.2 and PVE = 11.8. This QTL is located 1.4 MB away from qSUC-3-NC-2018.

The sucrose synthase gene *Glyma.09G073600* was highly expressed in seeds, followed by *Glyma.17G111400*, *Glyma.17G035800*, and *Glyma.08G043800* with moderated expression patterns in seeds. *Glyma.09G073600* and *Glyma.09G016600* are located close to qSUC-7-IL-2018, qRAF-1-IL-2018, and qRAF-2-IL-2018 on chr. 9. *Glyma.08G143500* is located close to qSUC-3-NC-2020, and *Glyma.05G003900* is positioned close to qSUC-5-IL-2018 and qSUC-

2-NC-2020 on chr. 5. These genes could be useful in gene editing technology or breeding programs to develop soybean cultivars with reduced amounts of RFOs and high amounts of sucrose, which is beneficial for human consumption and animal feed.

Further studies are needed to characterize these genes, identify their enzymes and protein products, and understand their roles in the sugar biosynthetic pathway in soybean.

6. Conclusions

In summary, we have identified 26 QTLs associated with the seed sugar contents and 57 candidate genes involved in the sucrose, raffinose, and stachyose biosynthetic pathway. Amongst these candidate genes, 16 were located less than 10 MB away from the QTL regions identified in this study.

On chr. 17, a cluster of four genes controlling the sugar pathway is collocated within 6 MB of two QTLs (*qSUC-10-NC-2018* and *qSTA-3-IL-2020*) that were identified in this study. Moreover, the raffinose synthase candidate gene *Glyma.06G175500* is 9.7MB away from qSUC-6-NC 2018 QTL on chr. 6. The invertase candidate gene *Glyma.02G016700* is located 3.6 and 0.2 MB away from qSUC-1-NC-2018 (R^2 = 47.9) and *qSUC-1-IL 2020* (R^2 = 3.6), respectively, on chr. 2. Moreover, the sucrose synthase candidate gene *Glyma.09G073600* and the raffinose synthase candidate gene *Glyma.09G016600* were found close to qSUC-7-IL-2018, qRAF-1-IL-2018, qRAF-2-IL-2018, and qRAF-1-IL-2018 on chr. 9.

Five QTL regions were commonly identified in the two environments, NC and IL, on chrs. 2, 5, 13, 17 and 20 ((qSUC-5-NC-2018 and qSUC-2-IL-2020), (qSUC-9-NC-2018, qSTA-1-NC-2018, and qSTA-1-IL-2020), (qSUC-12-NC-2018 and qSTA-4-IL-2020), (qSUC-10-NC-2018 and qSTA-3-IL-2020), and (qSUC-2-NC-2018 and qSUC-1-IL-2020)).

Five genes (*Glyma.09G073600, Glyma.08G143500, Glyma.17G111400, Glyma.05G003900,* and *Glyma.09G016600*) have SNPs and InDels between the Forrest and Williams 82 sequences. These SNPs could potentially explain the difference in sugar content between Forrest and Williams 82 cultivars.

Further studies are required to functionally characterize these genes so we can understand and validate their roles in the sugar biosynthetic pathway in soybean before they are used in breeding programs to produce soybean lines with high beneficial sucrose and low RFOs.

Supplementary Materials: The following supporting information can be downloaded at: https://www.mdpi.com/article/10.3390/plants12193498/s1, Table S1: Quantitative trait loci (QTL) that control sugars (sucrose, stachyose, and raffinose) contents in F × W82 RIL population in Spring Lake, NC in 2018; Table S2: Quantitative trait loci (QTL) that control sugars (sucrose, stachyose, and raffinose) contents in F × W82 RIL population in Carbondale, IL in 2020; Table S3: Comparison of the Williams 82 and Forrest cv. Sequences of the Glyma.09G073600, Glyma.08G143500, Glyma.17G111400, Glyma.17G035800, Glyma.09G016600 and Glyma.05G003900 candidate genes; Figure S1: Positions of QTL that control seed sucrose (qSUC), stachyose (qSTA), and raffinose (qRAF) contents on Chrs; Figure S2: Expression profiles of the sugars (sucrose, raffinose, and stachyose) pathway candidate genes in soybean based on RNAseq data available from RNAsequencing data; Figure S3. Physical positions corresponding to the *Glyma.17G037400* encoding for an invertase, *Glyma.17G045800* encoding for sucrose synthase, *Glyma.17G111400* encoding for raffinose synthase, and *Glyma.17G035800* encoding for UDP-D-glucose-4-epimerase, and the identified seed sugars QTL identified in this study on chr. 17 are shown.

Author Contributions: Conceptualization, K.M. and M.A.K.; methodology, D.K., J.Y., T.V., N.L., A.M., E.A., N.B. and M.E.; validation, M.A.K., K.M. and H.T.N.; formal analysis, D.K., J.Y. and N.B.; investigation, K.M. and M.A.K.; resources and data curation, K.M., M.A.K. and H.T.N.; writing—original draft preparation, D.K., M.A.K. and K.M.; review and editing, D.K., J.Y., N.B., N.L., T.V., M.A.K., K.M. and H.T.N.; supervision, M.A.K. and K.M.; project administration, M.A.K., K.M., and H.T.N. All authors have read and agreed to the published version of the manuscript.

Funding: This research was partially funded by the U.S. Department of Agriculture, Agricultural Research Service Project 6066-21220-014-000D. This project was also partially funded by the United Soybean Board, project # 2220-152-0104, and Southern Illinois University at Carbondale.

Data Availability Statement: The data presented in this study are available upon request from the corresponding author.

Acknowledgments: Technical support provided by Sandra Mosley is appreciated. Mention of trade names or commercial products in this publication is solely for the purpose of providing specific information and does not imply recommendation or endorsement by the United States Department of Agriculture (USDA). The USDA is an equal opportunity provider and employer.

Conflicts of Interest: The authors declare no conflict of interest.

References

1. Blackman, S.A.; Obendorf, R.L.; Leopold, A.C. Maturation Proteins and Sugars in Desiccation Tolerance of Developing Soybean Seeds. *Plant Physiol.* **1992**, *100*, 225–230. [CrossRef] [PubMed]
2. Hitz, W.D.; Carlson, T.J.; Kerr, P.S.; Sebastian, S.A. Biochemical and Molecular Characterization of a Mutation That Confers a Decreased Raffinosaccharide and Phytic Acid Phenotype on Soybean Seeds. *Plant Physiol.* **2002**, *128*, 650–660. [CrossRef] [PubMed]
3. Koch, K. Sucrose metabolism: Regulatory mechanisms and pivotal roles in sugar sensing and plant development. *Curr. Opin. Plant Biol.* **2004**, *7*, 235–246. [CrossRef] [PubMed]
4. Redekar, N.R.; Glover, N.M.; Biyashev, R.M.; Ha, B.-K.; Raboy, V.; Maroof, M.A.S. Genetic interactions regulating seed phytate and oligosaccharides in soybean (*Glycine max* L.). *PLoS ONE* **2020**, *15*, e0235120. [CrossRef]
5. Skoneczka, J.A.; Maroof, M.A.S.; Shang, C.; Buss, G.R. Identification of Candidate Gene Mutation Associated With Low Stachyose Phenotype in Soybean Line PI200508. *Crop Sci.* **2009**, *49*, 247–255. [CrossRef]
6. Horbowicz, M.; Obendorf, R.L. Seed desiccation tolerance and storability: Dependence on flatulence-producing oligosaccharides and cyclitols—Review and survey. *Seed Sci. Res.* **1994**, *4*, 385–405.
7. Sprenger, R.; Schlagenhaufer, R.; Kerb, R.; Bruhn, C.; Brockmöller, J.; Roots, I.; Brinkmann, U. Characterization of the glutathione S-transferase GSTT1 deletion: Discrimination of all genotypes by polymerase chain reaction indicates a trimodular genotype–phenotype correlation. *Pharmacogenet. Genom.* **2000**, *10*, 557–565. [CrossRef]
8. Pennycooke, J.C.; Jones, M.L.; Stushnoff, C. Down-regulating α-galactosidase enhances freezing tolerance in transgenic petunia. *Plant Physiol.* **2003**, *133*, 901–909.
9. ElSayed, A.I.; Rafudeen, M.S.; Golldack, D. Physiological aspects of raffinose family oligosaccharides in plants: Protection against abiotic stress. *Plant Biol.* **2014**, *16*, 1–8.
10. Keller, F.; Pharr, D.M. Metabolism of carbohydrates in sinks and sources: Galactosyl-sucrose oligosaccharides. In *Photoassimilate Distribution in Plants and Crops*; Routledge: London, UK, 1996; pp. 157–183.
11. Xue, H.; Chen, X.; Li, G. Involvement of phospholipid signaling in plant growth and hormone effects. *Curr. Opin. Plant Biol.* **2007**, *10*, 483–489. [CrossRef]
12. Okada, M.; Ye, K. Nuclear phosphoinositide signaling regulates messenger RNA export. *RNA Biol.* **2009**, *6*, 12–16. [PubMed]
13. Thole, J.M.; Nielsen, E. Phosphoinositides in plants: Novel functions in membrane trafficking. *Curr. Opin. Plant Biol.* **2008**, *11*, 620–631. [PubMed]
14. González, E.M.; Gordon, A.J.; James, C.L.; Arrese-lgor, C. The role of sucrose synthase in the response of soybean nodules to drought. *J. Exp. Bot.* **1995**, *46*, 1515–1523. [CrossRef]
15. Bellaloui, N.; Smith, J.R.; Gillen, A.M.; Ray, J.D. Effect of Maturity on Seed Sugars as Measured on Near-Isogenic Soybean (*Glycine max*) Lines. *Crop Sci.* **2010**, *50*, 1978–1987. [CrossRef]
16. Bellaloui, N.; Mengistu, A.; Fisher, D.K.; Abel, C.A. Soybean Seed Composition Constituents as Affected by Drought and Phomopsis in Phomopsis Susceptible and Resistant Genotypes. *J. Crop Improv.* **2012**, *26*, 428–453. [CrossRef]
17. Wijewardana, C.; Reddy, K.R.; Bellaloui, N. Soybean seed physiology, quality, and chemical composition under soil moisture stress. *Food Chem.* **2019**, *278*, 92–100. [CrossRef]
18. Kassem, M.A. *Soybean Seed Composition: Protein, Oil, Fatty Acids, Amino Acids, Sugars, Mineral Nutrients, Tocopherols, and Isoflavones*; Springer Nature: Berlin/Heidelberg, Germany, 2021.
19. Taiz, L. *Mineral Nutrition*, Plant Physiology ed.; Sinauer Associates Inc.: Sunderland, MA, USA, 1998.
20. Brown, A.V.; Conners, S.I.; Huang, W.; Wilkey, A.P.; Grant, D.; Weeks, N.T.; Cannon, S.B.; Graham, M.A.; Nelson, R.T. A new decade and new data at SoyBase, the USDA-ARS soybean genetics and genomics database. *Nucleic Acids Res.* **2021**, *49*, D1496–D1501. [CrossRef]
21. Salari, M.W.; Ongom, P.O.; Thapa, R.; Nguyen, H.T.; Vuong, T.D.; Rainey, K.M. Mapping QTL controlling soybean seed sucrose and oligosaccharides in a single family of soybean nested association mapping (SoyNAM) population. *Plant Breed.* **2021**, *140*, 110–122. [CrossRef]
22. Wang, T.L.; Domoney, C.; Hedley, C.L.; Casey, R.; Grusak, M.A. Can we improve the nutritional quality of legume seeds? *Plant Physiol.* **2003**, *131*, 886–891.
23. Arendt, E.K.; Zannini, E. *Cereal Grains for the Food and Beverage Industries*; Elsevier: Amsterdam, The Netherlands, 2013.
24. Avilés-Gaxiola, S.; Chuck-Hernández, C.; Serna Saldívar, S.O. Inactivation methods of trypsin inhibitor in legumes: A review. *J. Food Sci.* **2018**, *83*, 17–29.

25. Kerr, P.S.; Pearlstein, R.W.; Schweiger, B.J.; Becker-Manley, M.F.; Pierce, J.W. Nucleotide Sequences of Galactinol Synthase from Zucchini and Soybean. U.S. Patent US5648210A, 15 July 1997.
26. Frías, J.; Bakhsh, A.; Jones, D.; Arthur, A.; Vidal-Valverde, C.; Rhodes, M.; Hedley, C.L. Genetic analysis of the raffinose oligosaccharide pathway in lentil seeds. *J. Exp. Bot.* **1999**, *50*, 469–476. [CrossRef]
27. Hedley, C.L. *Carbohydrates in Grain Legume Seeds: Improving Nutritional Quality and Agronomic Characteristics*; CABI: Wallingford, UK, 2001.
28. Knizia, D.; Yuan, J.; Bellaloui, N.; Vuong, T.; Usovsky, M.; Song, Q.; Betts, F.; Register, T.; Williams, E.; Lakhssassi, N. The Soybean High Density 'Forrest'by 'Williams 82' SNP-Based Genetic Linkage Map Identifies QTL and Candidate Genes for Seed Isoflavone Content. *Plants* **2021**, *10*, 2029. [CrossRef] [PubMed]
29. Knizia, D.; Yuan, J.; Lakhssassi, N.; El Baze, A.; Cullen, M.; Vuong, T.; Mazouz, H.; Nguyen, H.; Kassem, M.A.; Meksem, K. QTL and Candidate Genes for Seed Tocopherol Content in 'Forrest'by 'Williams 82' Recombinant Inbred Line (RIL) Population of Soybean. *Plants* **2022**, *11*, 1258. [CrossRef] [PubMed]
30. Akond, M.; Liu, S.; Kantartzi, S.K.; Meksem, K.; Bellaloui, N.; Lightfoot, D.A.; Kassem, M.A. Quantitative trait loci underlying seed sugars content in "MD96-5722" by "Spencer" recombinant inbred line population of soybean. *Food Nutr. Sci.* **2015**, *6*, 964. [CrossRef]
31. Allen, G.C.; Flores-Vergara, M.; Kracynanski, S.; Kumar, S.; Thompson, W. A modified protocol for rapid DNA isolation from plant tissues using cetyltrimethylammonium bromide. *Nat. Protoc.* **2006**, *1*, 2320–2325.
32. Wu, X.; Vuong, T.D.; Leroy, J.A.; Grover Shannon, J.; Sleper, D.A.; Nguyen, H.T. Selection of a core set of RILs from Forrest× Williams 82 to develop a framework map in soybean. *Theor. Appl. Genet.* **2011**, *122*, 1179–1187. [CrossRef]
33. Wang, S.; Basten, C.; Zeng, Z. Windows QTL Cartographer 2.5_011. Department of Statistics, North Carolina State University: Raleigh, NC, USA, 2012. Available online: http://statgen.ncsu.edu/qtlcart/WQTLCart.htm (accessed on 28 August 2023).
34. Voorrips, R. MapChart: Software for the graphical presentation of linkage maps and QTLs. *J. Hered.* **2002**, *93*, 77–78. [CrossRef]
35. Patil, G.B.; Lakhssassi, N.; Wan, J.; Song, L.; Zhou, Z.; Klepadlo, M.; Vuong, T.D.; Stec, A.O.; Kahil, S.S.; Colantonio, V.; et al. Whole-genome re-sequencing reveals the impact of the interaction of copy number variants of the rhg1 and Rhg4 genes on broad-based resistance to soybean cyst nematode. *Plant Biotechnol. J.* **2019**, *17*, 1595–1611. [CrossRef]
36. Iftime, D.; Hannah, M.A.; Peterbauer, T.; Heyer, A.G. Stachyose in the cytosol does not influence freezing tolerance of transgenic Arabidopsis expressing stachyose synthase from adzuki bean. *Plant Sci.* **2011**, *180*, 24–30. [CrossRef]
37. González-Morales, S.I.; Chávez-Montes, R.A.; Hayano-Kanashiro, C.; Alejo-Jacuinde, G.; Rico-Cambron, T.Y.; de Folter, S.; Herrera-Estrella, L. Regulatory network analysis reveals novel regulators of seed desiccation tolerance in *Arabidopsis thaliana*. *Proc. Natl. Acad. Sci. USA* **2016**, *113*, E5232–E5241. [CrossRef]
38. Blöchl, A.; March, G.G.-d.; Sourdioux, M.; Peterbauer, T.; Richter, A. Induction of raffinose oligosaccharide biosynthesis by abscisic acid in somatic embryos of alfalfa (*Medicago sativa* L.). *Plant Sci.* **2005**, *168*, 1075–1082. [CrossRef]
39. Maughan, P.; Maroof, M.; Buss, G. Identification of quantitative trait loci controlling sucrose content in soybean (*Glycine max*). *Mol. Breed.* **2000**, *6*, 105–111. [CrossRef]
40. Kim, H.K.; Kang, S.T.; Oh, K.W. Mapping of putative quantitative trait loci controlling the total oligosaccharide and sucrose content of Glycine max seeds. *J. Plant Res.* **2006**, *119*, 533–538. [CrossRef] [PubMed]
41. Kim, H.-K.; Kang, S.-T.; Cho, J.-H.; Choung, M.-G.; Suh, D.-Y. Quantitative trait loci associated with oligosaccharide and sucrose contents in soybean (*Glycine max* L.). *J. Plant Biol.* **2005**, *48*, 106–112.
42. Mainali, H.R.; Vadivel, A.K.A.; Li, X.; Gijzen, M.; Dhaubhadel, S. Soybean cyclophilin GmCYP1 interacts with an isoflavonoid regulator GmMYB176. *Sci. Rep.* **2017**, *7*, 39550.
43. Liu, C.; Chen, H.; Yu, Q.; Gu, H.; Li, Y.; Tu, B.; Zhang, H.; Zhang, Q.; Liu, X. Identification of quantitative trait loci (QTLs) and candidate genes for seed sucrose and soluble sugar concentrations in soybean. *Crop Sci.* **2023**, *63*, 2976–2992. [CrossRef]
44. Zeng, A.; Chen, P.; Shi, A.; Wang, D.; Zhang, B.; Orazaly, M.; Florez-Palacios, L.; Brye, K.; Song, Q.; Cregan, P. Identification of quantitative trait loci for sucrose content in soybean seed. *Crop Sci.* **2014**, *54*, 554–564.
45. Cai, J.-S.; Feng, J.-Y.; Ni, Z.-J.; Ma, R.-H.; Thakur, K.; Wang, S.; Hu, F.; Zhang, J.-G.; Wei, Z.-J. An update on the nutritional, functional, sensory characteristics of soy products, and applications of new processing strategies. *Trends Food Sci. Technol.* **2021**, *112*, 676–689.
46. Sui, M.; Wang, Y.; Bao, Y.; Wang, X.; Li, R.; Lv, Y.; Yan, M.; Quan, C.; Li, C.; Teng, W. Genome-wide association analysis of sucrose concentration in soybean (*Glycine max* L.) seed based on high-throughput sequencing. *Plant Genome* **2020**, *13*, e20059.
47. Lee, J.S.; Kim, S.-M.; Kang, S. Fine mapping of quantitative trait loci for sucrose and oligosaccharide contents in soybean [Glycine max (L.) Merr.] using 180 K Axiom® SoyaSNP genotyping platform. *Euphytica* **2016**, *208*, 195–203.
48. Stombaugh, S.; Orf, J.H.; Jung, H.; Chase, K.; Lark, K.; Somers, D. Quantitative trait loci associated with cell wall polysaccharides in soybean seed. *Crop Sci.* **2004**, *44*, 2101–2106. [CrossRef]
49. Feng, C.; Morsy, M.; Giannoccaro, E.; Zhang, B.; Chen, P. Soybean seed sugar content and quantitative trait loci mapping. In *Plant Nutrition for Food Security, Human Health and Environmental Protection*; Fifteenth International Plant Nutrition Colloquium; Tsinghua University Press: Beijing, China, 2005.
50. Jaureguy, L.M. *Identification of Molecular Markers Associated with Seed Size, Protein and Sugar Content in Soybean*; University of Arkansas: Fayetteville, Arkansas, 2009.

51. Wang, X.; Jiang, G.-L.; Green, M.; Scott, R.A.; Song, Q.; Hyten, D.L.; Cregan, P.B. Identification and validation of quantitative trait loci for seed yield, oil and protein contents in two recombinant inbred line populations of soybean. *Mol. Genet. Genom.* **2014**, *289*, 935–949.
52. Dhungana, S.K.; Kulkarni, K.P.; Park, C.W.; Jo, H.; Song, J.T.; Shin, D.H.; Lee, J.D. Mapping quantitative trait loci controlling soybean seed starch content in an interspecific cross of 'Williams 82' (*Glycine max*) and 'PI 366121' (*Glycine soja*). *Plant Breed.* **2017**, *136*, 379–385.
53. Patil, G.; Vuong, T.D.; Kale, S.; Valliyodan, B.; Deshmukh, R.; Zhu, C.; Wu, X.; Bai, Y.; Yungbluth, D.; Lu, F. Dissecting genomic hotspots underlying seed protein, oil, and sucrose content in an interspecific mapping population of soybean using high-density linkage mapping. *Plant Biotechnol. J.* **2018**, *16*, 1939–1953. [PubMed]
54. Hu, L.; Wang, X.; Zhang, J.; Florez-Palacios, L.; Song, Q.; Jiang, G.-L. Genome-Wide Detection of Quantitative Trait Loci and Prediction of Candidate Genes for Seed Sugar Composition in Early Mature Soybean. *Int. J. Mol. Sci.* **2023**, *24*, 3167.
55. Silva, L.C.C.; da Matta, L.B.; Pereira, G.R.; Bueno, R.D.; Piovesan, N.D.; Cardinal, A.J.; God, P.I.V.G.; Ribeiro, C.; Dal-Bianco, M. Association studies and QTL mapping for soybean oil content and composition. *Euphytica* **2021**, *217*, 24. [CrossRef]

plants

MDPI

Article

Identification of Novel QTLs Associated with Frost Tolerance in Winter Wheat (*Triticum aestivum* L.)

Parisa Bolouri [1], Kamil Haliloğlu [1,*], Seyyed Abolghasem Mohammadi [2], Aras Türkoğlu [3], Emre İlhan [4], Gniewko Niedbała [5,*], Piotr Szulc [6] and Mohsen Niazian [7,*]

1 Department of Field Crops, Faculty of Agriculture, Ataturk University, 25240 Erzurum, Turkey
2 Department of Plant Breeding and Biotechnology, Faculty of Agriculture, University of Tabriz, Tabriz 5166616471, Iran
3 Department of Field Crops, Faculty of Agriculture, Necmettin Erbakan University, 42310 Konya, Turkey
4 Department of Molecular Biology and Genetics, Erzurum Technical University, 25240 Erzurum, Turkey
5 Department of Biosystems Engineering, Faculty of Environmental and Mechanical Engineering, Poznań University of Life Sciences, Wojska Polskiego 50, 60-627 Poznań, Poland
6 Department of Agronomy, Poznań University of Life Sciences, Dojazd 11, 60-632 Poznań, Poland
7 Field and Horticultural Crops Research Department, Kurdistan Agricultural and Natural Resources Research and Education Center, Agricultural Research, Education and Extension Organization (AREEO), Sanandaj 6616936311, Iran
* Correspondence: kamilh@atauni.edu.tr (K.H.); gniewko.niedbala@up.poznan.pl (G.N.); mniazian@ut.ac.ir (M.N.)

Abstract: Low temperature (cold) and freezing stress is a major problem during winter wheat growth. Low temperature tolerance (LT) is an important agronomic trait in winter wheat and determines the plants' ability to cope with below-freezing temperatures; thus, the development of cold-tolerant cultivars has become a major goal of breeding in various regions of the world. In this study, we sought to identify quantitative trait loci (QTL) using molecular markers related to freezing tolerance in winter. Thirty-four polymorphic markers among 425 SSR markers were obtained for the population, including 180 inbred lines of F_{12} generation wheat, derived from crosses (Norstar × Zagros) after testing with parents. LT_{50} is used as an effective selection criterion for identifying frost-tolerance genotypes. The progeny of individual F_{12} plants were used to evaluate LT50. Several QTLs related to wheat yield, including heading time period, 1000-seed weight, and number of surviving plants after overwintering, were identified. Single-marker analysis illustrated that four SSR markers with a total of 25% phenotypic variance determination were linked to LT50. Related QTLs were located on chromosomes 4A, 2B, and 3B. Common QTLs identified in two cropping seasons based on agronomical traits were two QTLs for heading time period, one QTL for 1000-seed weight, and six QTLs for number of surviving plants after overwintering. The four markers identified linked to LT_{50} significantly affected both LT_{50} and yield-related traits simultaneously. This is the first report to identify a major-effect QTL related to frost tolerance on chromosome 4A by the marker XGWM160. It is possible that some QTLs are closely related to pleiotropic effects that control two or more traits simultaneously, and this feature can be used as a factor to select frost-resistant lines in plant breeding programs.

Keywords: winter wheat; frost tolerance; LT_{50}; chromosome 4A; SSR marker; yield-related traits

Citation: Bolouri, P.; Haliloğlu, K.; Mohammadi, S.A.; Türkoğlu, A.; İlhan, E.; Niedbała, G.; Szulc, P.; Niazian, M. Identification of Novel QTLs Associated with Frost Tolerance in Winter Wheat (*Triticum aestivum* L.). *Plants* **2023**, *12*, 1641. https://doi.org/10.3390/plants12081641

Academic Editor: Abdelmajid Kassem

Received: 12 February 2023
Revised: 5 April 2023
Accepted: 7 April 2023
Published: 13 April 2023

1. Introduction

Winter wheat (*Triticum aestivum* L., 2n = 6x = 42, AABBDD) is a naturally formed allohexaploid species with seven groups of homoeologous chromosomes [1]. In Turkey, 49.37% of the total seed cultivated is wheat. As seed farming is relatively easy and suitable for mechanization, farmers often choose to cultivate these crops. According to TUIK data, Turkey's wheat cultivation area constitutes 3.2% of the world wheat cultivation area as of the 2019–2020 production season [2]. Temperature is an important environmental factor that affects wheat production [1]. Cold temperatures and frost are fundamental non-biological factors that reduce wheat production worldwide. Cold tolerance is a complex trait in wheat and includes morphological, physiological, biological, and hereditary elements [3]. Such stressors are frequent during the life of the plant and can reduce the yield of agricultural products [4,5].

The most important stage of plant development is the flowering induction stage, which occurs in many plant species based on the response to seasonal changes caused by the surrounding environment. The mechanism of cold tolerance in winter wheat has a fundamental relationship with the need for vernalization, which causes a delay in the transition from the vegetative to the reproductive stage; the result is cold tolerance [5].

Abiotic stresses such as frost stress are complex quantitative traits, where numerous stress-responsive genes take part to ensure the survival of plants [4]. It is possible that in a wide range of plant species, such responses are controlled by quantitative trait loci (QTL) [6]. The ability of plants to survive frost temperatures is critical for long-term survival. Although field survival is the ultimate measure of winter hardiness of a variety, field survival as a selection tool is unreliable because of variable levels of winter severity on different winter crops [7]. Recently, analyses of plants grown in fields and natural environments have revealed that most frost-responsive genes detected in the laboratory are also responsive to frost stress in these environments [8].

LT_{50}, the temperature predicted to be fatal to 50% of the plants, is typically used to quantify the frost tolerance of winter wheat cultivars. The way in which LT_{50} is measured varies among researchers [9]. Livingston [10] studied frost tolerance in different plant species (rye, wheat, barley, oat) and tested the plants after freezing. For this purpose, the frost-tolerance genotypes of barley and oats were subjected to −4, −7, −9, and −12 °C, and the more tolerant genotypes to −14, −16, and −18 °C. Plants were kept at each temperature for 2 h. Mahfoozi et al. [11] illustrated that the level of expression as well as the number of expressed proteins in plants grown in field conditions that experience cold acclimatization periods at sub-zero temperatures show much higher tolerance than plants acclimated to cold at low but above-zero temperatures (in controlled conditions).

To study different winter wheat cultivars, several biochemical, physiological, and morphological traits, such as plant height, heading time period, nutrient content, sucrose, glucose, raffinose, total sugar content, and the relationship between frost resistance genes, were investigated [12]. Studies have been conducted on protein factors such as transcription factors and protein kinases, which play a role in stress response and further regulation of gene expression [13]. Moreover, phenological, molecular, and metabolic analyses during vernalization have shown that there is a close relationship between the completion time of vernalization, the reduction of accumulation of metabolites, and the expression of frost-induced proteins [14].

The relationship between cold adaptation (acclimation) and other physiological events is important as cereals become tolerant to colder temperatures as a result of acclimatizing to cold exposure. In addition, adaptation activates different gene expression pathways [15]. Cold-activated genes may also be present in more than one gene locus. Response of plants to abiotic stress occurs via signal transduction [16]. Hannah et al. [17] reported that cold adaptation is the result of an increase or decrease in the expression of many genes. Thus, the controlling factor for frost tolerance depends on genetic (evolutionary) changes between vegetative and reproductive growth periods [18]. Therefore, the reaction of plants to low temperatures is primarily through activating metabolic pathways, followed by cell recognition (detection) and activation of genes that react to frost stress [19].

Wheat has both winter and spring growth phases, which are determined via vernalization (VRN) genes [20]. The genes involved in vernalization are located on the fifth chromosome and include homologous Vrn-A1, Vrn-A2, and Vrn-A3 genes [21]. The dominant or recessive alleles per locus lead to tolerance to frost or low temperatures in wheat with a winter growth type [18]. Freezing tolerance (Fr), which is linked to Vrn alleles, is an effective factor in gene stability. Fr-A_{11}, also known as Fr loci, are located on group

5 chromosomes in the loci of Fr-B_1 and Fr-D_1 [22,23]. Another locus, Fr-A_2, is located approximately 30 to 40 centimorgan (cM) to the vernalization locus Vrn-A_1 [24,25] and is also located on the fifth chromosome of *T. monococcum* [26]. According to Baga et al. [24] and Börner et al. [27], genetic mapping studies have shown which Fr and VRN genes are autonomous loci, and these loci are the main sources of variations observed in freezing tolerance [21,23]. Recent research has shown that allelic variations in the Vrn-A_1 locus significantly affect freezing tolerance. This QTL region associated with Fr-A_1 for freezing tolerance is due to the pleiotropic effect of Vrn-A_1 rather than being entirely dependent on the Fr-A_1 gene [19,28] and is located on the group 5 chromosomes [18]. Shindo et al. [29] also observed the importance of multiple locations on chromosome 2B, which controls heading time in wheat. In addition to the importance of chromosomes of the B genome in terms of frost tolerance, several studies related to the association mapping of wheat indicated that there were significant relationships between genetic markers and other agronomic traits, such as heading time, on the chromosomes of the B genome. Chromosome 4A is significantly associated with plant height, heading time, number of seeds per spike, spike length, spikelet number, and 1000-seed weight [24,30,31].

The selection of complex genetic traits, such as frost tolerance, can be simplified in plant breeding programs when associated markers are identified [32]. QTL and DNA markers, which are complex properties, can be used for indirect selection in selection programs with the aid of a marker [33]. Most phenotypes are quantitative in nature, and thus, significant variation for a trait of interest may be assigned to one or more loci (QTL). Identification and validation of QTLs requires associating them with one or more molecular markers. Knowing the location and the number of loci-wrapped traits, such as frost tolerance, increases the efficiency of selection of such agronomic traits. In recent years, molecular markers have been used as a useful complement to classical breeding techniques in the selection of quantitative traits, such as freezing tolerance in wheat. The aim of the present study is to identify gene or gene regions (QTLs) related to frost tolerance by SSR markers. QTLs linked to frost tolerance traits can be used in plant breeding programs for winter wheat.

2. Results

2.1. Evaluation of Yield-Related Traits

The mean time period from planting to 50% heading time in Zagros and Norstar was 181 and 195 days, respectively. For the F_{12} population, it was 185 days. The mean 1000-seed weight in Zagros and Norstar was 40.34 and 40.73 g, respectively. For the F_{12} population, the mean was 40.16 ± 0.28 g. The number of surviving plants after overwintering in Zagros and Norstar was 37.25 and 46.75 seedlings, respectively. The mean number of surviving plants after overwintering for the F_{12} population was 44.85 seedlings. Statistically significant correlations between frost tolerance (LT_{50}) were observed with heading time period (r = 0.154 *) and number of surviving plants after winter (r = 0.66). In addition, significant negative correlations were observed between winter survival and heading time period (r = −0.13) and 1000-seed weight (r = −0.223) (Table 1).

Table 1. Correlation between studied agronomic traits.

	Heading Time Period (HT)	1000-Seed Weight (1000—SW)	Number of Surviving Plants after Winter (WS)	LT_{50}
HT	1			
1000—SW	−0.097	1		
WS	−0.130	−0.223 **[1]	1	
LT_{50}	0.154 *[2]	0.084	0.66 *	1

[1,2] Significant at ** p = 0.01 and * p = 0.05 levels, respectively.

2.2. Evaluation of LT_{50} Values

The F_{12} population and their parents were examined using the freezing test, and LT_{50} values of the population varied in thermal range of −3 °C to −25 °C (ST1). The LT_{50} values of Zagros (Z) and Norstar (N) parents were determined as −4.5 °C and −24.34 °C, respectively. The Norstar variety showed maximum frost tolerance, with an LT_{50} value at −24.34 °C. The mean LT_{50} value for the F_{12} population showed a distribution close to the frost-sensitive Zagros parent; the mean LT_{50} value for the F_{12} population was −8.94 °C (Table 2) and was considered as the frost tolerance standard. In general, in the RIL population, 82 (45.5%) lines showed high tolerance, and 98 lines (54.44%) were susceptible to frost.

Table 2. The mean LT_{50} values in the recombinant lines (RILs) derived from Norstar × Zagros wheat cross.

LT_{50} (°C)	−1.5	−4.5	−7.5	−10.5	−13.5	−16.5	−19.5	−24.34
Genotype number	5	51	43	40	25	15	2	1

[1,2] Significant at ** $p = 0.01$ and * $p = 0.05$ levels, respectively.

2.3. Molecular Evaluation (Single Marker Analysis)

Thirty-four SSR markers were identified to be polymorphic between parents and thus were used for single-marker analysis [34]. Four QTL regions were found for the LT_{50} value (Table 3). The QTL regions were associated with XBarc101, XGWM340, XGWM160, and XGWM493 markers. In terms of variation related to the LT_{50} value, XGWM160 had the highest phenotypic variation (12%), located on chromosome 4A and linked to LT_{50}. The other three QTL regions expressed phenotypic variation from 2% to 7% (Table 3). These QTL regions were also located on chromosomes 2B and 3B, which were linked to LT_{50}. Additionally, QTL analysis (SMA) for yield-related traits showed that two marker loci (XGWM413 and XGWM165) were located on 1B and 5AL-12-~10 chromosomes, respectively, which were related to the heading time period. One marker locus (XGWM160) on 4A was associated with the 1000-seed weight. Six marker loci (XGWM340 on 3B, XBarc 154 on 7A, XBarc100 on 5AL-12-~10, XGWM501 on 2B, XGWM160 on 4A, and XWMC765 on 5D chromosomes) were associated with number of surviving plants after overwintering in the F_{12} population (Table 4).

Table 3. QTL analyses of SSR markers for LT_{50} in the recombinant lines (RILs) derived from Norstar × Zagros wheat cross.

Characteristic	Marker	Location	* %PV	% *p*-Value
LT_{50}	XBarc 101	2BL	7	0.002
	XGWM340	3B	2	0.036
	XGWM160	4A	12	0.000
	XGWM493	3B	4	0.004

* % PV: Phenotypic Variation.

Table 4. QTL analyses for the number of yield-related traits in the recombinant lines (RILs) derived from Norstar × Zagros wheat cross.

Characteristic	Marker	Location	% Phenotypic Variation (PV)	% *p*-Value
Heading time period	XBarc165	5AL-12~10	2	0.032
	XGWM413	1B	6	0.000
1000-seed weight	XGWM160	4A	16	0.000

Table 4. *Cont.*

Characteristic	Marker	Location	% Phenotypic Variation (PV)	% *p*-Value
Number of surviving plants after winter	XBarc 154	7A, D	5	0.011
	XBarc100	5AL-12~1	2	0.033
	XGWM501	2B	3	0.022
	XGWM340	3B	9	0.000
	XGWM160	4A	12	0.000
	XWMC765	5D	5	0.003

3. Discussion

Considering that low-temperature stress in late spring is a serious threat to winter wheat production, frost tolerance is one of the most important traits for wheat breeding programs. This trait is complex and controlled by QTLs. Although challenging, identification of these QTLs will greatly benefit agricultural development. Therefore, considering the importance of yield-related traits in winter wheat, such as heading time period, 1000-seed weight, and number of surviving plants after overwintering, as well as their effect on yield, many studies have focused on identifying the QTLs associated with these traits and characterizing the molecular control of these traits and their role in frost tolerance.

Different organs of winter wheat are different in terms of resistance to low-temperature stress, among which the leaves are the most sensitive to low-temperature stress. Although low-temperature stress during elongation and booting stages causes great damage to the young ear, reduces the number of distinct florets, and accelerates the degeneration of florets, it seems that genotypes with a longer heading time period are less damaged by frost stress. Because the biomass transferred to the sink organ decreases less in these genotypes, as a result, it does not reduce the grain yield [32,35–38]. Our studies revealed that there is a significant correlation between frost tolerance (LT50) and heading time period in the F12 population (r = 0.154 *). However, significant positive correlations ($p < 0.05$) were exhibited between heading time period and LT50, indicating a high response to selection of these traits (Table 1). Therefore, lines with a larger heading time period were also more tolerant to low temperatures and frost. Several QTLs with pleiotropic effects are involved in controlling this trait. There may be genetic linkage between frost tolerance and late maturity (winter growth habit) [32,35–38]. Our results indicate that with an increasing period of heading time, the expression of genes related to frost tolerance strongly increases. Heo et al. [39] reported that transcription factors are better expressed with extended periods of heading time and could be promising candidates for identifying the molecular mechanisms and fitness of freezing tolerance. In addition, the shortening of the day length in autumn leads to the induction of the FT1/VRN3 gene upstream of the key gene of springing VRN1 [40]. Downregulation of Cor/Lea (cold-responsive or cold-regulated/Late-embryogenesis-abundant) genes and frost tolerance under controlled conditions was reported by Fowler et al. [9]. There is strong evidence that the Cor/Lea gene can contribute to freezing tolerance [41].

Low temperature (LT) represents a critical environmental factor in determining winter survival (WS) of winter wheat species. This means that during the acclimation process, highly tolerant varieties accumulate dehydrin proteins and transcripts earlier and at a higher level than less tolerant varieties.

We identified four QTL linked to LT_{50} that controlled frost tolerance on three wheat chromosomes (B3, 2BL, and 4A), which described a total of 25% of phenotypic variation for LT50. In addition, a QTL identified on chromosome 4A had a major effect on LT_{50}, and this accounted for 12% of the total phenotypic variance. For this purpose, it is possible to use QTLs with major effects linked to LT_{50} with a *p*-Value close to zero for selection in breeding programs. The segregation ratio of the percentage of tolerant F12 lines in the population shows that the resistance may be conditioned by more than one gene. Baga et al. [24] reported the importance of QTL regions on chromosomes 1D, 2A, 2B, 6D, and 7B in frost tolerance. However, the results of other studies indicate the existence of QTLs linked to frost resistance on chromosomes 5B, 5D, 5A, 2D, 2A, and 4B [40,42,43]. A previous study conducted by Sutka [44] emphasized the importance of QTLs linked to frost tolerance on chromosome 2B, which is consistent with our results. Traits related to yield, such as the number of surviving plants and frost resistance, were identified in chromosome 2B in our results. On the other hand, Kruse et al. [45] identified one QTL on chromosomes 5A and 4B associated with freezing tolerance by using 155 recombinant inbred lines with 663 molecular markers in $F_{2:5}$ lines in bread wheat. Numerous studies have indicated the location of the freezing tolerance genes mapped to homologous 5th group chromosomes [23,32,44], showing the importance of chromosome 5A in frost tolerance [22,32,44,46]. Previously, the study by Ballesta et al. [47] found that at least one of the 175 SNP markers was related to the drought tolerance index, which explained up to 6% of the phenotypic changes. Forty-five SNPs were associated with more than one tolerance index (up to four agronomic traits). Most linkages were located on chromosome 4A, supporting the hypothesis that this chromosome plays a key role in drought tolerance and should be used for wheat improvement. In the present study, our results show for first time that among QTLs linked to frost tolerance, a major effect QTL (12%) was identified with the aid of GWM 160 SSR marker on chromosome 4A, which indicates the importance of this locus on chromosome 4A. QTL have been verified by genome-wide association studies using a diverse panel of 276 winter wheat genotypes of one QTL on chromosomes 3A, 3D, 4A, and 7D [48]. Although Galiba et al. [49] and Juhasz et al. [50] emphasized that genes controlling osmotic regulation and proline content are mainly located on chromosomes 5D and 5A, the contribution of other chromosomes, especially 4A in frost tolerance, cannot be ignored. Based on our results, a major QTL on chromosome 4A is linked with both frost resistance and number of surviving plants after overwintering. This indicates the importance of chromosome 4A in frost tolerance in winter wheat. Single marker analysis was preferred due to its simplicity and specific conditions to determine QTLs.

4. Materials and Methods

4.1. Plant Material and Mapping Population

The Norstar variety is a cold-tolerant (LT_{50} −22.25 °C) variety developed in Saskatchewan, Canada, in the 1980s and was used as the maternal parent. The other parent, Zagros, is an early and summer wheat variety that was developed by the International Agricultural Research Center for Dry Areas (ICARDA) as highly sensitive to cold (LT_{50} −3.5 °C) [36]. The mapping population consisted of 182 recombinant inbred lines (RILs) (F_{12}), derived from a cross between Norstar and Zagros; two parental lines were used as genetic material.

The F_{12} generation seeds were planted in a greenhouse with temperature conditions of 20 °C and a 10/14 h (D/N) photoperiod. After 5 weeks, when the growing seedlings reached the 3- to 4-leaf stage [51], the plants were transplanted to a freezing chamber for vernalizing at 2 ± 0.5 °C for 6 weeks. The vernalized seedlings were then prepared for a freezing test.

4.2. Freezing Tolerance Screening

Freezing tolerance was evaluated from −3 °C to −25 °C at −2 °C increments in 12 test ranges after cold acclimation according to Fowler et al. [12] and Mahfoozi et al. [51] (Figure 1) (ST1). Vernalized seedlings of the genotypes were placed in plastic pots and wetted and kept at 2–4 °C for 2 days for the adaptation assay freezing test. At the end of this period, seedlings were transferred to a programmable freezer for array freezing tests (Figure 1C,D) [32,52] and kept for 1 h at each temperature point. After the freezing test stage, samples (pots) were kept in a growth chamber with a 10/14 h (D/N) photoperiod at 4 °C for 24 h. Samples were then allowed to regrow in the greenhouse with a 10/14 h (D/N) photoperiod at 20 °C for 3 weeks, and LT_{50} was recorded for the entire population. The final LT_{50} was calculated after probity transformation. Accordingly, if at least 5 out of 10 plants survived, the degree of frost tolerance was considered for that genotype (LT_{50} value).

Figure 1. Stages of freezing test; (**A**) Cultivation of genotypes in the greenhouse, (**B**) Preparing seedlings for frost exposure (freezing test), (**C**) Preparing seedlings for transfer to the freezing test machine, (**D**) Freezing test machine.

4.3. Field Experiment

The F_{12} RILs were planted at the Research Farm of Ataturk University for 2 years (2014–2016) in a randomized complete design (RCBD) with two replications. Thereafter, the following yield-related parameters were determined: heading time, time period between the sowing date and the time when almost half of the spikes of each row of plants emerged from the flag leaf sheath, 1000-seed weight, number of surviving plants after winter, number of plants that emerged after winter. This experiment was performed to identify any probable correlation between these traits and LT_{50} and any possible common QTL governing both traits.

4.4. Genotyping

Nuclear DNA was extracted from young leaves of wheat plants of individual F_{12} ten-day seedlings germinated from seeds of each genotype, as previously described [53]. DNA samples were examined with 0.8% agarose gel electrophoresis for quality and then quantified by a Nanodrop device. DNA samples were diluted to 20 ng/µL concentration. A set of 425 SSRs from the Wheat database (XBARC, XCFA, XCFD, XGWM, XWMC, and XWMS) involve the 21 chromosomes of wheat. Wheat microsatellites (SSR) were chosen from http://www.graingenes.org (accessed on 5 January 2022). For molecular analysis, parental-line polymorphisms were assessed by 425 SSR primer pairs distributed on all wheat chromosomes (ST2). Thirty-four polymorphic SSR primer pairs were used for genotyping the F_{12} population. QTL mapping was performed based on single mapping. PCR was performed to amplify the sequence in the SSR molecular markers (94 °C for 1 min (one cycle); 94 °C for 20 s, 50–62 °C for 35 s, 72 °C for 45 s (35–38 cycles), and final extension at 72 °C for 45 s (one cycle, then hold at 4 °C indefinitely) [54]. The PCR products were then loaded onto 6% polyacrylamide gels. Bands were separated by electrophoresis at 100 V containing 0.5 μg/mL ethidium bromide for approximately 2 h using 0.5 × TBE buffer, along with a DNA ladder, and examined under ultraviolet light. Finally, the gels were photographed using a digital camera (Model Nikon Coolpix500, Nikon, Japan) under UV light [55,56].

4.5. Data Analysis

Before analyzing the obtained phenotypic data, a normality test was performed with the SPSS program and the *Shapiro–Wilk* method was used for non-parametric analysis. The SAS program (SAS Institute, Inc, NC, USA. http://www.sas.com (accessed on 5 January 2022)) was used for variation analysis. LT_{50} values were determined by probit regression analysis from the SAS program [57]. QTL analysis was performed to identify the QTL associated with frost tolerance in winter wheat using 182 plants of an F_{12} population derived from crosses between two bread wheat genotypes using the MAPMANAGER-QTX20 program based on the values of Single Marker Analysis (SMA) [32]. The percentage of phenotypic variation explained by markers was calculated based on R-square regression analysis based on SMA using MAPMANAGER-QTX20 software [32]. Due to fewer polymorphic markers, no maps were constructed, and the QTL analysis used the SMA method.

5. Conclusions

Wheat culture is strongly affected by types of stress, such as cold and freezing. Therefore, the generation of cold-tolerant cultivars is one of the essential challenges to genetics and breeders. SSR markers, due to high efficiency, are useful for the detection of QTLs related to abiotic stress, such as cold. In this study, some primers were also linked to yield-related traits, such as number of plants surviving after overwintering, 1000-seed weight in field conditions, and frost tolerance, which suggests pleiotropic effects. Therefore, in a genotype with greater 1000-seed weight and a greater number of plants surviving after overwintering, the activity of frost tolerance genes is prolonged, and the expression of these structural genes is increased. In addition, QTL-rich regions on chromosome 4A were detected, supporting the hypothesis that this chromosome has a key role to play in frost tolerance and should be exploited for wheat improvement. In addition, the traits LT50, 1000-seed weight, and number of surviving plants after winter were located in the 4A genome and have been associated with frost tolerance. This suggests that a set of gene loci on a set of wheat chromosomes plays a role in the degree of frost tolerance. Other SSR markers and gene expression mechanisms should be investigated.

Supplementary Materials: The following supporting information can be downloaded at: https://www.mdpi.com/article/10.3390/plants12081641/s1, Table S1: Freezing test LT_{50} values. Table S2: 425- SSR primers for QTL analysis among RIL accessions.

Author Contributions: Conceptualization, P.B., K.H. and S.A.M.; methodology, K.H., S.A.M., A.T., G.N. and M.N.; software, S.A.M.; validation, S.A.M., G.N., P.S. and M.N.; formal analysis, K.H. and S.A.M.; investigation, K.H. and A.T.; resources, K.H., A.T. and E.İ.; data curation, K.H. and A.T.; writing—original draft preparation, P.B, K.H., S.A.M., A.T., E.İ., G.N., P.S. and M.N.; writing—review and editing, K.H., S.A.M., A.T., E.İ., G.N., P.S. and M.N.; visualization, K.H., S.A.M. and A.T.; supervision, K.H., S.A.M. and G.N.; project administration, A.T. and M.N. All authors have read and agreed to the published version of the manuscript.

Funding: This research received no external funding.

Data Availability Statement: Data are contained within the article and Supplementary Materials.

Acknowledgments: Plant materials were provided by the Tabriz University Breeding Center, Iran. We would like to thank M. Moghaddam for use of the freezing chamber in the Department of Plant Breeding and Biotechnology, Faculty of Agriculture, University of Tabriz, Tabriz, Iran. We would also like to thank Peter Poczai from the Botany Unit, Finnish Museum of Natural History, University of Helsinki, Helsinki, Finland for review of this article.

Conflicts of Interest: The authors have no conflict of interest to declare.

References

1. Salamini, F.; Özkan, H.; Brandolini, A.; Schäfer-Pregl, R.; Martin, W. Genetics and geography of wild cereal domestication in the near east. *Nat. Rev. Genet.* **2002**, *3*, 429–441. [CrossRef]
2. TUIK. Herbal Product Statistics. 2018. Available online: http://www.tuik.gov.tr (accessed on 9 August 2018).
3. Wąsek, I.; Dyda, M.; Gołębiowska, G.; Tyrka, M.; Rapacz, M.; Szechyńska-Hebda, M.; Wędzony, M. Quantitative trait loci and candidate genes associated with freezing tolerance of winter triticale (× *Triticosecale* Wittmack). *J. Appl. Genet.* **2022**, *63*, 15–33. [CrossRef]
4. Kidokoro, S.; Shinozaki, K.; Yamaguchi-Shinozaki, K. Transcriptional regulatory network of plant cold-stress responses. *Trends Plant Sci.* **2022**, *27*, 922–935. [CrossRef]
5. Chun, J.; Yu, X.; Griffith, M. Genetic studies of antifreeze proteins and their correlation with winter survival in wheat. *Euphytica* **1998**, *102*, 219–226. [CrossRef]
6. Limin, A.; Fowler, D. Inheritance of cold hardiness in *Triticum aestivum* × *synthetic* hexaploid wheat crosses. *Plant Breed.* **1993**, *110*, 103–108. [CrossRef]
7. Gusta, L.; O'connor, B.; Gao, Y.P.; Jana, S. A re-evaluation of controlled freeze-tests and controlled environment hardening conditions to estimate the winter survival potential of hardy winter wheats. *Can. J. Plant Sci.* **2001**, *81*, 241–246. [CrossRef]
8. Nagano, A.J.; Kawagoe, T.; Sugisaka, J.; Honjo, M.N.; Iwayama, K.; Kudoh, H. Annual transcriptome dynamics in natural environments reveals plant seasonal adaptation. *Nat. Plants* **2019**, *5*, 74–83. [CrossRef]
9. Fowler, D.; Chauvin, L.; Limin, A.; Sarhan, F. The regulatory role of vernalization in the expression of low-temperature-induced genes in wheat and rye. *Theor. Appl. Genet.* **1996**, *93*, 554–559. [CrossRef]
10. Livingston, D.P., III. The second phase of cold hardening: Freezing tolerance and fructan isomer changes in winter cereal crowns. *Crop Sci.* **1996**, *36*, 1568–1573. [CrossRef]
11. Mahfoozi, S.; Majdi, M.; Janmohammadi, M.; Sasani, S.; Tavakol-Afshari, R.; Hosseini-Salekdeh, G. Developmental control of cold tolerance in wheat (*Triticum aestivum*). *Plant Prod. Genet.* **2019**, *2*, 53–68.
12. Fowler, D.; Gusta, L.; Tyler, N. Selection for winterhardiness in wheat. III. Screening methods. *Crop Sci.* **1981**, *21*, 896–901. [CrossRef]
13. Liu, B.; Wang, X.Y.; Cao, Y.; Arora, R.; Zhou, H.; Xia, Y.P. Factors affecting freezing tolerance: A comparative transcriptomics study between field and artificial cold acclimations in overwintering evergreens. *Plant J.* **2020**, *103*, 2279–2300. [CrossRef]
14. Bhattacharya, A. *Effect of Low-Temperature Stress on Germination, Growth, and Phenology of Plants: A Review*; Physiological Processes in Plants Under Low Temperature Stress: Berlin/Heidelberg, Germany, 2022; pp. 1–106.
15. Fowler, S.; Thomashow, M.F. Arabidopsis transcriptome profiling indicates that multiple regulatory pathways are activated during cold acclimation in addition to the CBF cold response pathway. *Plant Cell* **2002**, *14*, 1675–1690. [CrossRef] [PubMed]
16. Heidarvand, L.; Maali Amiri, R. What happens in plant molecular responses to cold stress? *Acta Physiol. Plant.* **2010**, *32*, 419–431. [CrossRef]
17. Hannah, M.A.; Heyer, A.G.; Hincha, D.K. A global survey of gene regulation during cold acclimation in *Arabidopsis thaliana*. *PLoS Genet.* **2005**, *1*, e26. [CrossRef] [PubMed]
18. Limin, A.E.; Fowler, D.B. Low-temperature tolerance and genetic potential in wheat (*Triticum aestivum* L.): Response to photoperiod, vernalization, and plant development. *Planta* **2006**, *224*, 360–366. [CrossRef]
19. Denesik, T.J. Quantitative Expression Analysis of Four Low-Temperature-Tolerance-Associated Genes during Cold Acclimation in Wheat (*Triticum aestivum* L.). Master's Thesis, University of Saskatchewan, Saskatoon, SK, Canada, 2007.
20. Law, C.; Worland, A. Genetic analysis of some flowering time and adaptive traits in wheat. *New Phytol.* **1997**, *137*, 19–28. [CrossRef]
21. Snape, J.; Butterworth, K.; Whitechurch, E.; Worland, A. Waiting for fine times: Genetics of flowering time in wheat. In *Wheat in a Global Environment*; Springer: Berlin/Heidelberg, Germany, 2001; pp. 67–74.
22. Sutka, J.; Galiba, G.; Vagujfalvi, A.; Gill, B.S.; Snape, J.W. Physical mapping of the Vrn-A1 and Fr1 genes on chromosome 5A of wheat using deletion lines. *Theor. Appl. Genet.* **1999**, *99*, 199–202. [CrossRef]
23. Tóth, B.; Galiba, G.; Fehér, E.; Sutka, J.; Snape, J.W. Mapping genes affecting flowering time and frost resistance on chromosome 5B of wheat. *Theor. Appl. Genet.* **2003**, *107*, 509–514. [CrossRef]
24. Båga, M.; Chodaparambil, S.V.; Limin, A.E.; Pecar, M.; Fowler, D.B.; Chibbar, R.N. Identification of quantitative trait loci and associated candidate genes for low-temperature tolerance in cold-hardy winter wheat. *Funct. Integr.* **2007**, *7*, 53–68. [CrossRef]
25. Vagujfalvi, A.; Galiba, G.; Cattivelli, L.; Dubcovsky, J. The cold-regulated transcriptional activator Cbf3 is linked to the frost-tolerance locus Fr-A2 on wheat chromosome 5A. *Mol. Genet. Genom.* **2003**, *269*, 60–67. [CrossRef] [PubMed]
26. Vagujfalvi, A.; Galiba, G.; Dubcovsky, J.; Cattivelli, L. Two loci on wheat chromosome 5A regulate the differential cold-dependent expression of the cor14b gene in frost-tolerant and frost-sensitive genotypes. *Mol. Gen. Genet.* **2000**, *263*, 194–200. [CrossRef] [PubMed]
27. Börner, A.; Schumann, E.; Fürste, A.; Cöster, H.; Leithold, B.; Röder, M.; Weber, W. Mapping of quantitative trait loci determining agronomic important characters in hexaploid wheat (*Triticum aestivum* L.). *Theor. Appl. Genet.* **2002**, *105*, 921–936. [CrossRef] [PubMed]

28. Zhu, J.; Pearce, S.; Burke, A.; See, D.R.; Skinner, D.Z.; Dubcovsky, J.; Garland-Campbell, K. Copy number and haplotype variation at the VRN-A1 and central FR-A2 loci are associated with frost tolerance in hexaploid wheat. *Theor. Appl. Genet.* **2014**, *127*, 1183–1197. [CrossRef]
29. Shindo, C.; Tsujimoto, H.; Sasakuma, T. Segregation analysis of heading traits in hexaploid wheat utilizing recombinant inbred lines. *Heredity* **2003**, *90*, 56–63. [CrossRef]
30. Fowler, D.; N'Diaye, A.; Laudencia-Chingcuanco, D.; Pozniak, C. Quantitative trait loci associated with phenological development, low-temperature tolerance, grain quality, and agronomic characters in wheat (*Triticum aestivum* L.). *PLoS ONE* **2016**, *11*, e0152185. [CrossRef]
31. Liu, L.; Wang, L.; Yao, J.; Zheng, Y.; Zhao, C. Association mapping of six agronomic traits on chromosome 4A of wheat (*Triticum aestivum* L.). *Mol. Plant Breed.* **2010**, *1*, 1–10. [CrossRef]
32. Sofalian, O.; Mohammadi, S.A.; Aharizad, S.; Moghaddam, M.; Shakiba, M.R. Mapping of QTLs for frost tolerance and heading time using SSR markers in bread wheat. *Afr. J. Biotechnol.* **2009**, *8*, 20.
33. Collard, B.C.; Jahufer, M.; Brouwer, J.; Pang, E.C.K. An introduction to markers, quantitative trait loci (QTL) mapping and marker-assisted selection for crop improvement: The basic concepts. *Euphytica* **2005**, *142*, 169–196. [CrossRef]
34. Hyne, V.; Kearsey, M. QTL analysis: Further uses of 'marker regression. *Theor. Appl. Genet.* **1995**, *91*, 471–476. [CrossRef]
35. Brule Babel A, Fowler D: Genetic control of cold hardiness and vernalization requirement in winter wheat. *Crop Sci.* **1988**, *28*, 879–884. [CrossRef]
36. Fowler, D.; Limin, A.; Ritchie, J. Low-temperature tolerance in cereals: Model and genetic interpretation. *Crop Sci.* **1999**, *39*, 626–633. [CrossRef]
37. Limin, A.; Fowler, D. Developmental traits affecting low-temperature tolerance response in near-isogenic lines for the vernalization locus Vrn-A1 in wheat (*Triticum aestivum* L. em Thell). *Ann. Bot.* **2002**, *89*, 579–585. [CrossRef]
38. Liu, F.; Xu, W.; Song, Q.; Tan, L.; Liu, J.; Zhu, Z.; Fu, Y.; Su, Z.; Sun, C. Microarray-assisted fine-mapping of quantitative trait loci for cold tolerance in rice. *Mol. Plant* **2013**, *6*, 757–767. [CrossRef] [PubMed]
39. Heo, J.Y.; Feng, D.; Niu, X.; Mitchell-Olds, T.; Van Tienderen, P.H.; Tomes, D.; Schranz, M.E. Identification of quantitative trait loci and a candidate locus for freezing tolerance in controlled and outdoor environments in the overwintering crucifer *Boechera stricta*. *Plant Cell Environ.* **2014**, *37*, 2459–2469. [CrossRef] [PubMed]
40. Case, A.J.; Skinner, D.Z.; Garland-Campbell, K.A.; Carter, A.H. Freezing tolerance-associated quantitative trait loci in the Brundage× Coda wheat recombinant inbred line population. *Crop Sci.* **2014**, *54*, 982–992. [CrossRef]
41. Xu, K.; Zhao, Y.; Gu, J.; Zhou, M.; Gao, L.; Sun, R.X.; Wang, W.W.; Zhang, S.H.; Yang, X.J. Proteomic analysis reveals the molecular mechanism underlying the cold acclimation and freezing tolerance of wheat (*Triticum aestivum* L.). *Plant Sci.* **2022**, *318*, 111242. [CrossRef]
42. Maruyama, K.; Sakuma, Y.; Kasuga, M.; Ito, Y.; Seki, M.; Goda, H.; Shimada, Y.; Yoshida, S.; Shinozaki, K.; Yamaguchi-Shinozaki, K. Identification of cold—Inducible downstream genes of the Arabidopsis DREB1A/CBF3 transcriptional factor using two microarray systems. *Plant J.* **2004**, *38*, 982–993. [CrossRef]
43. Taleei, A.; Mirfakhraee, R.; Mardi, M.; Zali, A.; Mahfouzi, C. QTL markers associated with low temperature tolerance in winter wheat. *Int. J. Biol. Sci.* **2010**, *2*, 39–47.
44. Sutka, J. Genes for frost resistance in wheat. *Euphytica* **2001**, *119*, 69–177. [CrossRef]
45. Kruse, E.B.; Carle, S.W.; Wen, N.; Skinner, D.Z.; Murray, T.D.; Garland-Campbell, K.A.; Carter, A.H. Genomic regions associated with tolerance to freezing stress and snow mold in winter wheat. *G3 Genes Genomes Genet.* **2017**, *7*, 775–780. [CrossRef] [PubMed]
46. Sutka, J.; Snape, J.W. Location of a gene for frost resistance on chromosome 5A of wheat. *Euphytica* **1989**, *42*, 41–44. [CrossRef]
47. Ballesta, P.; Mora, F.; Del Pozo, A. Association mapping of drought tolerance indices in wheat: QTL-rich regions on chromosome 4A. *Sci. Agric.* **2019**, *77*, e20180153. [CrossRef]
48. Soleimani, B.; Lehnert, H.; Babben, S.; Keilwagen, J.; Koch, M.; Arana-Ceballos, F.A.; Chesnokov, Y.; Pshenichnikova, T.; Schondelmaier, J.; Ordon, F. Genome wide association study of frost tolerance in wheat. *Sci. Rep.* **2022**, *12*, 5275. [CrossRef] [PubMed]
49. Galiba, G.; Kerepesi, I.; Snape, J.W.; Sutka, J. Location of a gene regulating cold-induced carbohydrate production on chromosome 5A of wheat. *Theor. Appl. Genet.* **1997**, *95*, 265–270. [CrossRef]
50. Juhász, Z.; Boldizsár, Á.; Nagy, T.; Kocsy, G.; Marincs, F.; Galiba, G.; Bánfalvi, Z. Pleiotropic effect of chromosome 5A and the mvp mutation on the metabolite profile during cold acclimation and the vegetative/generative transition in wheat. *BMC Plant Biol.* **2015**, *15*, 1–13. [CrossRef]
51. Zadoks, J.C.; Chang, T.T.; Konzak, C.F. A decimal code for the growth stages of cereals. *Weed Res.* **1974**, *14*, 415–421. [CrossRef]
52. Mahfoozi, S.; Limin, A.; Fowler, D. Influence of vernalization and photoperiod responses on cold hardiness in winter cereals. *Crop Sci.* **2001**, *41*, 1006–1011. [CrossRef]
53. Zeinalzadehtabrizi, H.; Hosseinpour, A.; Aydin, M.; Haliloglu, K. A modified genomic DNA extraction method from leaves of sunflower for PCR based analyzes. *J. Biodivers. Environ. Sci.* **2015**, *7*, 222–225.
54. Haliloğlu, K.; Türkoğlu, A.; Öztürk, A.; Niedbała, G.; Niazian, M.; Wojciechowski, T.; Piekutowska, M. Genetic Diversity and Population Structure in Bread Wheat Germplasm from Türkiye Using iPBS-Retrotransposons-Based Markers. *Agronomy* **2023**, *13*, 255. [CrossRef]

55. Haliloglu, K.; Turkoglu, A.; Tan, M.; Poczai, P. SSR-Based Molecular Identification and Population Structure Analysis for Forage Pea (*Pisum sativum* var. arvense L.) Landraces. *Genes* **2022**, *13*, 1086. [CrossRef] [PubMed]
56. Özkan, G.; Haliloğlu, K.; Türkoğlu, A.; Özturk, H.I.; Elkoca, E.; Poczai, P. Determining genetic diversity and population structure of common bean (*Phaseolus vulgaris* L.) landraces from Türkiye using SSR markers. *Genes* **2022**, *13*, 1410. [CrossRef] [PubMed]
57. Skinner, D.; Garland-Campbell, K. The relationship of LT50 to prolonged freezing survival in winter wheat. *Can. J. Plant Sci.* **2008**, *88*, 885–889. [CrossRef]

plants

MDPI

Communication

Genetic Analyses of Seed Longevity in *Capsicum annuum* L. in Cold Storage Conditions

Mian Abdur Rehman Arif [1,*], Pasquale Tripodi [2], Muhammad Qandeel Waheed [1], Irfan Afzal [3], Sibylle Pistrick [4], Gudrun Schütze [4] and Andreas Börner [4,*]

1 Nuclear Institute for Agriculture and Biology, Faisalabad 38000, Pakistan
2 Research Centre for Vegetable and Ornamental Crops, Council for Agricultural Research and Economics (CREA), 84098 Pontecagnano Faiano, Italy
3 Seed Physiology Lab, Department of Agronomy, University of Agriculture, Faisalabad 38000, Pakistan
4 Leibniz Institute of Plant Genetics and Crop Plant Research, Corrensstr. 3, 06466 Seeland, Germany
* Correspondence: m.a.rehman.arif@gmail.com (M.A.R.A.); boerner@ipk-gatersleben.de (A.B.); Tel.: +92-3335521394 (M.A.R.A.); +49-394825229 (A.B.)

Abstract: Seed longevity is the most important trait in the genebank management system. No seed can remain infinitely viable. There are 1241 accessions of *Capsicum annuum* L. available at the German Federal ex situ genebank at IPK Gatersleben. *C. annuum* (*Capsicum*) is the most economically important species of the genus *Capsicum*. So far, there is no report that has addressed the genetic basis of seed longevity in *Capsicum*. Here, we convened a total of 1152 *Capsicum* accessions that were deposited in Gatersleben over forty years (from 1976 to 2017) and assessed their longevity by analyzing the standard germination percentage after 5–40 years of storage at −15/−18 °C. These data were used to determine the genetic causes of seed longevity, along with 23,462 single nucleotide polymorphism (SNP) markers covering all of the 12 *Capsicum* chromosomes. Using the association-mapping approach, we identified a total of 224 marker trait associations (MTAs) (34, 25, 31, 35, 39, 7, 21 and 32 MTAs after 5-, 10-, 15-, 20-, 25-, 30-, 35- and 40-year storage intervals) on all the *Capsicum* chromosomes. Several candidate genes were identified using the blast analysis of SNPs, and these candidate genes are discussed.

Keywords: genetics; candidate genes; GWAS; *Capsicum*; genebanks; seed longevity; cold storage

check for updates

Citation: Arif, M.A.R.; Tripodi, P.; Waheed, M.Q.; Afzal, I.; Pistrick, S.; Schütze, G.; Börner, A. Genetic Analyses of Seed Longevity in *Capsicum annuum* L. in Cold Storage Conditions. *Plants* **2023**, *12*, 1321. https://doi.org/10.3390/plants12061321

Academic Editor: Fengxia Liu

Received: 26 January 2023
Revised: 6 March 2023
Accepted: 13 March 2023
Published: 14 March 2023

1. Introduction

Seeds are considered the building blocks of genebanks, which came in to being to preserve plant genetic resources and avoid the risk of extinction and of genetic erosion [1]. Seed storage in the genebanks also ensures the preservation of allelic (or genic) combinations in germplasm collections [2], thus serving as the raw material to breed new cultivars [3]. The success of plant genetic resources stored in genebanks was recently realized for wheat [4] when several genebank lines of wheat from Mexican genebank were crossed with several elite cultivars developed at CIMMYT [4] to produce a large number of different pre-breeding germplasm sets and were distributed to resource-poor countries including India and Pakistan. The resultant germplasm revealed considerable success in providing favorable alleles for disease resistance [5], salinity tolerance [6], nitrogen-use efficiency [7], nematode resistance [8], Karnal bunt resistance [9] and drought tolerance [10]. More recent evaluations of Mexican wheat landraces coupled with genetic mapping have also revealed their latent potential toward food security in the upcoming decades [11].

Global food supply relies on the availability of viable seeds [12]. To maintain their germinability, the genetic resources stored in the form of seeds need to be regularly evaluated [13]. A drop in their germination beyond a certain threshold indicates that regeneration is required [14]. This renders "seed longevity" the single most important trait in the genebank management system [15]. Seed longevity refers to the time period during which a seed remains viable and capable of producing healthy seedlings [16,17]. Research on seed longevity is of extreme significance to genebank management [12]. No seed can remain infinitely viable. Conditions during seed production, crop harvesting, post-harvest conditions and, later, the storage conditions determine seed viability [18]. Seed viability, however, is also variable among species and even between varieties, indicating that the genetic component also plays an important role in determining seed longevity [3].

Vegetable seeds constitute ~19,000 accessions of the IPK germplasm collection [19]. According to the genebank information system (GBIS) of the IPK (https://gbis.ipk-gatersleben.de/gbis2i/faces/index.jsf, accessed on 20 October 2022), there were 1241 accessions of *Capsicum annuum* L. available. *C. annuum* is the most economically important species of the genus *Capsicum* [20]. A recent report has shed considerable light on its evolution and trade history in addition to mapping genes related to its plant architecture, fruit quality and flower-related traits [21]. Genetic analyses of seed longevity, however, in *Capsicum* are non-existent. Recently, however, the molecular mechanisms involved in seed longevity in different *Capsicum* species and varieties were illustrated. Less domesticated species (*C. chinense* and *C. frutescens*) exhibited higher germination rates and longevity after AA. Differential gene expression analyses exhibited that *aspartic protease in guard cell 1 (ASPG1)* and *homeobox protein 25 (HB25)* expression were higher in long-lived accessions. In addition, a positive correlation between the amount of lignin and seed viability was demonstrated [20].

Among the two most common techniques to investigate the genetic components of a trait, association mapping (AM) is advantageous to conventional linkage mapping technique because AM does not require the generation of a defined "population". AM utilizes unrelated accessions or collections of germplasm [3]. AM has been performed on a range of crop species with respect to seed longevity analysis, including *Arabidopsis* [22], rice [23], barley [24] and wheat [14]. Most of the studies that analyzed seed longevity involved the lab-based AA or CD tests, the results of which in major crop plants such as wheat have remained under debate in comparison to long-term cold-storage aging [14]. Moreover, seed longevity assessment under lab conditions is costly due to the growing of a plant for one complete season, harvesting it and subsequently storing it, followed by experimental protocols that involve seed aging at high temperatures and high relative humidity [15]. Here, we report on the molecular genetic analyses of seed longevity in *Capsicum annuum* L. by using the genebank germination data generated over a period of 40 years and employing the AM protocol.

2. Results

2.1. Standard Germination after Various Storage Periods

The standard germination percentage after various storage periods varied considerably. For example, the germination % after 1–5 years of storage was considerably high with a mean value (±standard deviation) of 86.08 ± 14.82%, whereas the germination % after 36–40 years of storage was 64.49 ± 23.87% (Figure 1). The germination % after 6–10 years dropped sharply to 71.63 ± 26.84%. However, mean germination % after 11–15 and 16–20 years of storage remained 80.33 ± 18.15% and 80.09 ± 21.14%, respectively. Likewise, germination % during the periods of 21–25 and 26–30 years of storage dropped minimally, with mean values of 77.64 ± 18.30% and 77.73 ± 15.92%, respectively. Afterward, there was some decline in survival after 31–35 years of storage when germination% was 70.06 ± 18.28%.

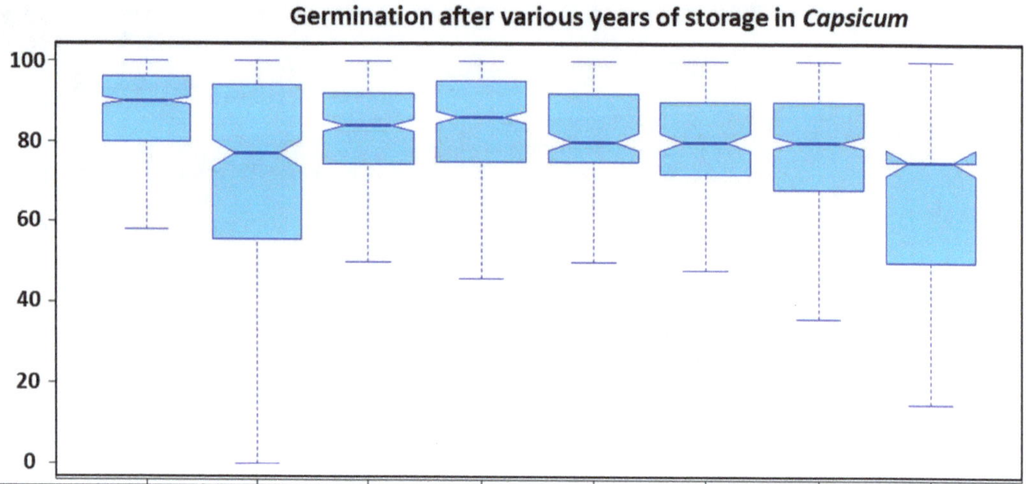

	1–5 years	6–10 years	11–15 years	16–20 years	21–25 years	26–30 years	31–35 years	36–40 years
N	813	320	389	507	164	170	605	140
Mean	86.08 ± 14.82	71.63 ± 26.84	80.33 ± 18.15	80.09 ± 21.14	77.64 ± 18.30	77.73 ± 15.92	70.06 ± 18.28	64.49 ± 23.87
Range	0-100	0-100	6-100	0-100	0-100	0-100	0-100	0-100

Figure 1. Mean (±standard deviation) germination percentages (seed survival) after various years of storage in *Capsicum*, where N = sample size.

2.2. *Genome-Wide Association (GWA) Mapping*

Association mapping was carried out separately for each storage interval. We identified a total of 34 significant maker trait associations (MTAs) (including 10 highly significant MTAs) for the longevity of *Capsicum* seeds stored from 1 to 5 years (Table S1, Figure 2). These MTAs were located on chromosomes 1 (6 MTAs), 2 (2 MTAs), 4 (2 MTAs), 5 (3 MTAs), 6 (2 MTAs), 7 (4 MTAs), 9 (4 MTAs), 10 (6 MTAs), 11 (3 MTAs) and 12 (2 MTAs), which explained 1.6–7.6% phenotypic variance. For the storage period of 6–10 years, 25 significant MTAs were detected on chromosomes 1 (1 MTA), 2 (3 MTAs), 3 (2 MTAs), 4 (1 MTA), 5 (2 MTAs), 6 (2 MTAs), 7 (3 MTAs), 8 (1 MTA), 9 (8 MTAs), 10 (1 MTA) and 12 (1 MTA). These MTAs explained 3.3 to 6.6% phenotypic variance. Likewise, 31 significant MTAs (including 1 highly significant MTA) were detected for germination after the storage period of 11–15 years, and these MTAs were located on chromosomes 1 (5 MTAs), 2 (1 MTA), 3 (2 MTAs), 4 (4 MTAs), 5 (1 MTA), 6 (1 MTA), 7 (1 MTA), 8 (2 MTAs), 9 (2 MTAs), 10 (7 MTAs), 11 (4 MTAs) and 12 (2 MTAs). These MTAs were responsible for 3.0 to 8.0% phenotypic variation.

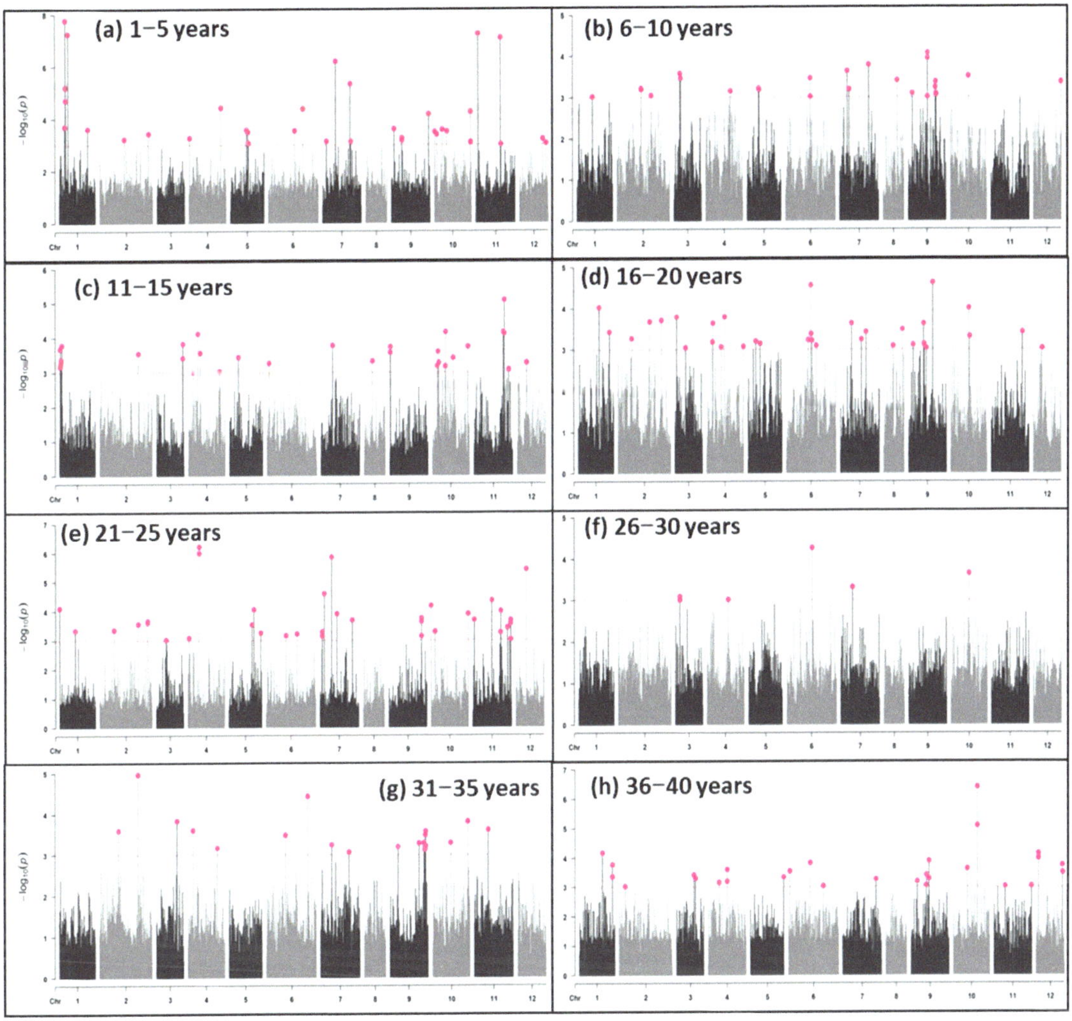

Figure 2. The plot of a genome-wide scan (GWA analysis) of SNP markers associated with seed longevity over various periods of storage ((**a**) after 1–5 years of storage, (**b**) after 6–10 years of storage, (**c**) after 11–15 years of storage, (**d**) after 16–20 years of storage, (**e**) after 21–25 years of storage, (**f**) after 26–30 years of storage, (**g**) after 31–35 years of storage and (**h**) after 6–10 years of storage) in *Capsicum* accessions. The chromosomes are shown on the *x*-axis, the genome-wide scan −log10 (*p* values) is shown on the *y*-axis, and the significantly associated SNPs are highlighted in pink.

For the longevity after 16–20 years of storage, another 35 significant MTAs (including 2 highly significant MTAs) were found. These were exhibited on chromosomes 1 (2 MTAs), 2 (3 MTA), 3 (2 MTAs), 4 (5 MTAs), 5 (2 MTAs), 6 (5 MTAs), 7 (3 MTAs), 8 (1 MTA), 9 (7 MTAs), 10 (2 MTAs), 11 (1 MTA) and 12 (1 MTA), which explained between 2.3 and 4.5% differences in longevity. The highest number of MTAs (39 significant MTAs including 5 highly significant MTAs) for longevity were detected after 21–25 years of storage. Here, the chromosomes involved were 1 (2 MTAs), 2 (4 MTA), 3 (1 MTA), 4 (3 MTAs), 5 (4 MTAs), 6 (2 MTAs), 7 (6 MTAs), 9 (3 MTAs), 10 (5 MTAs), 11 (8 MTAs) and 12 (1 MTA), and the variation explained was 9.7–23.9%. On the contrary, the least number of MTAs (7 significant MTAs including 2 highly significant MTAs) were detected for the storage period of 26–30 years. These were located on chromosomes 2 (3 MTAs), 4 (1 MTA), 6 (1 MTA), 7 (1 MTA) and 10 (1 MTA). The phenotypic variance explained, however, was 8.0–14.3%.

In the case of storage for 31–35 years, 8 different chromosomes [chromosome 2 (2 MTAs), 3 (1 MTA), 4 (2 MTAs), 6 (2 MTAs), 7 (2 MTAs), 9 (9 MTAs), 10 (2 MTAs) and 11 (1 MTA)] carried 21 significant MTAs, and the variance explained was 1.8–3.9%. Finally, another 32 MTAs (including 2 highly significant MTAs) were discovered for longevity after storage for 36–40 years, which involved all the *Capsicum* chromosomes except chromosome 8. These MTAs were located on chromosomes 1 (3 MTAs), 2 (1 MTA), 3 (2 MTAs), 4 (5 MTAs), 5 (1 MTA), 6 (4 MTAs), 7 (1 MTA), 9 (5 MTAs), 10 (3 MTAs), 11 (2 MTAs) and 12 (6 MTAs), which were responsible for a 8.6–30.2% difference in longevity. Thus, 224 MTAs (including 22 highly significant MTAs) collectively were detected on all the *Capsicum* chromosomes for longevity after various years of storage.

3. Discussion

3.1. Variation in Germination over Various Periods of Storage

Seed deterioration depends on many factors encompassing the environmental and genetic components [14]. This was exhibited in wheat when seeds from the multiplication year of 1974 were stored and tested 34 years later, and germination % varied from 0 to 100%. Likewise, a huge variation in seed survival was witnessed albeit after storage below freezing temperatures ($-15/-18$ °C) in *Capsicum*. A direct comparison among the results of different intervals is not possible because of the involvement of dissimilar accessions tested during each interval. However, some accessions were found to be common across different intervals. For example, there were 171, 261, 419, 163, 108, 318 and 54 accessions from 1- to 5-year intervals that were common in the 6–10-year, 11–15-year, 16–20-year, 21–25-year, 26–30-year, 31–35-year and 36–40-year intervals, respectively. ANOVA results indicated that germination % among common accessions was significantly different in all such cases (data not shown). No reports exist in which such comparisons were made in any crop. It is, however, known that different factors are involved in different aging procedures such as accelerated aging (AA) and controlled deterioration (CD) methods [3]. A comparison between natural aging (seeds stored at 0 ± 1 °C at constant moisture contents of 8 ± 2%) and the deterioration of fresh seeds after AA and CD yielded different results in wheat [14].

3.2. GWA Analyses and Candidate Genes

The seed lots were handled in the same way (from seed sowing to harvest and post-harvest treatments) and were maintained at the IPK genebank since then. The differential behavior in the investigated material was thought to be due to differences in the genetic build of these accessions. Our analyses identified a total of 224 MTAs for 8 different years of storage periods in which chromosome 9 carried the highest number of MTAs (38 MTAs), followed by chromosome 10 (27 MTAs), followed by chromosome 4 (23 MTAs) (Table S2). Chromosome 7 carried 21 MTAs, whereas chromosomes 1, 6 and 11 carried 19 MTAs each. On the other hand, chromosome 2 carried 16 MTAs, and 13 MTAs were located on each of chromosomes 3 and 5. Finally, 12 and 4 MTAs were located on chromosomes 12 and 8, respectively. Thus, this is the very first report of the GWA of seed longevity in *Capsicum*; no comparison with previous studies can be made. Blast analyses of the reported SNPs identified several candidate genes for longevity (Table S3). Of the 220 associated SNPs, 167 SNPs successfully provided hits with certain candidate genes. These 167 SNPs could further be divided into 5 groups based on the function they perform (Figure 3, Table S3). The first group included ten genes which were involved mainly in growth- and development-related processes. The other group constituted 72 genes which were mainly enzymes that were either specifically produced under (both biotic and abiotic) stress or produced under normal conditions. The third group constituted 10 genes that were mainly transcription factors. The fourth group included 18 genes that were mainly transporter genes, whereas the fifth group constituted 48 genes that were mainly uncharacterized and/or hypothetical proteins.

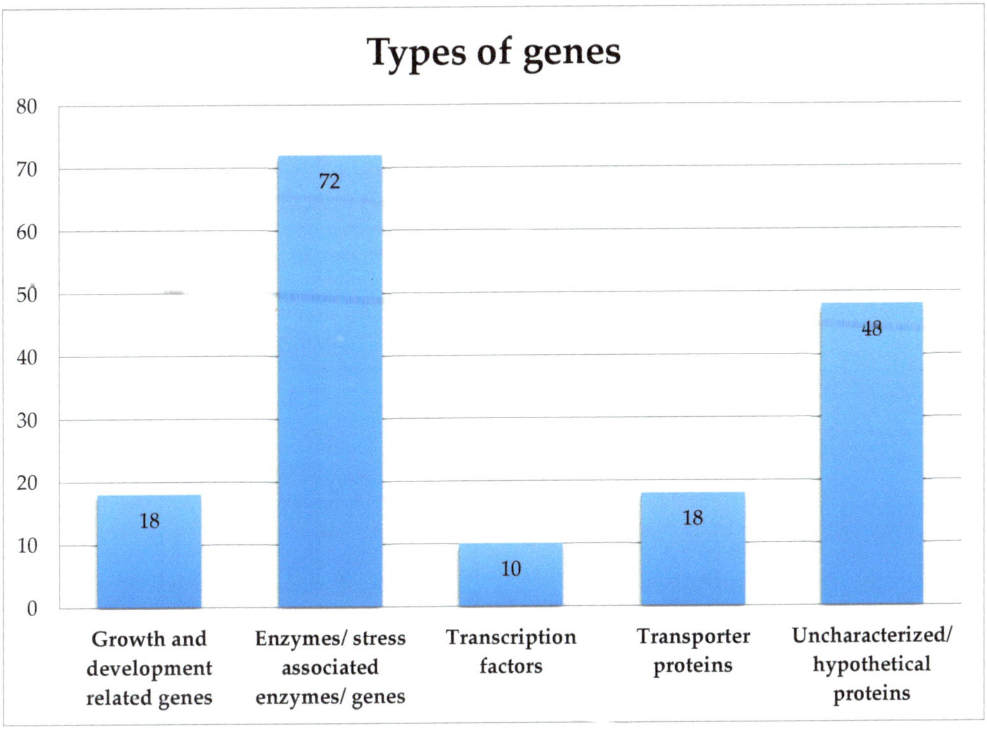

Figure 3. Types of candidate genes linked with the longevity-associated SNPs.

In the following, we provide some details (chromosome by chromosome) for the candidate genes (Table S4) that are linked with SNPs that explain >4% phenotypic variation or are reported to be associated with longevity in other crops.

On chromosome 1, we identified subtilisin-like protease 4 (associated with SNP S1_7853500), 3-oxoacyl-[acyl-carrier-protein] (ACPs) synthase I (associated with SNPs S1_1430591 and S1_1435099), phosphatidylinositol 4-kinase gamma 2 (associated with SNP S1_1921287) and B3-domain-containing transcription factor NGA1 isoform X2 (associated with SNP S1_133318). Subtilisin-like proteases (subtilases) are serine proteases that play specific roles in plant development and signaling cascades. Several subtilases are specifically induced following pathogen infection or under stress [25]. They are also identified as the S-nitrosylation target in potato S-nitrosylation candidates in the potato–*Phytophthora infestans* system [26]. On the other hand, ACPs are a central cofactor for de novo fatty acid synthesis, acyl chain modification and chain-length termination during lipid biosynthesis in living organisms. Different ACP isoforms have been found to be responsible for the biosynthesis of fatty acids and lipids for specific purposes in plants [27]. In addition, ACP has also been identified as a candidate gene for resistance against different insects [thrips, orange (OWBM) and yellow (YWBM) wheat blossom midges] in wheat [28]. Similarly, phosphatidylinositol 4-kinase gamma 2 is known to play a role in the phosphorylation of phosphatidylinositol (PI) to PI 4-phosphate, which is one of the key reactions in the production of phosphoinositides, which are lipid regulators of several cellular functions [29]. NGA transcription factors are involved mainly in developing pistils; they are also involved in regulating the shape and size of lateral organs such as leaves and petals and the regulation of seed size [30].

On chromosome 2, the candidate genes causing significant variation toward seed survival include putative *leucine-rich repeat receptor-like protein kinase* (associated with SNP *S2_59895267*), putative *uroporphyrinogen decarboxylase, chloroplastic-like* (associated with SNP *S2_117264835*), putative *aspartic proteinase nepenthesin-2-like* (associated with SNP *S2_157010230*) and *THO complex subunit 2-like* (associated with SNP s*S2_165845447* and *S2_165845425*). *Aspartic proteinase nepenthesin-2-like* was reported during the periods of 11–15 years and 20–25 years of storage. The nepenthesin aspartic proteases, which are produced by specialized cells in the lower part of the pitchers, are aimed primarily at the digestion of prey trapped by the plant [31]. *Aspartic protease in guard cell 1* has recently been reported as a candidate gene for longevity in *Capsicum* [20]. *Aspartic proteases* mobilize seed-storage proteins and play a crucial role in the germination process and seed longevity [32]. Likewise, the THO complex that is encoded by *THO complex subunit 2-like* is a key component in the co-transcriptional formation of messenger ribonucleo-particles that are competent to be exported from the nucleus (unknown precise function). The THO complex is also involved in mRNA processing and its transport from the nucleus. It also plays a role in small interfering RNA-dependent processes in plants [33,34]. The importance of the *THO complex subunit 2-like* is also evident from the fact that it was identified as a candidate gene for longevity during the periods of 1–5 years and 21–25 years of storage and on multiple chromosomes (chromosome 2 and 11).

The candidate genes for longevity on chromosome 3 include mitochondrial NADH dehydrogenase (ubiquinone) flavoprotein 2 (associated with SNP S3_145194851), chalcone synthase 1B (associated with SNP S3_7697933) and ethylene-responsive transcription factor 4 (associated with SNP S3_12004386). Both NADH dehydrogenase and ethylene-responsive transcription factor have previously been reported as candidate genes for seed dormancy/pre-harvest sprouting (PHS) (in wheat) [35] and longevity (in wheat and barley) [13,36], respectively. Chalcone synthase (CHS) is a crucial rate-limiting enzyme in the flavonoid biosynthetic pathway that catalyzes the condensation of malonyl-CoA and ρ-coumaroyl-CoA to produce naringenin chalcone, which serves as the precursor of a variety of flavonoid derivatives. These flavonoids are involved in the response to and protection of plants from abiotic and biotic stress, including ultraviolet radiation, temperature, humidity and pathogenic attack [37,38]

The most important candidate genes on chromosome 4 include *early nodulin-93 isoform X2* (associated with SNP *S4_15009085*), *transcription factor TGA7* (associated with SNP *S4_209765879*), *3-ketoacyl-CoA synthase 19* (associated with SNP *S4_124426*) and *vacuolar protein-sorting-associated protein 8 homolog* (associated with SNPs *S4_28351186* and *S4_28351189*). *Early nodulin* has been identified as a candidate gene for longevity, dormancy and PHS in wheat [3,35]. The *3-ketoacyl-CoA synthase* is involved in lateral organ development and cuticular wax synthesis in *Medicago truncatula* [39]. The *TGA* family of transcription factors plays important roles in the systemic acquired resistance (SAR) in plants. However, despite its important roles in plant immunity, the molecular mechanism for the DNA binding of *TGA7* remains unclear [40]. Vacuolar protein-sorting-associated proteins (Vps) are part of the Endosomal Sorting Complex Required for Transport (ESCRT), which performs the topologically unique membrane bending and scission reaction away from the cytoplasm [41].

Scarecrow-like protein 30 (associated with SNP *S5_226924789*), ankyrin repeat-contain ing protein 2A and ITN1 (associated with SNPs *S5_238134029* and *S5_237978262*, respectively), protein ACCELERATED CELL DEATH 6 (ACD6) (associated with SNP *S5_237978236*), GRF1-interacting factor 1 (associated with SNP S5_26823246), histone acetyltransferase HAC1-like (associated with SNP *S5_14262255*) and putative LRR receptor-like serine/threonine-protein kinase-like (associated with SNP S5_24462479) were among the candidate genes on chromosome 5. Scarecrow-like protein is a transcription factor belonging to the GRAS family. It regulates root growth and the cell cycle and also mediates resistance to environmental stresses [42]. Recently, 85 ankyrin repeat-containing protein (ANK) genes in *C. annuum* were identified. Our ANK loci on chromosome 5 (SNPs: *S5_237978262* and *S5_238134029*) could correspond to any of the CaANK35-CaANK51 genes mapped at the distal end of chromosome 5 on the *C. annuum* L. genome [43]. ANKs have also been identified as candidate genes against insect (OWBM and YWBM) resistance in wheat [28]. ACD6 is a multipass membrane protein with an ankyrin domain that acts in a positive feedback loop with the defense signal salicylic acid (SA) [44]. GRFs are a class of plant-specific proteins involved in the regulation of stem and leaf development that act mainly as positive regulators of cell proliferation [45]. Histone acetyltransferase HAC1-like encode for histone acetyltransferases that play a crucial role in the control of cell fate and influence cell cycle progression, plant responses to environmental conditions, and gene interactions [46,47]. LRR receptor-like serine/threonine-protein kinase-like was also reported for the longevity-associated SNP (S2_59895267) on chromosome 2 and functions in protein phosphorylation and the transmembrane receptor protein tyrosine kinase signaling pathway. It is an integral component of the plasma membrane, where it functions as an ATP-binding site and is expressed in the flowering stage and plant embryo stage in flowers or seeds [48]. It has also been detected as a candidate for seed longevity in wheat [13].

Among the candidate genes on chromosome 6, the most important were *ribose-phosphate pyrophosphokinase 1-like* (associated with SNP *S6_196787696*), *11S globulin seed storage protein* (associated with SNP *S6_1150637*), *beta-galactosidase* (associated with SNP *S6_213563914*) and putative *ATP synthase subunit O, mitochondrial-like* (associated with SNP *S6_182246997*). *ATP synthase subunit O, mitochondrial-like* was also associated with SNPs on chromosomes 7 (SNP: *S7_157986416*) and 9 (SNP: *S9_35876497*). *Ribose-phosphate pyrophosphokinase,* which is also known as *phosphoribosyldiphosphate synthetase* (PRPP), catalyzes the biosynthesis of PRPP. PRPP is a precursor for the synthesis of pyrimidine, purine, pyridine nucleotides, tryptophan and histidine [49]. Plant seed storage proteins function as the major nitrogen source for the developing plant. The 11S-type globulins are non-glycosylated proteins which form hexameric structures. They are the proteins required for the development or growth of seeds [50]. *Beta-galactosidase* is considered to be an important regulator involved in fruit ripening in *Capsicum* [51]. The reduced form of the *mitochondrial ATP synthase* holoenzyme leads to wide-ranging defects in energy-demanding cellular processes. Hence, it is required to protect plants from various stresses such as heat [52].

Important genes on chromosome 7 include pentatricopeptide repeat-containing protein-mitochondrial-like (associated with SNP S7_42146593), NADP-dependent glyceraldehyde-3-phosphate dehydrogenase (associated with SNP S7_8830422), COP9 signalosome complex (CSN) subunit 8-like (associated with SNP S7_244455030), 1-acyl-sn-glycerol-3-phosphate acyltransferase (associated with SNP S7_245349068) and protein phosphatase 2C 27 (associated with SNP S7_13703435). Here, pentatricopeptide repeat-containing protein was associated with longevity after 11–15 years, 21–25 years and 26–30 years of storage. Pentatricopeptide repeat-containing protein [members of the pentatricopeptide repeat (PPR) protein] family are sequence-specific RNA-binding proteins that play crucial roles in organelle RNA metabolism [53]. In addition, PPR is also involved in YBWM resistance in wheat [28]. PPRs have also been identified as candidate genes that are involved in seed vigor under low-temperature conditions in rapeseed [54]. Glyceraldehyde-3-phosphate dehydrogenase is used in a variant of glycolysis that conserves energy as NADPH rather than as ATP [55]. CSN is an evolutionarily conserved multiprotein complex that regulates many aspects of plant development [56]. In addition, the glycerol-3-phosphate acyltransferase gene plays a pivotal role in cold resistance in a variety of plant species [57], whereas a Type 2C protein phosphatase, CaADIP1 (Capsicum annuum ABA and Drought-Induced Protein phosphatase 1), is known to be expressed on leaves on treatment with ABA, drought and NaCl treatments [58].

Putative *cysteine synthase* (associated with SNP *S8_140480324*) and *syntaxin-32-like* (associated with SNP *S8_119314147*) were the candidate genes on chromosome 8. The former is an enzyme responsible for the formation of cysteine from *O*-acetyl-serine and hydrogen sulfide with the concomitant release of acetic acid [59], and the latter is reported to be involved in host defense responses against pathogen attack [60].

The most important candidate genes on chromosome 9 include salicylate O-methyltran sferase-like (SAMT) (associated with SNPs S9_91765981 and S9_91766032), solute carrier family 35 member F1-like (associated with SNP S9_649935), cell division cycle protein 48-like protein (associated with SNP S9_5554728), superoxide dismutase (associated with SNP S9_93034919) and eukaryotic translation initiation factor isoform 4G-1-like isoform 1 (associated with SNP S9_270289444). SAMT regulates the SA signaling pathway and catalyzes the methylation of SA with *S*-adenosyl-L-methionine as the methyl donor to form methyl salicylate. SAMT appears to play an important role in plant response to drought stress by modulating the SA-signaling pathway [61]. Solute carrier family 35 member F1-like is akin to osmotin-like protein (OSML81). OSMLs belong to the thumatin-like protein family and are known to play a role in seed longevity in wheat and barley [35,36]. Cell division cycle proteins are known to be involved in cell division, growth processes and seed longevity [13]. Likewise, superoxide dismutase and eukaryotic translation initiation factor isoform 4G-1-like isoform 1 are also reported to be candidate genes for seed longevity [14] and PHS [35] in wheat.

On chromosome 10, the most important candidate genes include putative *histone H3.3-like* (associated with SNP *S10_374592* and multiple seed storage durations) and *high mobility group B protein 6* (associated with SNP *S10_16302829*). *Histone H3.3-like* is a candidate gene for longevity in wheat [3], whereas the gene encoding the *"high-mobility group B protein 6"* is a *WRKY transcription factor* involved in the nucleosome/chromatin assembly [62].

Chromosome 11 carries longevity genes such as putative flavin-containing monooxygenase 1-like (FMO) (associated with SNP S11_152967951), protein YAE1 isoform X3 (associated with SNP S11_257151051) and putative F-box protein-like (associated with SNP S11_257213134). FMOs are oxidoreductases and possess remarkable diversity and functionality in the oxygenation reactions, which are crucial steps within hormone metabolism, pathogen resistance, signaling and chemical defense [63]. YAE1 proteins are essential for growth under aerobic conditions and may provide protection from damage due to reactive oxygen species [64]. CaF-box is known to be expressed mainly in stems and seeds, and the transcript is markedly up-regulated in response to cold stress, ABA and SA) treatment, and down-regulated under osmotic and heavy metal stress [65]. However, its role in seed longevity is unknown. Finally, on chromosome 12, the most important candidate gene was tonoplast dicarboxylate transporter (associated with SNPs S12_2792525, S12_2792536 and S12_2792561), and it is known to play an important role in malate and citrate transport (organic acid metabolism) [66].

4. Materials and Methods

4.1. Materials

A total of 1152 *Capsicum* accessions was convened in this investigation. These were deposited in the IPK-Gatersleben over a number of years (from 1976 to 2017) (Table S5) and kept in glass containers (Figure 4). A large proportion of them, however, were deposited during the years 1976 (165 accessions), 1977 (130 accessions) and 1978 (115 accessions). Of these accessions, 1137 were of *Capsicum annuum* L., 14 were *C. annuum* var. *glabriusculum*, and 1 was *C. frutescens*. These accessions were stored at below-freezing temperatures (1976–2010 −15°C; since 2011 −18 °C). Their details can be accessed through the Gatersleben genebank information system by providing the ID (identity number) of the accessions.

Figure 4. Storage of various *Capsicum* accessions [CAP1104, CAP1106, CAP1108, CAP1164, CAP1165, CAP1168, CAP1188, CAP1189 and CAP1190 (for details see Table S5) using the genebank ID number] in glass containers kept at below-freezing temperatures in the IPK genebank.

4.2. Standard Germination Tests

We performed 3103 germination assays on 1152 *Capsicum* accessions from 1990 to 2022. Not all accessions were assessed after each year. For easy understanding, we divided the germination assays into 8 intervals: (1) germination assays performed between 1 and 5 years of storage (812 tests), (2) germination assays performed between 6 and 10 years of storage (320 tests), (3) germinations assays performed between 11 and 15 years of storage (389 tests), (4) germinations assays performed between 16 and 20 years of storage (506 tests), (5) germination assays performed between 21and 25 years of storage (163 tests), (6) germination assays performed between 26 and 30 years of storage (169 tests), (7) germination assays performed between 31 and 35 years of storage (604 tests) and (8) germinations assays performed between 36 and 40 years of storage (140 tests) (Table S6). Stored seeds in genebanks over many years are precious materials, and hence, only a limited quantity could be made available for research. Because of that, one single replicate of 50 seeds of each accession was retrieved from the glass containers and germinated on round filter paper with glass covers on Jacobsen Apparatus at 25 $\pm$ 2 °C and 23 $\pm$ 2 °C during the day and night, respectively. The germination percentages were recorded on the eighth day according to International Seed Testing Association (ISTA) protocols.

4.3. Genotyping

For the purpose of genotyping, 100 mg fresh leaf tissue that was collected from individual plants upon germination was used for DNA extraction. DNeasy Plant Mini Kit (QIAGEN, Düsseldorf, Germany) or the Sbeadex maxi plant kit (LGC Genomics, London, UK) was used to extract the highest-quality DNA, the quantity and quality parameters of which were determined using both spectrometry (ND-1000; NanoDrop, ThermoScientific, Wlatham, MA, USA) and fluorometry (Qubit 2.0 Fluorometer, Invitrogen, Carlsbad, CA, USA) methods. Samples with 260/280 and 230/260 ratios ranging between 1.8 to 2.2 and 1.8 to 2.0, respectively, and with a less-than-twofold deviation between fluorimetric and spectrophotometric readings were subjected to genotyping-by-sequencing (GBS). Genotyping was carried out using an Illumina HiSeq2500 platform generating 1 $\times$ 107-bp single-end reads version 3 chemistry (Illumina, San Diego, CA, USA), which resulted in the generation of 23,462 single nucleotide polymorphism (SNP) markers covering all the 12 *Capsicum* chromosomes. Other relevant details including the details of bioinformatics techniques and tools are available from Tripodi et al. [21].

4.4. Genome-Wide Association (GWA) Analyses

We performed the GWA analyses by utilizing the data of 23,462 high-quality SNP markers [21] and the data of standard germination tests that were obtained as mentioned above. The MLM (mixed linear model) option implemented in the *TASSEL v5.2.43* [67] software was used. A pre-requisite of this model is the provision of the population structure (Q-matrix) or principal component analysis (PCA) matrix and a kinship (K-matrix) matrix. These matrices are used as covariates in the MLM model to avoid false positives during analyses. The PCA matrix and the K-matrix could be generated through *TASSEL v5.2.43*. We ran each of the analyses using the PC = 3, PC = 4 and PC = 5 options for correct estimates. We found that 3, 4 or 5 PCs yielded 90–95% similar marker trait association with slight variation in the *p*-values of the associated SNPs. Thus, we kept PC = 5 for the final analysis. We claimed the SNPs in significant association with longevity that gave a *p*-value of 0.001 ($-$log10 value of 3). In addition, highly significant *p*-values were calculated by taking the reciprocal of the number of markers [13]. Thus, SNPs with *p*-values of 4.26×10^{-5} were considered to be highly significantly associated.

4.5. Blast Analysis

In order to look for the candidate genes linked with the associated SNPs, sequence retrieval of each significant or highly significant SNP was performed. This was achieved by retrieving a raw sequence of 301 nucleotides considering 150 bases upstream and 150 bases downstream of each candidate SNP. The sequence was generated from the reference genome *C. annuum* CM334 [68] version 1.6 using samtools faidx [69] via a blast analysis of all the retrieved sequences of associated SNPs. These sequences were used as a query in the NCBI BLASTX (https://blast.ncbi.nlm.nih.gov/Blast.cgi?PROGRAM=blastx&PAGE_TYPE=BlastSearch&LINK_LOC=blasthome, accessed on 3 January 2023) research tool database for functional gene annotations. The topmost hits with the smallest *E*-value and a high percentage of query coverage were reported as potential candidate genes.

5. Conclusions

To conclude, we presented the very first comprehensive genetic analyses of seed longevity in *Capsicum* using the real-time data after long-term cold storage and the untapped natural genetic diversity. Several candidate genes have been reported for seed longevity in *Capsicum*. Some of them have already been reported for longevity in wheat, barley or other crops, whereas others are novel. Our findings lay the foundation for the comprehensive future studies of seed longevity in *Capsicum*.

Supplementary Materials: The following supporting information can be downloaded at: https://www.mdpi.com/article/10.3390/plants12061321/s1, Table S1: SNPs significantly associated with seed longevity after various years of storage (Gr5, Gr10, Gr15, Gr20, Gr25, Gr30, Gr35 and Gr40 = germination after 1–5, 6–10, 11–15, 16–20,21–25, 26–30, 31–35 and 36–40 years of storage, respectively). Highly significantly associated SNPs are highlighted in yellow. Table S2: Total SNPs significantly or highly significantly associated SNPs (in brackets) after various years of storage. Table S3: Blast hit results with SNPs (including chromosome and position of the SNP), types of genes based on function with candidate gene annotations. Table S4: Blast hit results with SNPs (including chromosome and position of the SNP), candidate gene annotations, probable function and/or association with any other trait (where known). Duplicate genes are highlighted. Table S5: Genebank codes, origin (where known) and initial year of deposit in IPK genebank of the 1152 *Capsicum* accessions used. Table S6: Total number of germination assays performed after various years of storage. References [70–113] are cited in the supplementary materials.

Author Contributions: Conceptualization, M.A.R.A. and A.B.; methodology, M.A.R.A. and P.T.; software, M.A.R.A. and P.T.; formal analysis, M.A.R.A., P.T. and M.Q.W.; investigation, S.P. and G.S.; resources, A.B.; data curation, M.A.R.A., P.T. and A.B.; writing—original draft preparation, M.A.R.A.; writing—review and editing, P.T., M.Q.W., I.A. and A.B.; visualization, M.A.R.A.; supervision, A.B. All authors have read and agreed to the published version of the manuscript.

Funding: This research received no external funding.

Data Availability Statement: Not applicable.

Acknowledgments: The authors thank the whole IPK genebank staff for performing germination tests over many years.

Conflicts of Interest: The authors declare no conflict of interest.

References

1. Linington, S.; Pritchard, H. Genebanks. In *Encyclopaedia of Biodiversity*; Levin, S.A., Ed.; Academic Press: San Francisco, CA, USA, 2001.
2. Fu, Y.B. The vulnerability of plant genetic resources conserved ex situ. *Crop Sci.* **2017**, *57*, 2314–2328. [CrossRef]
3. Arif, M.R.; Nagel, M.; Neumann, K.; Kobiljski, B.; Lohwasser, U.; Börner, A. Genetic studies of seed longevity in hexaploid wheat using segregation and association mapping approaches. *Euphytica* **2012**, *186*, 1–13. [CrossRef]
4. Singh, S.; Jighly, A.; Sehgal, D.; Burgueño, J.; Joukhadar, R.; Singh, S.; Sharma, A.; Vikram, P.; Sansaloni, C.; Govindan, V. Direct introgression of untapped diversity into elite wheat lines. *Nat. Food* **2021**, *2*, 819–827. [CrossRef]
5. Saleem, K.; Shokat, S.; Waheed, M.Q.; Arshad, H.M.I.; Arif, M.A.R. A GBS-Based GWAS Analysis of Leaf and Stripe Rust Resistance in Diverse Pre-Breeding Germplasm of Bread Wheat (*Triticum aestivum* L.). *Plants* **2022**, *11*, 2363. [CrossRef] [PubMed]
6. Akram, S.; Ghaffar, M.; Wadood, A.; Shokat, S.; Hameed, A.; Waheed, M.Q.; Arif, M.A.R. A GBS-based genome-wide association study reveals the genetic basis of salinity tolerance at the seedling stage in bread wheat (*Triticum aestivum* L.). *Front. Genet.* **2022**, *13*, 1–19. [CrossRef] [PubMed]
7. Sharma, A.; Arif, M.A.; Shamshad, M.; Rawale, K.S.; Brar, A.; Burgueño, J.; Shokat, S.; Kaur, R.; Vikram, P.; Srivastava, P. Preliminary dissection of grain yield and related traits at differential nitrogen levels in diverse pre-breeding wheat germplasm through association mapping. *Mol. Biotechnol.* **2022**, *65*, 116–130. [CrossRef]
8. Dababat, A.; Arif, M.A.R.; Toktay, H.; Atiya, O.; Shokat, S.; Gul, E.; Imren, M.; Singh, S. A GWAS to identify the cereal cyst nematode (*Heterodera filipjevi*) resistance loci in diverse wheat prebreeding lines. *J. Appl. Genet.* **2021**, *62*, 93–98. [CrossRef]
9. Singh, S.; Sehgal, D.; Kumar, S.; Arif, M.; Vikram, P.; Sansaloni, C.; Fuentes-Dávila, G.; Ortiz, C. GWAS revealed a novel resistance locus on chromosome 4D for the quarantine disease Karnal bunt in diverse wheat pre-breeding germplasm. *Sci. Rep.* **2020**, *10*, 5999. [CrossRef]
10. Singh, S.; Vikram, P.; Sehgal, D.; Burgueño, J.; Sharma, A.; Singh, S.K.; Sansaloni, C.P.; Joynson, R.; Brabbs, T.; Ortiz, C. Harnessing genetic potential of wheat germplasm banks through impact-oriented-prebreeding for future food and nutritional security. *Sci. Rep.* **2018**, *8*, 12527. [CrossRef]
11. Suhalia, A.; Sharma, A.; Kaur, S.; Sarlach, R.S.; Shokat, S.; Singh, S.; Arif, M.A.R.; Singh, S. Characterization of Mexican wheat landraces for drought and salt stress tolerance potential for future breeding. *Cereal Res. Commun.* **2022**, 1–12. [CrossRef]
12. Arif, M.A.R.; Börner, A. Mapping of QTL associated with seed longevity in durum wheat (*Triticum durum* Desf.). *J. Appl. Genet.* **2019**, *60*, 33–36. [CrossRef] [PubMed]
13. Arif, M.A.R.; Börner, A. An SNP based GWAS analysis of seed longevity in wheat. *Cereal Res. Commun.* **2020**, *48*, 149–156. [CrossRef]
14. Arif, M.A.R.; Nagel, M.; Lohwasser, U.; Börner, A. Genetic architecture of seed longevity in bread wheat (*Triticum aestivum* L.). *J. Biosci.* **2017**, *42*, 81–89. [CrossRef] [PubMed]
15. Arif, M.A.R.; Afzal, I.; Börner, A. Genetic Aspects and Molecular Causes of Seed Longevity in Plants—A Review. *Plants* **2022**, *11*, 598. [CrossRef] [PubMed]
16. Sano, N.; Rajjou, L.; North, H.M.; Debeaujon, I.; Marion-Poll, A.; Seo, M. Staying alive: Molecular aspects of seed longevity. *Plant Cell Physiol.* **2016**, *57*, 660–674. [CrossRef]
17. Barton, L.V. *Seed Preservation and Longevity*; Leonard Hill: London, UK, 1961.
18. McDonald, M. Seed deterioration: Physiology, repair and assessment. *Seed Sci. Technol.* **1999**, *27*, 177–237.
19. Börner, A.; Khlestkina, E.K.; Chebotar, S.; Nagel, M.; Arif, M.A.R.; Neumann, K.; Kobiljski, B.; Lohwasser, U.; Röder, M.S. Molecular markers in management of ex situ PGR–A case study. *J. Biosci.* **2012**, *37*, 871–877. [CrossRef]
20. Bissoli, G.; Bono, M.; Martínez-Almonacid, I.; Moreno-Peris, E.; Renard, J.; Espinosa, A.; Naranjo, M.Á.; Yenush, L.; Fita, A.; Serrano, R. Seed coat lignification level is crucial in *Capsicum* spp. seed longevity. *Physiol. Plant.* **2022**, *174*, e13600. [CrossRef]
21. Tripodi, P.; Rabanus-Wallace, M.T.; Barchi, L.; Kale, S.; Esposito, S.; Acquadro, A.; Schafleitner, R.; van Zonneveld, M.; Prohens, J.; Diez, M.J. Global range expansion history of pepper (*Capsicum* spp.) revealed by over 10,000 genebank accessions. *Proc. Natl. Acad. Sci. USA* **2021**, *118*, e2104315118. [CrossRef]
22. Renard, J.; Niñoles, R.; Martínez-Almonacid, I.; Gayubas, B.; Mateos-Fernández, R.; Bissoli, G.; Bueso, E.; Serrano, R.; Gadea, J. Identification of novel seed longevity genes related to oxidative stress and seed coat by genome-wide association studies and reverse genetics. *Plant Cell Environ.* **2020**, *43*, 2523–2539. [CrossRef]
23. Lee, J.-S.; Velasco-Punzalan, M.; Pacleb, M.; Valdez, R.; Kretzschmar, T.; McNally, K.L.; Ismail, A.M.; Cruz, P.C.S.; Sackville Hamilton, N.R.; Hay, F.R. Variation in seed longevity among diverse Indica rice varieties. *Ann. Bot.* **2019**, *124*, 447–460. [CrossRef] [PubMed]

24. Nagel, M.; Kranner, I.; Neumann, K.; Rolletschek, H.; Seal, C.E.; Colville, L.; Fernández-Marín, B.; Börner, A. Genome-wide association mapping and biochemical markers reveal that seed ageing and longevity are intricately affected by genetic background and developmental and environmental conditions in barley. *Plant Cell Environ.* **2015**, *38*, 1011–1022. [CrossRef] [PubMed]
25. Figueiredo, A.; Monteiro, F.; Sebastiana, M. Subtilisin-like proteases in plant–pathogen recognition and immune priming: A perspective. *Front. Plant Sci.* **2014**, *5*, 739. [CrossRef] [PubMed]
26. Abramowski, D.; Arasimowicz-Jelonek, M.; Izbiańska, K.; Billert, H.; Floryszak-Wieczorek, J. Nitric oxide modulates redox-mediated defense in potato challenged with Phytophthora infestans. *Eur. J. Plant Pathol.* **2015**, *143*, 237–260. [CrossRef]
27. Li, M.-J.; Wang, X.-J.; Su, L.; Bi, Y.-P.; Wan, S.-B. Characterization of Five Putative Acyl Carrier Protein (ACP) Isoforms from Developing Seeds of *Arachis hypogaea* L. *Plant Mol. Biol. Rep.* **2010**, *28*, 365–372. [CrossRef]
28. Arif, M.A.R.; Waheed, M.Q.; Lohwasser, U.; Shokat, S.; Alqudah, A.M.; Volkmar, C.; Börner, A. Genetic insight into the insect resistance in bread wheat exploiting the untapped natural diversity. *Front. Genet.* **2022**, *13*, 89. [CrossRef] [PubMed]
29. Balla, A.; Tuymetova, G.; Barshishat, M.; Geiszt, M.; Balla, T. Characterization of type II phosphatidylinositol 4-kinase isoforms reveals association of the enzymes with endosomal vesicular compartments. *J. Biol. Chem.* **2002**, *277*, 20041–20050. [CrossRef]
30. Salava, H.; Thula, S.; Sánchez, A.S.; Nodzyński, T.; Maghuly, F. Genome Wide Identification and Annotation of NGATHA Transcription Factor Family in Crop Plants. *Int. J. Mol. Sci.* **2022**, *23*, 7063. [CrossRef]
31. Kadek, A.; Mrazek, H.; Halada, P.; Rey, M.; Schriemer, D.C.; Man, P. Aspartic protease nepenthesin-1 as a tool for digestion in hydrogen/deuterium exchange mass spectrometry. *Anal. Chem.* **2014**, *86*, 4287–4294. [CrossRef]
32. Shen, W.; Yao, X.; Ye, T.; Ma, S.; Liu, X.; Yin, X.; Wu, Y. Arabidopsis aspartic protease ASPG1 affects seed dormancy, seed longevity and seed germination. *Plant Cell Physiol.* **2018**, *59*, 1415–1431. [CrossRef]
33. Jimeno, S.; Aguilera, A. The THO complex as a key mRNP biogenesis factor in development and cell differentiation. *J. Biol.* **2010**, *9*, 6. [CrossRef]
34. Francisco-Mangilet, A.G.; Karlsson, P.; Kim, M.H.; Eo, H.J.; Oh, S.A.; Kim, J.H.; Kulcheski, F.R.; Park, S.K.; Manavella, P.A. THO 2, a core member of the THO/TREX complex, is required for micro RNA production in Arabidopsis. *Plant J.* **2015**, *82*, 1018–1029. [CrossRef] [PubMed]
35. Arif, M.R.; Neumann, K.; Nagel, M.; Kobiljski, B.; Lohwasser, U.; Börner, A. An association mapping analysis of dormancy and pre-harvest sprouting in wheat. *Euphytica* **2012**, *188*, 409–417. [CrossRef]
36. Nagel, M.; Vogel, H.; Landjeva, S.; Buck-Sorlin, G.; Lohwasser, U.; Scholz, U.; Börner, A. Seed conservation in ex situ genebanks—Genetic studies on longevity in barley. *Euphytica* **2009**, *170*, 5–14. [CrossRef]
37. Hou, Q.; Li, S.; Shang, C.; Wen, Z.; Cai, X.; Hong, Y.; Qiao, G. Genome-wide characterisation of chalcone synthase genes in Chinese cherry and functional characterisation of CpCHS1 under drought stress. *Front. Plant Sci.* **2022**, 3054.
38. Fini, A.; Brunetti, C.; di Ferdinando, M.; Ferrini, F.; Tattini, M. Stress-induced flavonoid biosynthesis and the antioxidant machinery of plants. *Plant Signal. Behav.* **2011**, *6*, 709–711. [CrossRef]
39. Yang, T.; Li, Y.; Liu, Y.; He, L.; Liu, A.; Wen, J.; Mysore, K.S.; Tadege, M.; Chen, J. The 3-ketoacyl-CoA synthase WFL is involved in lateral organ development and cuticular wax synthesis in *Medicago truncatula*. *Plant Mol. Biol.* **2021**, *105*, 193–204. [CrossRef]
40. Shi, X.; Che, Z.; Xu, G.; Ming, Z. Crystal structure of transcription factor TGA7 from Arabidopsis. *Biochem. Biophys. Res. Commun.* **2022**, *637*, 322–330. [CrossRef]
41. Schmidt, O.; Teis, D. The ESCRT machinery. *Curr. Biol.* **2012**, *22*, R116–R120. [CrossRef]
42. Wang, X.; Luo, Y.; Shi, L.; Pang, P.; Gao, G. Analysis of expression characteristics of scarecrow-like gene Stsl-1 elicited by exogenous hormone and *Ralstonia solanacearum* infection in potato. *Int. J. Agric. Biol.* **2019**, *22*, 201–208.
43. Lopez-Ortiz, C.; Peña-Garcia, Y.; Natarajan, P.; Bhandari, M.; Abburi, V.; Dutta, S.K.; Yadav, L.; Stommel, J.; Nimmakayala, P.; Reddy, U.K. The ankyrin repeat gene family in *Capsicum* spp: Genome-wide survey, characterization and gene expression profile. *Sci. Rep.* **2020**, *10*, 4044. [CrossRef] [PubMed]
44. Zhang, Z.; Shrestha, J.; Tateda, C.; Greenberg, J.T. Salicylic acid signaling controls the maturation and localization of the Arabidopsis defense protein ACCELERATED CELL DEATH6. *Mol. Plant* **2014**, *7*, 1365–1383. [CrossRef] [PubMed]
45. Omidbakhshfard, M.A.; Proost, S.; Fujikura, U.; Mueller-Roeber, B. Growth-regulating factors (GRFs): A small transcription factor family with important functions in plant biology. *Mol. Plant* **2015**, *8*, 998–1010. [CrossRef] [PubMed]
46. Friedmann, D.R.; Marmorstein, R. Structure and mechanism of non-histone protein acetyltransferase enzymes. *FEBS J.* **2013**, *280*, 5570–5581. [CrossRef] [PubMed]
47. Dekker, F.J.; van den Bosch, T.; Martin, N.I. Small molecule inhibitors of histone acetyltransferases and deacetylases are potential drugs for inflammatory diseases. *Drug Discov. Today* **2014**, *19*, 654–660. [CrossRef]
48. Hanada, K.; Sawada, Y.; Kuromori, T.; Klausnitzer, R.; Saito, K.; Toyoda, T.; Shinozaki, K.; Li, W.-H.; Hirai, M.Y. Functional compensation of primary and secondary metabolites by duplicate genes in *Arabidopsis thaliana*. *Mol. Biol. Evol.* **2011**, *28*, 377–382. [CrossRef]
49. Hove-Jensen, B.; Andersen, K.R.; Kilstrup, M.; Martinussen, J.; Switzer, R.L.; Willemoës, M. Phosphoribosyl diphosphate (PRPP): Biosynthesis, enzymology, utilization, and metabolic significance. *Microbiol. Mol. Biol. Rev.* **2017**, *81*, e00040-16. [CrossRef]
50. Hayashi, M.; Mori, H.; Nishimura, M.; Akazawa, T.; Hara-Nishimura, I. Nucleotide sequence of cloned cDNA coding for pumpkin 11-S globulin β subunit. *Eur. J. Biochem.* **1988**, *172*, 627–632. [CrossRef]
51. Ogasawara, S.; Abe, K.; Nakajima, T. Pepper β-galactosidase 1 (PBG1) plays a significant role in fruit ripening in bell pepper (*Capsicum annuum*). *Biosci. Biotechnol. Biochem.* **2007**, *71*, 309–322. [CrossRef]

52. Liu, T.; Arsenault, J.; Vierling, E.; Kim, M. Mitochondrial ATP synthase subunit d, a component of the peripheral stalk, is essential for growth and heat stress tolerance in *Arabidopsis thaliana*. *Plant J.* **2021**, *107*, 713–726. [CrossRef]
53. Yan, J.; Yao, Y.; Hong, S.; Yang, Y.; Shen, C.; Zhang, Q.; Zhang, D.; Zou, T.; Yin, P. Delineation of pentatricopeptide repeat codes for target RNA prediction. *Nucleic Acids Res.* **2019**, *47*, 3728–3738. [CrossRef] [PubMed]
54. Luo, T.; Zhang, Y.; Zhang, C.; Nelson, M.N.; Yuan, J.; Guo, L.; Xu, Z. Genome-wide association mapping unravels the genetic control of seed vigor under low-temperature conditions in rapeseed (*Brassica napus* L.). *Plants* **2021**, *10*, 426. [CrossRef]
55. Rosenberg, L.L.; Arnon, D.I. The preparation and properties of a new glyceraldehyde-3-phosphate dehydrogenase from photosynthetic tissues. *J. Biol. Chem.* **1955**, *217*, 361–371. [CrossRef]
56. Pacurar, D.I.; Pacurar, M.L.; Lakehal, A.; Pacurar, A.M.; Ranjan, A.; Bellini, C. The Arabidopsis Cop9 signalosome subunit 4 (CSN4) is involved in adventitious root formation. *Sci. Rep.* **2017**, *7*, 628. [CrossRef] [PubMed]
57. Li, X.; Liu, P.; Yang, P.; Fan, C.; Sun, X. Characterization of the glycerol-3-phosphate acyltransferase gene and its real-time expression under cold stress in Paeonia lactiflora Pall. *PLoS ONE* **2018**, *13*, e0202168. [CrossRef] [PubMed]
58. Lim, C.W.; Lee, S.C. Pepper protein phosphatase type 2C, CaADIP1 and its interacting partner CaRLP1 antagonistically regulate ABA signalling and drought response. *Plant Cell Environ.* **2016**, *39*, 1559–1575. [CrossRef]
59. Hatzfeld, Y.; Maruyama, A.; Schmidt, A.; Noji, M.; Ishizawa, K.; Saito, K. β-Cyanoalanine synthase is a mitochondrial cysteine synthase-like protein in spinach and Arabidopsis. *Plant Physiol.* **2000**, *123*, 1163–1172. [CrossRef]
60. Rabuma, T.; Gupta, O.P.; Yadav, M.; Chhokar, V. Integrative RNA-Seq analysis of *Capsicum annuum* L.-*Phytophthora capsici* L. pathosystem reveals molecular cross-talk and activation of host defence response. *Physiol. Mol. Biol. Plants* **2022**, *28*, 171–188. [CrossRef]
61. Wang, G.; Li, Q.; Wang, C.; Jin, C.; Ji, J.; Guan, C. A salicylic acid carboxyl methyltransferase-like gene LcSAMT from *Lycium chinense*, negatively regulates the drought response in transgenic tobacco. *Environ. Exp. Bot.* **2019**, *167*, 103833. [CrossRef]
62. Sgarbi, C.; Malbrán, I.; Saldúa, L.; Lori, G.A.; Lohwasser, U.; Arif, M.A.R.; Börner, A.; Yanniccari, M.; Castro, A.M. Mapping Resistance to Argentinean Fusarium (*Graminearum*) Head Blight Isolates in Wheat. *Int. J. Mol. Sci.* **2021**, *22*, 13653. [CrossRef]
63. Thodberg, S.; Jakobsen Neilson, E.H. The "green" FMOs: Diversity, functionality and application of plant flavoproteins. *Catalysts* **2020**, *10*, 329. [CrossRef]
64. Paul, V.D.; Mühlenhoff, U.; Stümpfig, M.; Seebacher, J.; Kugler, K.G.; Renicke, C.; Taxis, C.; Gavin, A.-C.; Pierik, A.J.; Lill, R. The deca-GX3 proteins Yae1-Lto1 function as adaptors recruiting the ABC protein Rli1 for iron-sulfur cluster insertion. *eLife* **2015**, *4*, e08231. [CrossRef] [PubMed]
65. Chen, R.; Guo, W.; Yin, Y.; Gong, Z.-H. A novel F-box protein CaF-box is involved in responses to plant hormones and abiotic stress in pepper (*Capsicum annuum* L.). *Int. J. Mol. Sci.* **2014**, *15*, 2413–2430. [CrossRef] [PubMed]
66. Liu, R.; Li, B.; Qin, G.; Zhang, Z.; Tian, S. Identification and functional characterization of a tonoplast dicarboxylate transporter in tomato (*Solanum lycopersicum*). *Front. Plant Sci.* **2017**, *8*, 186. [CrossRef]
67. Bradbury, P.J.; Zhang, Z.; Kroon, D.E.; Casstevens, T.M.; Ramdoss, Y.; Buckler, E.S. TASSEL: Software for association mapping of complex traits in diverse samples. *Bioinformatics* **2007**, *23*, 2633–2635. [CrossRef]
68. Kim, S.; Park, J.; Yeom, S.-I.; Kim, Y.-M.; Seo, E.; Kim, K.-T.; Kim, M.-S.; Lee, J.M.; Cheong, K.; Shin, H.-S. New reference genome sequences of hot pepper reveal the massive evolution of plant disease-resistance genes by retroduplication. *Genome Biol.* **2017**, *18*, 210. [CrossRef]
69. Li, H.; Handsaker, B.; Wysoker, A.; Fennell, T.; Ruan, J.; Homer, N.; Marth, G.; Abecasis, G.; Durbin, R. The sequence alignment/map format and SAMtools. *Bioinformatics* **2009**, *25*, 2078–2079. [CrossRef]
70. Driesen, E.; Van den Ende, W.; De Proft, M.; Saeys, W. Influence of Environmental Factors Light, CO_2, Temperature, and Relative Humidity on Stomatal Opening and Development: A Review. *Agronomy* **2020**, *10*, 1975. [CrossRef]
71. Zhou, Y.; Mumtaz, M.A.; Zhang, Y.; Shu, H.; Hao, Y.; Lu, X.; Cheng, S.; Zhu, G.; Wang, Z. Response of Anthocyanin Accumulation in Pepper (*Capsicum annuum*) Fruit to Light Days. *Int. J. Mol. Sci.* **2022**, *23*, 8357. [CrossRef]
72. Afzal, A.J.; Wood, A.J.; Lightfoot, D.A. Plant receptor-like serine threonine kinases: Roles in signaling and plant defense. *Mol. Plant Microbe Interact.* **2008**, *21*, 507–517. [CrossRef]
73. Mayorga-Gómez, A.; Nambeesan, S.U. Temporal expression patterns of fruit-specific α-EXPANSINS during cell expansion in bell pepper (*Capsicum annuum* L.). *BMC Plant Biol.* **2020**, *20*, 241. [CrossRef] [PubMed]
74. Elder, G.H.; Roberts, A.G. Uroporphyrinogen decarboxylase. *J. Bioenerg. Biomembr.* **1995**, *27*, 207–214. [CrossRef] [PubMed]
75. Kang, H.; Hwang, I. Vacuolar Sorting Receptor-Mediated Trafficking of Soluble Vacuolar Proteins in Plant Cells. *Plants* **2014**, *3*, 392–408. [CrossRef] [PubMed]
76. Wang, J.-P.; Xu, Y.-P.; Munyampundu, J.-P.; Liu, T.-Y.; Cai, X.-Z. Calcium-dependent protein kinase (CDPK) and CDPK-related kinase (CRK) gene families in tomato: Genome-wide identification and functional analyses in disease resistance. *Mol. Genet. Genom.* **2015**, *291*, 661–676. [CrossRef]
77. Iyamu, I.; Al-Hamashi, A.; Huang, R. A Pan-Inhibitor for Protein Arginine Methyltransferase Family Enzymes. *Biomolecules* **2021**, *11*, 854. [CrossRef]
78. Manohar, M.; Shigaki, T.; Hirschi, K.D. Plant cation/H+ exchangers (CAXs): Biological functions and genetic manipulations. *Plant Biol.* **2011**, *13*, 561–569. [CrossRef]

79. Aitouguinane, M.; El Alaoui-Talibi, Z.; Rchid, H.; Fendri, I.; Abdelkafi, S.; El-Hadj, M.D.O.; Boual, Z.; Dubessay, P.; Michaud, P.; Le Cerf, D.; et al. A Novel Sulfated Glycoprotein Elicitor Extracted from the Moroccan Green Seaweed *Codium decorticatum* Induces Natural Defenses in Tomato. *Appl. Sci.* **2022**, *12*, 3643. [CrossRef]
80. Jin, J.F.; He, Q.Y.; Li, P.F.; Lou, H.Q.; Chen, W.W.; Yang, J.L. Genome-Wide Identification and Gene Expression Analysis of Acyl-Activating Enzymes Superfamily in Tomato (*Solanum lycopersicum*) under Aluminum Stress. *Front. Plant Sci.* **2021**, *12*, 754147. [CrossRef]
81. Bick, J.A.; Setterdahl, A.T.; Knaff, D.B.; Chen, Y.; Pitcher, L.H.; Zilinskas, B.A.; Leustek, T. Regulation of the plant-type 5′-adenylyl sulfate reductase by oxidative stress. *Biochemistry* **2001**, *40*, 9040–9048. [CrossRef]
82. Yang, S.; Huang, L.; Song, J.; Liu, L.; Bian, Y.; Jia, B.; Wu, L.; Xin, Y.; Wu, M.; Zhang, J.; et al. Genome-Wide Analysis of DA1-Like Genes in Gossypium and Functional Characterization of GhDA1-1A Controlling Seed Size. *Front. Plant Sci.* **2021**, *12*, 647091. [CrossRef]
83. Kim, J.H.; Tsukaya, H. Regulation of plant growth and development by the growth-regulating factor and grf-interacting factor duo. *J. Exp. Bot.* **2015**, *66*, 6093–6107. [CrossRef] [PubMed]
84. Hahn, A.; Vonck, J.; Mills, D.J.; Meier, T.; Kühlbrandt, W. Structure, mechanism, and regulation of the chloroplast ATP synthase. *Science* **2018**, *360*, eaat4318. [CrossRef] [PubMed]
85. Su, H.-G.; Zhang, X.-H.; Wang, T.-T.; Wei, W.-L.; Wang, Y.-X.; Chen, J.; Zhou, Y.-B.; Chen, M.; Ma, Y.-Z.; Xu, Z.-S.; et al. Genome-Wide Identification, Evolution, and Expression of GDSL-Type Esterase/Lipase Gene Family in Soybean. *Front. Plant Sci.* **2020**, *11*, 726. [CrossRef] [PubMed]
86. Ruegger, M.; Dewey, E.; Gray, W.M.; Hobbie, L.; Turner, J.; Estelle, M. The TIR1 protein of *Arabidopsis* functions in auxin response and is related to human SKP2 and yeast Grr1p. *Genes Dev.* **1998**, *12*, 198–207. [CrossRef]
87. Nick, P.; Heuing, A.; Ehmann, B. Plant chaperonins: A role in microtubule-dependent wall formation? *Protoplasma* **2000**, *211*, 234–244. [CrossRef]
88. Sarnowski, T.J.; Ríos, G.; Jásik, J.; Swiezewski, S.; Kaczanowski, S.; Li, Y.; Kwiatkowska, A.; Pawlikowska, K.; Kozbiał, M.; Kozbiał, P.; et al. SWI3 Subunits of Putative SWI/SNF Chromatin-Remodeling Complexes Play Distinct Roles during *Arabidopsis* Development. *Plant Cell* **2005**, *17*, 2454–2472. [CrossRef]
89. Hove-Jensen, B. Mutation in the phosphoribosylpyrophosphate synthetase gene (prs) that results in simultaneous requirements for purine and pyrimidine nucleosides, nicotinamide nucleotide, histidine, and tryptophan in *Escherichia coli*. *J. Bacteriol.* **1988**, *170*, 1148–1152. [CrossRef]
90. Guo, C.; Guo, L.; Li, X.; Gu, J.; Zhao, M.; Duan, W.; Ma, C.; Lu, W.; Xiao, K. TaPT2, a high-affinity phosphate transporter gene in wheat (*Triticum aestivum* L.), is crucial in plant Pi uptake under phosphorus deprivation. *Acta Physiol. Plant.* **2014**, *36*, 1373–1384. [CrossRef]
91. Curien, G.; Giustini, C.; Montillet, J.-L.; Mas-Y-Mas, S.; Cobessi, D.; Ferrer, J.-L.; Matringe, M.; Grechkin, A.; Rolland, N. The chloroplast membrane associated ceQORH putative quinone oxidoreductase reduces long-chain, stress-related oxidized lipids. *Phytochemistry* **2016**, *122*, 45–55. [CrossRef]
92. Kaczmarska, A.; Pieczywek, P.M.; Cybulska, J.; Zdunek, A. Structure and functionality of Rhamnogalacturonan I in the cell wall and in solution: A review. *Carbohydr. Polym.* **2021**, *278*, 118909. [CrossRef]
93. Hayama, R.; Yang, P.; Valverde, F.; Mizoguchi, T.; Furutani-Hayama, I.; Vierstra, R.D.; Coupland, G. Ubiquitin carboxyl-terminal hydrolases are required for period maintenance of the circadian clock at high temperature in Arabidopsis. *Sci. Rep.* **2019**, *9*, 17030. [CrossRef] [PubMed]
94. Garcıa-Pineda, E.; Castro-Mercado, E.; Lozoya-Gloria, E. Gene expression and enzyme activity of pepper (*Capsicum annuum* L.) ascorbate oxidase during elicitor and wounding stress. *Plant Sci.* **2004**, *166*, 237–243. [CrossRef]
95. Oomen, R.J.; Doeswijk-Voragen, C.H.; Bush, M.S.; Vincken, J.P.; Borkhardt, B.; Van Den Broek, L.A.; Visser, R.G. In muro fragmentation of the rhamnogalacturonan I backbone in potato (*Solanum tuberosum* L.) results in a reduction and altered location of the galactan and arabinan side-chains and abnormal periderm development. *Plant J.* **2002**, *30*, 403–413. [CrossRef] [PubMed]
96. Vogel, G.; Fiehn, O.; Jean-Richard-Dit-Bressel, L.; Boller, T.; Wiemken, A.; Aeschbacher, R.A.; Wingler, A. Trehalose metabolism in Arabidopsis: Occurrence of trehalose and molecular cloning and characterization of trehalose-6-phosphate synthase homologues. *J. Exp. Bot.* **2001**, *52*, 1817–1826. [CrossRef] [PubMed]
97. Bonner, E.R.; Cahoon, R.E.; Knapke, S.M.; Jez, J.M. Molecular basis of cysteine biosynthesis in plants: Structural and functional analysis of *O*-acetylserine sulfhydrylase from *Arabidopsis thaliana*. *J. Biol. Chem.* **2005**, *280*, 38803–38813. [CrossRef]
98. Sanders, S.L.; Weil, P. Identification of Two Novel TAF Subunits of the Yeast *Saccharomyces cerevisiae* TFIID Complex. *J. Biol. Chem.* **2000**, *275*, 13895–13900. [CrossRef]
99. Dubos, C.; Stracke, R.; Grotewold, E.; Weisshaar, B.; Martin, C.; Lepiniec, L. MYB transcription factors in Arabidopsis. *Trends Plant Sci.* **2010**, *15*, 573–581. [CrossRef]
100. Baek, H.J.; Kang, Y.K.; Roeder, R.G. Human Mediator Enhances Basal Transcription by Facilitating Recruitment of Transcription Factor IIB during Preinitiation Complex Assembly. *J. Biol. Chem.* **2006**, *281*, 15172–15181. [CrossRef]
101. Lopez-Ortiz, C.; Dutta, S.K.; Natarajan, P.; Peña-Garcia, Y.; Abburi, V.; Saminathan, T.; Nimmakayala, P.; Reddy, U.K. Genome-wide identification and gene expression pattern of ABC transporter gene family in *Capsicum* spp. *PLoS ONE* **2019**, *14*, e0215901. [CrossRef]

102. Sun, X.-L.; Yu, Q.-Y.; Tang, L.-L.; Ji, W.; Bai, X.; Cai, H.; Liu, X.-F.; Ding, X.-D.; Zhu, Y.-M. GsSRK, a G-type lectin S-receptor-like serine/threonine protein kinase, is a positive regulator of plant tolerance to salt stress. *J. Plant Physiol.* **2012**, *170*, 505–515. [CrossRef]
103. Bonza, M.C.; Morandini, P.; Luoni, L. At-ACA8 encodes a plasma membrane-localized calcium-ATPase of Arabidopsis with a calmodulin-binding domain at the N terminus. *Plant Physiol.* **2000**, *123*, 1495–14506. [CrossRef] [PubMed]
104. Bar, M.; Aharon, M.; Benjamin, S.; Rotblat, B.; Horowitz, M.; Avni, A. AtEHDs, novel Arabidopsis EH-domain-containing proteins involved in endocytosis. *Plant J.* **2008**, *55*, 1025–1038. [CrossRef]
105. Cross, R.L.; Müller, V. The evolution of A-, F-, and V-type ATP synthases and ATPases: Reversals in function and changes in the H+/ATP coupling ratio. *FEBS Lett.* **2004**, *576*, 1–4. [CrossRef] [PubMed]
106. Martínez, O.; Arce-Rodríguez, M.; Hernández-Godínez, F.; Escoto-Sandoval, C.; Cervantes-Hernández, F.; Hayano-Kanashiro, C.; Ordaz-Ortiz, J.; Reyes-Valdés, M.; Razo-Mendivil, F.; Garcés-Claver, A.; et al. Transcriptome Analyses Throughout Chili Pepper Fruit Development Reveal Novel Insights into the Domestication Process. *Plants* **2021**, *10*, 585. [CrossRef] [PubMed]
107. Wang, J.; Sun, D.; Wang, M.; Cheng, A.; Zhu, Y.; Mao, S.; Ou, X.; Zhao, X.; Huang, J.; Gao, Q.; et al. Multiple functions of heterogeneous nuclear ribonucleoproteins in the positive single-stranded RNA virus life cycle. *Front. Immunol.* **2022**, *13*, 989298. [CrossRef] [PubMed]
108. Jasiński, M.; Stukkens, Y.; Degand, H.; Purnelle, B.; Marchand-Brynaert, J.; Boutry, M. A plant plasma membrane ATP binding cassette–type transporter is involved in antifungal terpenoid secretion. *Plant Cell* **2001**, *13*, 1095–1107. [CrossRef]
109. Piotrowski, M.; Janowitz, T.; Kneifel, H. Plant C-N Hydrolases and the Identification of a Plant *N*-Carbamoylputrescine Amidohydrolase Involved in Polyamine Biosynthesis. *J. Biol. Chem.* **2003**, *278*, 1708–1712. [CrossRef]
110. Fischer, C.; DeFalco, T.; Karia, P.; Snedden, W.A.; Moeder, W.; Yoshioka, K.; Dietrich, P. Calmodulin as a Ca^{2+}-Sensing Subunit of Arabidopsis Cyclic Nucleotide-Gated Channel Complexes. *Plant Cell Physiol.* **2017**, *58*, 1208–1221. [CrossRef]
111. Chung, Y.S.; Lee, Y.G.; Silva, R.R.; Park, S.; Park, M.Y.; Lim, Y.P.; Kim, C. Potential SNPs related to microspore culture in *Raphanus sativus* based on a single-marker analysis. *Can. J. Plant Sci.* **2018**, *98*, 1072–1083. [CrossRef]
112. Available online: https://www.arabidopsis.org/servlets/TairObject?name=AT2G41630&type=locus (accessed on 3 January 2023).
113. Zhang, Q.; Liu, H. Functioning mechanisms of Shugoshin-1 in centromeric cohesion during mitosis. *Essays Biochem.* **2020**, *64*, 289–297. [CrossRef]

plants

MDPI

Article

QTL Analysis for Bread Wheat Seed Size, Shape and Color Characteristics Estimated by Digital Image Processing

Mian Abdur Rehman Arif [1,†], Evgenii G. Komyshev [2,3,4,†], Mikhail A. Genaev [2,3,4], Vasily S. Koval [2,4], Nikolay A. Shmakov [2,3,4], Andreas Börner [5,*] and Dmitry A. Afonnikov [2,3,4,*]

1 Nuclear Institute for Agriculture and Biology, Faisalabad 38000, Pakistan
2 Institute of Cytology and Genetics, Siberian Branch of the Russian Academy of Sciences, 630090 Novosibirsk, Russia
3 Faculty of Natural Sciences, Novosibirsk State University, 630090 Novosibirsk, Russia
4 Kurchatov Genomics Center, Institute of Cytology and Genetics, Siberian Branch of the Russian Academy of Sciences, 630090 Novosibirsk, Russia
5 Leibniz Institute of Plant Genetics and Crop Plant Research, 06466 Seeland, Germany
* Correspondence: boerner@ipk-gatersleben.de (A.B.); ada@bionet.nsc.ru (D.A.A.); Tel.: +49-394825229 (A.B.); +7-(383)-363-49-63 (D.A.A.)
† These authors contributed equally to this work.

Abstract: The size, shape, and color of wheat seeds are important traits that are associated with yield and flour quality (size, shape), nutritional value, and pre-harvest sprouting (coat color). These traits are under multigenic control, and to dissect their molecular and genetic basis, quantitative trait loci (QTL) analysis is used. We evaluated 114 recombinant inbred lines (RILs) in a bi-parental RIL mapping population (the International Triticeae Mapping Initiative, ITMI/MP) grown in 2014 season. We used digital image analysis for seed phenotyping and obtained data for seven traits describing seed size and shape and 48 traits of seed coat color. We identified 212 additive and 34 pairs of epistatic QTLs on all the chromosomes of wheat genome except chromosomes 1A and 5D. Many QTLs were overlapping. We demonstrated that the overlap between QTL regions was low for seed size/shape traits and high for coat color traits. Using the literature and KEGG data, we identified sets of genes in *Arabidopsis* and rice from the networks controlling seed size and color. Further, we identified 29 and 14 candidate genes for seed size-related loci and for loci associated with seed coat color, respectively.

Keywords: wheat; seed size; seed shape; seed coat color; phenotyping; candidate genes; QTLs

Citation: Arif, M.A.R.; Komyshev, E.G.; Genaev, M.A.; Koval, V.S.; Shmakov, N.A.; Börner, A.; Afonnikov, D.A. QTL Analysis for Bread Wheat Seed Size, Shape and Color Characteristics Estimated by Digital Image Processing. *Plants* **2022**, *11*, 2105. https://doi.org/10.3390/plants11162105

Academic Editor: Abdelmajid Kassem

Received: 22 July 2022
Accepted: 10 August 2022
Published: 12 August 2022

1. Introduction

Bread wheat (*Triticum aestivum* L.) is a major staple crop. Millions of people depend on its production (https://www.fao.org/faostat/en/#data, accessed on 20 January 2022). This has led to an ongoing search for and study of genes that control wheat yield traits. Some of them are the characteristics of wheat seeds (size and shape) which have been shown to be related to seed weight [1–4]. Seed size and shape have also been shown to be related to flour quality and composition: small kernels can contribute to enhancing the bread-making quality of flour while having a detrimental effect on the milling yield [5]. To find genes that control these traits in wheat, QTL analysis is used. This analysis makes it possible to identify sets of markers that are associated with seed size or shape traits. Studies have shown that seed size and shape in wheat are controlled by a large number of loci located on almost all chromosomes [6–13]. Identification of these loci combined with molecular analysis can identify genes that are involved in controlling seed weight or size [14–19]. Based on genetic and molecular studies in both the model organism *Arabidopsis thaliana* and cereals, it is now established that seed weight is affected by multiple molecular and genetic aspects that lead to dynamic changes in cell division, expansion, and differentiation during seed development. Several important biological pathways contribute to seed weight, such as ubiquitination, phytohormones, G-proteins, photosynthesis, epigenetic modifications, and microRNAs [20,21]. Knowledge of the pathways controlling seed development in well-studied organisms allows the prioritization of candidate genes controlling these traits in wheat as well for their further study by molecular methods [15].

Another important characteristic of wheat seeds is the color of the shell. It characterizes the pigments and metabolites it contains. Purple and blue coloring of seeds is determined by the presence of anthocyanins. A yellowish color may be due to the presence of carotenoids. A reddish brown or dark brown coloration of the seeds is due to the presence of flavonoids such as proanthocyanidins and phlobaphenes [22]. Genetic control of color formation in both seeds and other plant organs is carried out by genes encoding the enzymes involved in pigment biosynthesis as well as regulatory genes [23]. For a number of pigments, these genes have been well studied; however, for some pigments, the molecular mechanisms of biosynthesis are still poorly understood [24].

The presence of pigments in the seed coat affects various technological properties of the seed [25] and is associated with antioxidant properties [26]. Therefore, varieties and lines with diverse seed coloration are of active interest in the food industry [27,28]. Seed shell color in wheat is also associated with important characteristics such as germination ability and pre-harvest sprouting (PHS). Red seeds are less susceptible to PHS [29]. QTL searches for seed color and PHS resistance are often simultaneously performed [30,31].

Recently, genotyping technologies have made great progress and include diversity array technology, genotyping-by-sequencing [32], and SNPs [33,34]. High-throughput genotyping can achieve high-density marker mapping [35]. This allows more QTLs to be obtained and, as a result, more accurately establish the molecular mechanisms controlling important plant phenotypic traits [8,32,33,36,37].

In the present work, we performed a SNP-based QTL search for seven traits of seed size/shape and 48 traits of shell color evaluated on the basis of digital image analysis on a set of 114 recombinant inbred lines (RILs) of the "International Triticeae Mapping Initiative" mapping population (ITMI/MP) and their parental plants.

2. Results

2.1. Analysis of the Seed Traits in ITMI Population

Figure 1 shows the distribution of six of the fifty-five seed characteristics in the ITMI population. Three of them characterize size (sL, sW, sA), and three characterize color (Lab_mL, Lab_ma, Lab_mb). The distributions were bell-shaped, and the hypothesis of normality was not rejected for the characteristics of seed length and area (Shapiro–Wilks test, $p < 0.05$), but for the width and color characteristics (Figure 1). Overall, the hypothesis of normality was not rejected based on this test for 22 of the 55 traits.

In order to visualize the distribution of genotypes in the space of the considered traits, we performed principal component analysis for the traits of shape/size (all seven traits) (Figure 2), color (12 traits of average values of color components in four-color spaces) of seeds independently (Figure 3) and all these 19 traits simultaneously (Figure S1). From the diagram of the principal components in the size/shape feature space, we see that the first component characterizes the roundness of the seeds and is correlated with circularity. The second component characterizes seed size (most related to width and area). The most characteristic genotypes are ITMI_082 (the most rounded seeds), ITMI_075 (large area), ITMI_048 (small area), and ITMI_111 and Synthetic_W7984 (most elongated). The second parental genotype is located in this diagram on the far-right side of the diagram, close to the *X*-axis, i.e., it has a rounded seed shape. It is difficult to distinguish any noticeable clusters in this diagram: it is a cloud with some distant genotypes. Notably, this cloud is more sparce in the upper half-plane (PC2 > 0) and compact in the lower half-plane (PC2 < 0).

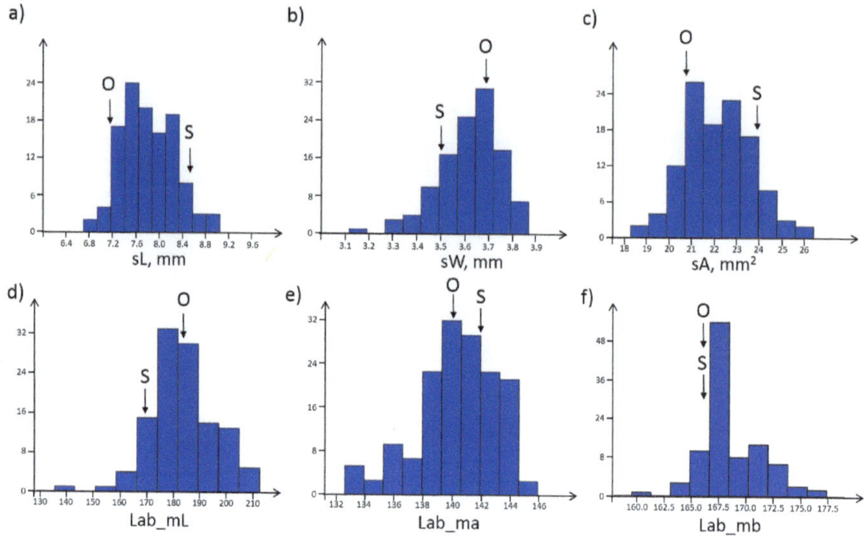

Figure 1. Distribution of six characteristics of seed size and color. The *X*-axis is the value of the characteristic, and the *Y*-axis is the frequency in the sample. (**a**) seed length, sL; (**b**) seed width, sW; (**c**) area of seed projection in the image, sA; (**d**) intensity of component L of Lab color space (lightness); (**e**) intensity of component a of Lab color space (redness); (**f**) intensity of component b of Lab color space (yellowness). The arrows show the characteristic values for the parental genotypes Opata (O) and Synthetic (S).

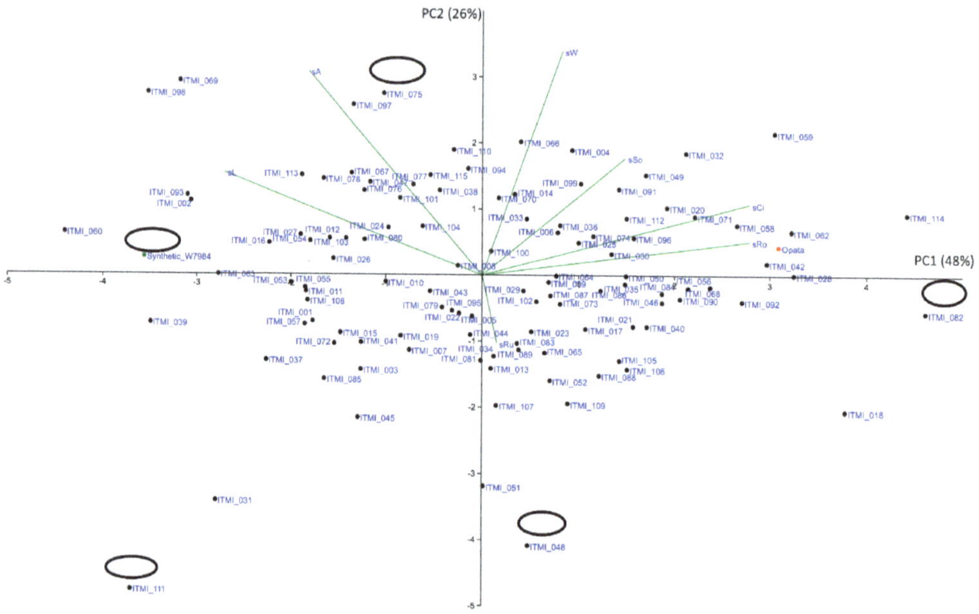

Figure 2. PCA biplot of seed size and shape of ITMI/MP performed using seed 7 coat traits. Ellipses represent seed size and shape for some contrast genotypes in the same scale. Parent genotypes are shown by green (Synthetic_W7984) and red (Opata) dots. PC1, PC2 axes denote principal components 1 and 2, percentage of variance explained shown in parentheses.

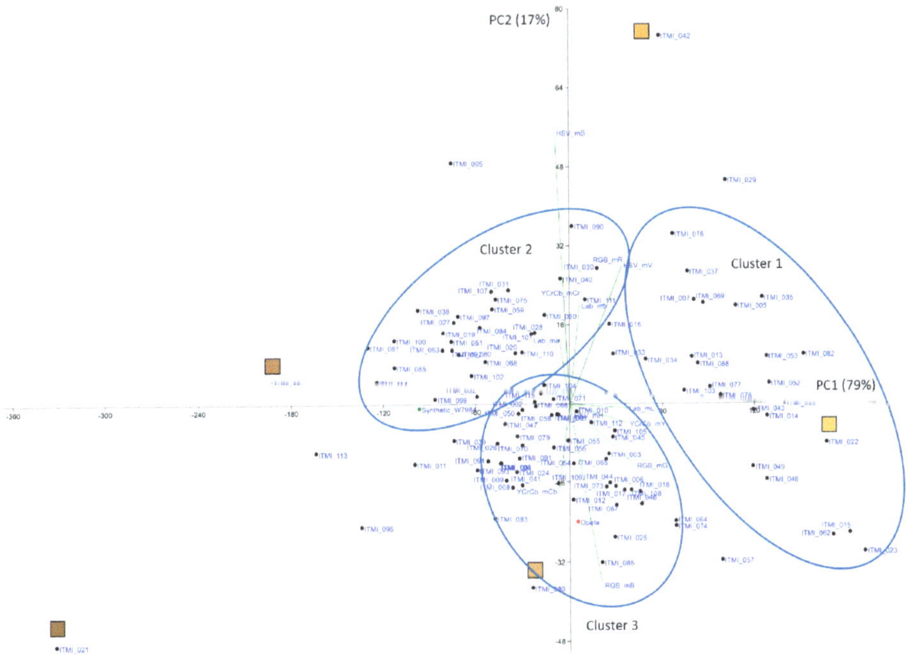

Figure 3. PCA biplot of seed coat color of ITMI/MP performed using seed coat traits (mean values for 12 color components of four-color spaces). Color bars represent coat color for some contrast genotypes. Parent genotypes shown by green (Synthetic_W7984) and red (Opata) dots. PC1 and PC2 axes denote principal components 1 and 2. Explained percentage of variance shown in parentheses. Three clusters of genotypes shown by ellipses.

The diagram of the principal components in the color feature space shows that the first component primarily characterizes the lightness of the shell (correlates with Lab_mL and YCrCb_mY). The second component characterizes seed color saturation and reddish shade (positively correlates with HSV_mS, RGB_mR, and Lab_ma characteristics). The most characteristic genotypes are ITMI_2 (the lightest shell), ITMI_042 (the most saturated color), ITMI_088 (the palest shell hue), and ITMI_021 and ITMI_087 (the darkest shell hue). Three clusters can be distinguished on the plot (Figure 3). Seeds from plants of the first cluster have a lighter color (large values of PC1), with a large part of them having a more reddish color (positive values of PC2). Seeds from plants of the second cluster have darker (negative values of PC1) and more reddish color (positive values of PC2). Seeds from plants of the second cluster have less reddish color (positive values of PC2 and PC1 values dispersed about 0 value). Interestingly, parent genotypes Synthetic_W7984 and Opata fall into distinct clusters on the plot (Cluster 2 and 3, respectively).

The diagram of the principal components in the seven size/shape and 12 color feature space (Figure S1) shows that the first component characterizes the color of the seed shell: the negative values are characteristics of reddish color (PC1 positively correlates with Lab_mb and negatively correlates with Lab_ma). The second component characterizes seed size/shape (positively correlated with roundness, sRo, and circularity, sCi and negatively correlates with area, sA, and length, sL). No clear clusters were detected in this plot for genotypes.

2.2. *QTL Analysis*

Genetic analysis of the three characteristics of seed size (sL, sW, and sA), four characteristics of shape (sCi, sRo, sRu, and sSo), and 48 characteristics of color (12 characteristics each of RGB, HSV, L *a *b, and YCrCb) revealed a total of 20, 22 and 170 QTLs (212 in total) (Figure 4, Table S1), correspondingly, on all the chromosomes of wheat genome except chromosomes 1A and 5D. The number of QTLs varied from one (characteristics: HSV_mS, HSV_dCH_2, HSV_dCS_2, and HSV_dCV_2) to ten (characteristic: sA) for one single trait. The majority of the traits yielded three (13 characters) to four QTLs (19 characters).

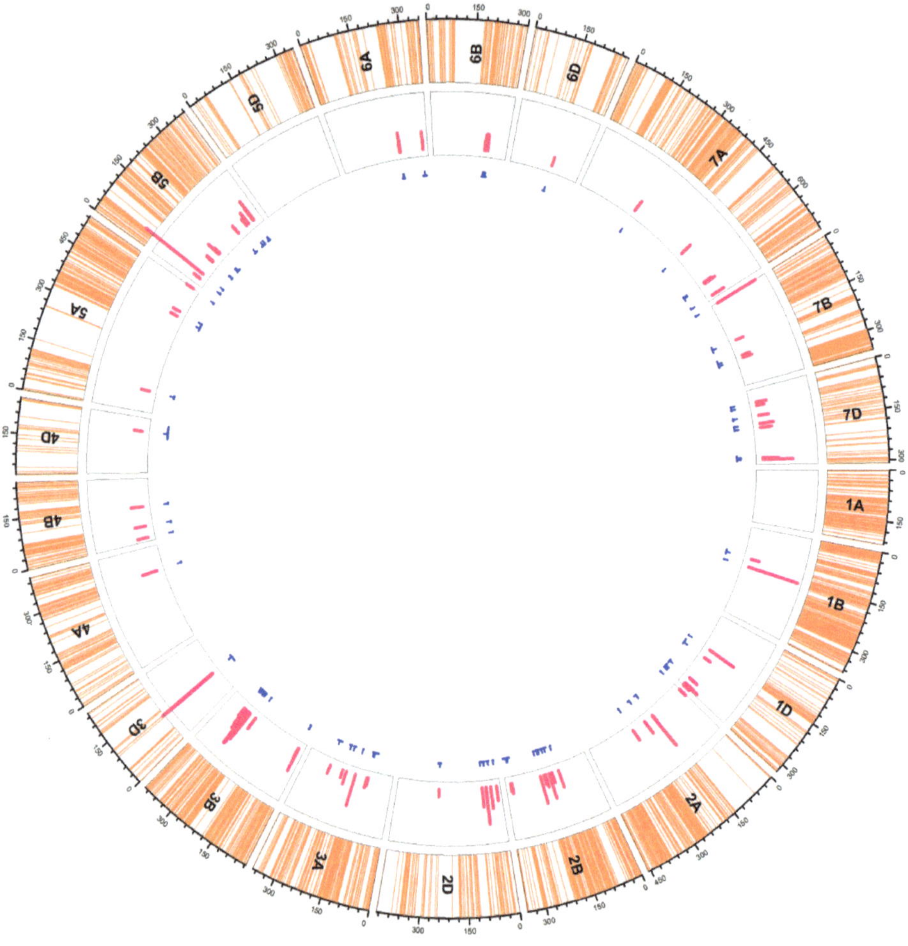

Figure 4. Distribution of additive QTLs (blue lines in the inner circle). Light orange lines in the outer track indicate the SNP positions on each chromosome. Pink bars in the second circle indicate the LOD values of QTLs. The blue lines under the track circle indicate the confidence interval of QTLs with small vertical lines pointing to the peak position of QTL. For details, see Table S1.

Among chromosomes, the highest number of QTLs was observed on chromosome 3B (46 QTLs), followed by chromosomes 3D and 6B with 34 and 27 QTLs, correspondingly. Chromosome 5B carried 15 QTLs, and the chromosome 2B carried 14 QTLs, whereas chromosome 7A carried 12 QTLs. This was followed by chromosome 1D with 11 QTLs. Chromosomes 3A and 7D carried nine QTLs each, and chromosome 6A carried seven QTLs. Chromosomes 6A and 2D carried seven and six QTLs, respectively. Five QTLs resided on each of chromosomes 2A and 4B, whereas four QTLs resided on each of chromosomes 5A and 7B. On the other hand, two QTLs were detected on each of chromosomes 1B and 6D. Finally, chromosomes 4A and 4D carried one QTL each. In terms of groups, group 3 chromosomes carried the highest number of QTLs (89), whereas group 4 chromosomes carried the least number of QTLs (five). Group 6 chromosomes carried 36 QTLs, whereas each of group 2 and 7 chromosomes carried 25 QTLs. On the other hand, group 5 chromosomes carried 19 QTLs, and group 1 chromosomes carried 13 QTLs.

Additionally, we were able to detect 34 pairs of epistatic QTLs controlling at least 22 characters in our RILs, with five characters under the influences of more than one pair of epistatic QTLs (Figure 5, Table S2). These QTLs involved all the wheat chromosomes except chromosomes 1A, 4A, 4B, and 6A. The most frequently involved chromosome was 3D (12 times), followed by chromosomes 3A (11 times), and 3B (nine times). Chromosome 2D was involved six times, whereas chromosomes 5B and 5D were involved four times each. Chromosomes 1D, 2B, and 7A were involved three times each. Two times involvement was observed for chromosomes 1B, 2A, 4D, 6D, and 6D, whereas the chromosomes 5A, 7B, and 7D were only involved one time.

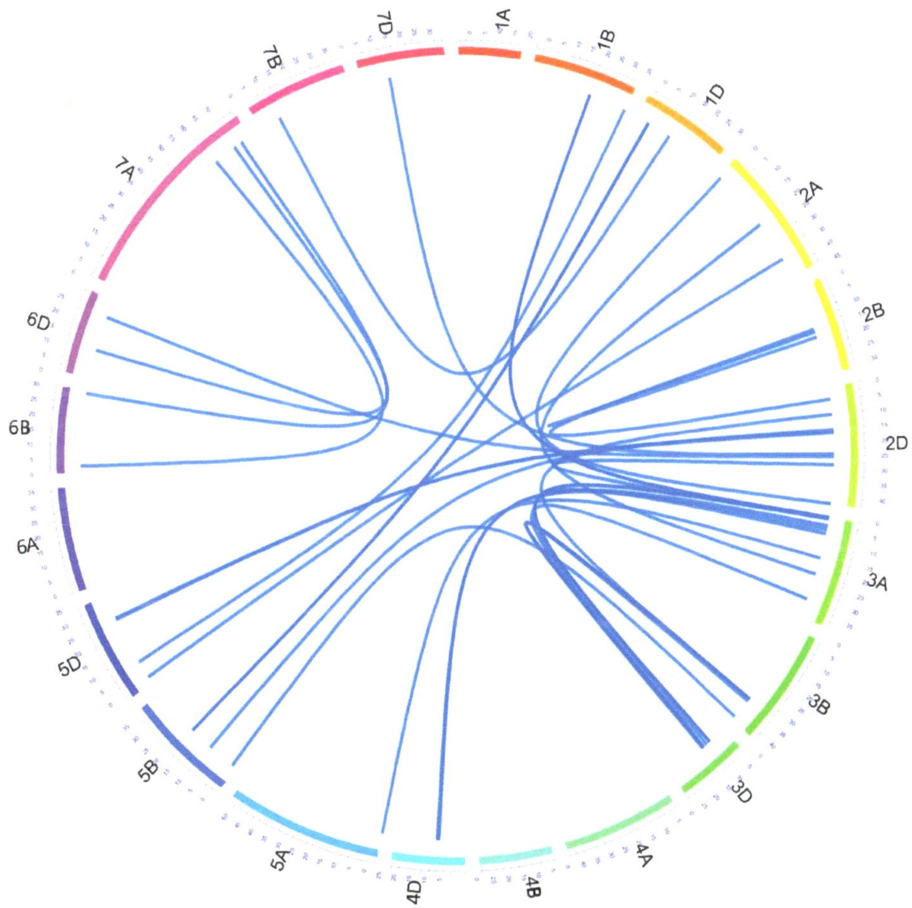

Figure 5. Epistasis QTL network in the ITMI/MP. Outer circular plot represents the hexaploid genome arranged in chromosomes (chrs) 1–21 (1A–7D) in clockwise direction. Numbers on colored outer circle represents cM on respective chrs. Blue-colored connections represent epistasis QTLs controlling different traits. For details, see Table S2.

2.3. Analysis of the Similarity of Traits by QTL Location

We observed remarkable overlap between QTL locations for different traits. For example, two QTLs related to seed shape, *Q.sCi-2B* [c] (circularity) and *Q.sSo-2B* [c] (solidity), were located in the same position 129 of chromosome 2B. The 3B chromosome has loci with multiple traits associations: position 39.179 (two traits of size), position 298.179 (seven color traits), position 299.179 (twelve color traits), position 300.179 (two color traits), position 306.179 (ten color traits), position 308.179 (two color traits), position 311.179 (two color traits and one shape trait, rugosity), and position 324.179 (three color traits). This is not surprising because our parameters estimated from images represent various quantifications of the same biological seed property (i.e., seed weight, pigment concentrations in the coat, etc.). This suggests that the set of our characteristics is degenerating and that many of them, in fact, are controlled by the same genes.

To evaluate the similarity of various traits under analysis, we hierarchically clustered them by the degree of the overlap of QTL locations (Figure 6). The tree diagram demonstrates several interesting features. Firstly, size/shape characteristics (right part of the tree) are clearly separated from the color traits (with the exception of rugosity, sRu). Secondly, some traits with a small QTL number (one to two) are also separated from other traits. Thirdly, a remarkable number of traits related to yellowness form a large cluster. Finally, traits related to the seed lightness (Lab_mL, HSV_mV, and YCrCb_mY) fall in the same cluster, and their QTLs are highly overlapped. Other color traits are irregularly dispersed on the tree within the large cluster of color traits.

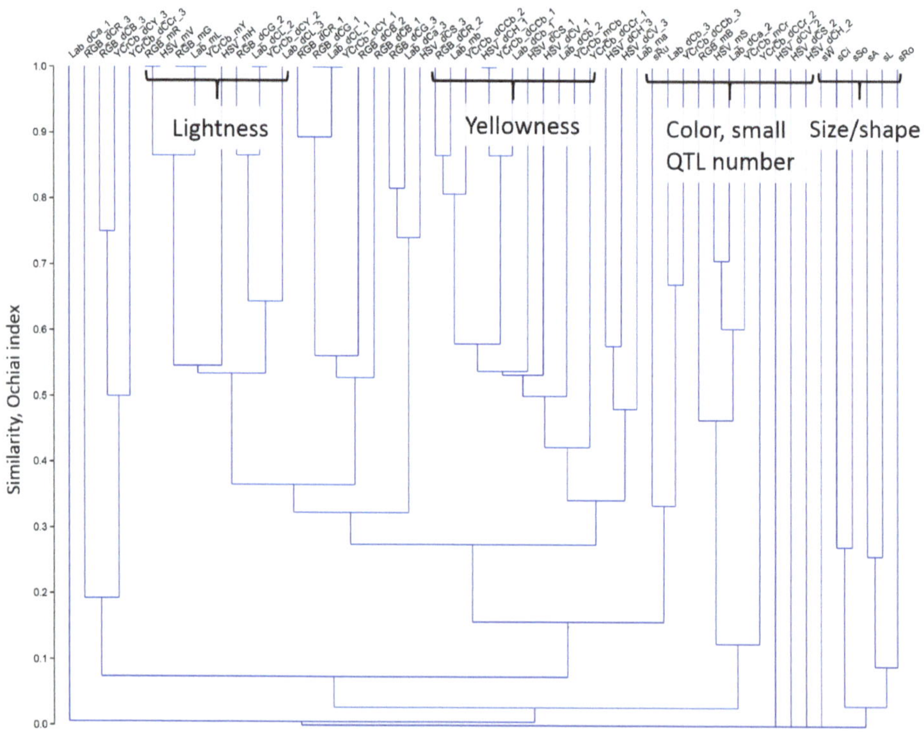

Figure 6. The similarity tree for seed traits obtained by the degree of the overlap between their QTL locations. The vertical axis represents the similarity measure based on the Ochiai index (*Y*-axis). Leaves correspond to seed traits described in [1] (for trait abbreviation, see Section 4.2. *Quantitative Characteristics of Seed Shape, Size and Color*). Groups of traits with strong overlapping of the QTL locations are shown by curly brackets.

2.4. Gene Prioritization

A search for orthologous groups for the eight pathways of pigment biosynthesis and their precursors identified 307 KEGG orthologs involved in these processes (Table S3). A review of the literature [18,19,38] identified 155 *Arabidopsis* and 42 rice genes involved in the molecular processes of seed development (Table S3). Of these genes, 193 were found to have sequence identifiers in the KEGG database, and 109 of them were associated with KEGG orthologous groups (Table S4).

For prioritization of genes, we used 48 highly significant QTLs with LOD > 3 for which marker positions were identified in the wheat genomic sequence (Table S5). On this basis, we identified 2787 unique genes localized to marker-limited sites. Of these, 1422 genes associated with seed size/shape, and 1365 genes associated with seed color. After filtering by expression level, 823 genes associated with seed color remained (Table S6). For these sets of genes, we performed KEGG orthogroup assignment using BlastKOALA and KofamKOALA services. For 464 genes associated with size trait loci and 321 genes associated with color traits, such orthogroups were found.

For 29 genes from the seed size-related loci, we found a match within the orthogroup list obtained based on the analysis of the literature data (Table 1). Eleven genes identified in this way belong to regulatory proteins (transcription factors EREBP, HD-ZIP, and MYBP; loci on chromosomes 3A, 2B, 2D, and 7D). Six genes belong to translation initiation factors (ELF2C; loci on chromosomes 2B and 7D). Five genes relate to enzymes associated with ubiquitination processes (loci on chromosomes 2D). Four genes have chitinase activity (locus on chromosome 7D), two genes with cytokin dehydrogenase activity (chromosomes 3A and 7D), and one aarF domain-containing kinase gene (chromosome 7D).

Table 1. List of candidate genes from QTLs associated with seed size/shape. Columns of the table contain QTL name (QTL), chromosome and position in cm (Chr/Pos), gene ID, KEGG orthogroup ID, KEGG orthogroup description, and EC number, if provided.

QTL	Chr/Pos	Gene ID	KO ID	Description	EC
Q.sA-3A	3A/155	*TraesCS3A03G0787100*	K09286	EREBP; EREBP-like factor	-
		TraesCS3A03G0782400	K09338	HD-ZIP; homeobox-leucine zipper protein	-
		TraesCS3A03G0763900	K00279	CKX; cytokinin dehydrogenase	EC:1.5.99.12
Q.sSo-4A	4A/305	*TraesCS4A03G1100100*	K19045	BB; E3 ubiquitin-protein ligase BIG BROTHER and related proteins	EC:2.3.2.27
Q.sA-2B.2	2B/208	*TraesCS2B03G0313000*	K09286	EREBP; EREBP-like factor	-
Q.sCi-2B [c]; sSo-2B [c]	2B/129	*TraesCS2B03G1115600*	K11593	ELF2C, AGO; eukaryotic translation initiation factor 2C	-
		TraesCS2B03G1114800	K11593	ELF2C, AGO; eukaryotic translation initiation factor 2C	-
		TraesCS2B03G1105600	K09422	MYBP; transcription factor MYB, plant	-
		TraesCS2B03G1109700	K09286	EREBP; EREBP-like factor	-
		TraesCS2B03G1116700	K11593	ELF2C, AGO; eukaryotic translation initiation factor 2C	-
		TraesCS2B03G1104200	K11593	ELF2C, AGO; eukaryotic translation initiation factor 2C	-
		TraesCS2B03G1109900	K09286	EREBP; EREBP-like factor	-
		TraesCS2B03G1106500	K09338	HD-ZIP; homeobox-leucine zipper protein	-
Q.sW-2D	2D/74	*TraesCS2D03G0133000*	K09422	MYBP; transcription factor MYB, plant	-
		TraesCS2D03G0143000	K09602	OTUB1; ubiquitin thioesterase protein OTUB1	EC:3.4.19.12
		TraesCS2D03G0143400	K09602	OTUB1; ubiquitin thioesterase protein OTUB1	EC:3.4.19.12

Table 1. *Cont.*

QTL	Chr/Pos	Gene ID	KO ID	Description	EC
Q.sCi-2D [g]; Q.sRo-2D [g]	2D/58	*TraesCS2D03G0107400*	K09602	ubiquitin thioesterase protein OTUB1	EC:3.4.19.12
		TraesCS2D03G0107900	K09602	ubiquitin thioesterase protein OTUB1	EC:3.4.19.12
sSo-7D.2	7D/141	*TraesCS7D03G1008800*	K09286	EREBP; EREBP-like factor	-
		TraesCS7D03G0987700	K09338	HD-ZIP; homeobox-leucine zipper protein	-
		TraesCS7D03G0983800	K09286	EREBP; EREBP-like factor	-
		TraesCS7D03G0972400	K08869	ADCK, ABC1; aarF domain-containing kinase	-
		TraesCS3A03G0763900	K00279	CKX; cytokinin dehydrogenase	EC:1.5.99.12
sL-7D [bb]; sRo-7D [bb]	7D/287	*TraesCS7D03G1260500*	K20547	CHIB; basic endochitinase B	EC:3.2.1.14
		TraesCS7D03G1260300	K20547	CHIB; basic endochitinase B	EC:3.2.1.14
		TraesCS7D03G1260400	K20547	CHIB; basic endochitinase B	EC:3.2.1.14
		TraesCS7D03G1286900	K11593	ELF2C, AGO; eukaryotic translation initiation factor 2C	-
		TraesCS7D03G1260600	K20547	CHIB; basic endochitinase B	EC:3.2.1.14
		TraesCS7D03G1287400	K11593	ELF2C, AGO; eukaryotic translation initiation factor 2C	-

For genes from the loci associated with seed coat color, 14 found a match with the orthogroups of the metabolic pathways of the KEGG database related to pigment biosynthesis (Table 2). Eight genes were involved in the phenylpropanoid biosynthesis (loci on chromosomes 3A, 3B, 6A, and 6B). Two genes were involved in the carotenoid biosynthesis pathway (loci on chromosomes 2A and 6A), and one gene each was involved in flavone and flavonol biosynthesis, flavonoid biosynthesis, tryptophan metabolism, and terpenoid backbone biosynthesis.

Table 2. List of candidate genes from QTLs associated with seed shell color. Columns of the table contain QTL name (QTL), chromosome and position in cm (Chr/Pos), gene ID, KEGG orthogroup ID, KEGG orthogroup description, EC number, KEGG pathway ID and description.

Trait	Chr/Pos	Gene ID	KO ID	KO Description	EC	KEGG Pathway ID	KEGG Pathway Description
Q.YCrCb_dCCr_1-2A.3	2A/196	*TraesCS6A03G0725700*	K09840	NCED; 9-cis-epoxycarotenoid dioxygenase	EC:1.13.11.51	map00906	Carotenoid biosynthesis
		TraesCS2A03G0158600	K22772	FG2; flavonol-3-O-glucoside L-rhamnosyltransferase	EC:2.4.1.159	map00944	Flavone and flavonol biosynthesis
		TraesCS6A03G0953500	K13065	HCT; shikimate O-hydroxycinnamoyltransferase	EC:2.3.1.133	map00941	Flavonoid biosynthesis
		TraesCS2A03G0099800	K13066	COMT; caffeic acid 3-O-methyltransferase/ acetylserotonin O-methyltransferase	EC:2.1.1.68; 2.1.1.4	map00380	Tryptophan metabolism
Q.Lab_dCb_3-3A [i*]	3A/195	*TraesCS3A03G0925800*	K01904	4CL; 4-coumarate–CoA ligase	EC:6.2.1.12	map00940	Phenylpropanoid biosynthesis
		TraesCS3A03G0925900	K01904	4CL; 4-coumarate–CoA ligase	EC:6.2.1.12	map00940	Phenylpropanoid biosynthesis
Q.RGB_dCB_1-3B.1	3B/269.179	*TraesCS3B03G1115600*	K12355	REF1; coniferyl-aldehyde dehydrogenase	EC:1.2.1.68	map00940	Phenylpropanoid biosynthesis
HSV_dCH_1-3B [m**]	306.179	*TraesCS3B03G1278200*	K01904	4CL; 4-coumarate–CoA ligase	EC:6.2.1.12	map00940	Phenylpropanoid biosynthesis

Table 2. *Cont.*

Trait	Chr/Pos	Gene ID	KO ID	KO Description	EC	KEGG Pathway ID	KEGG Pathway Description
HSV_dCH_3-6A [t,***]	6A/246	*TraesCS6A03G0725000*	K09843	CYP707A; (+)-abscisic acid 8′-hydroxylase	EC:1.14.14.137	map00906	Carotenoid biosynthesis
		TraesCS6A03G0741000	K00021	HMGCR; hydroxymethylglutaryl-CoA reductase (NADPH)	EC:1.1.1.34	map00900	Terpenoid backbone biosynthesis
Q.YCrCb_dCCr_1-6A	6A/340	*TraesCS6A03G0953500*	K13065	HCT; shikimate O-hydroxycinnamoyltransferase	EC:2.3.1.133	map00940	Phenylpropanoid biosynthesis
		TraesCS2A03G0163500	K00430	peroxidase	EC:1.11.1.7	map00940	Phenylpropanoid biosynthesis
		TraesCS2A03G0164200	K00430	peroxidase	EC:1.11.1.7	map00940	Phenylpropanoid biosynthesis
Lab_dCL_2-6B [x,****]	6B/220	*TraesCS6B03G0367700*	K00430	peroxidase	EC:1.11.1.7	map00940	Phenylpropanoid biosynthesis

* Co-located QTL: Q.YCrCb_dCCb_3-3A [i]. ** Co-located QTLs: HSV_dCH_3-3B [m]; HSV_dCS_1-3B [m]; HSV_dCV_1-3B [m]; HSV_dCV_3-3B [m]; Lab_dCb_1-3B [m]; Lab_dCb_2-3B [m]; Lab_mb-3B [m]; RGB_dCR_2-3B [m]; YCrCb_dCCb_1-3B [m]; YCrCb_dCCb_2-3B [m]; YCrCb_dCCr_1-3B [m]; YCrCb_mCb-3B [m]. *** Co-located QTL: HSV_dCV_3-6A [t]; Lab_ma-6A [t]. **** Co-located QTLs: Lab_dCL_3-6B [x]; Lab_mb-6B [x]; RGB_dCG_2-6B [x]; RGB_dCR_2-6B [x]; YCrCb_dCCb_2-6B [x]; YCrCb_dCY_2-6B [x]; YCrCb_dCY_3-6B [x].

3. Discussion

3.1. Using Digital Image Analysis for QTL Identification

Based on the analysis of digital images, we identified QTLs associated with quantitative seed characteristics in wheat. With the development of modern phenotyping technologies [39,40], such approaches are increasingly being used [7–11,41]. Modern digital cameras and image processing algorithms have made great progress; they allow us to estimate even small differences in the color characteristics of seeds, their shape and size with high accuracy. In addition, these approaches have one interesting feature: the use of a large number of quantitative characteristics that are essentially derived from the same biological trait of the plant. For example, seed shape and size could be described as the sets of elliptic Fourier components [41] or virtual curves [42,43]. Components of various digital spaces [44,45] represent seed coat color. In the case when quantitative traits are derived from the same biological trait, we can assume that they will be associated with the same loci. Williams and Sorrels [11] used QTL for a set of seed size and shape characteristics derived from the developed seed image in two projections (elliptic Fourier components) as well as length, width, and thousand-kernel weight (TKW). They studied two populations, one of which, SynOpDH, was derived from crosses of the same parents used to obtain the ITMI population, and the other Cayuga × Caledonia was a doubled-haploid mapping population (C × C). Thirty-one loci were identified for the SynOpDH population which controlled from one to four traits per locus. Thirty loci were identified for the C × C population which also controlled from one to four traits per locus.

From our results (Figure 6), it is apparent that many color trait loci overlap between each other but not with the QTLs of size/shape (Table S1; chromosomes 3B, 3D, and 6B). The exception is rugosity which reflects the roughness of the shell and is probably associated with color distortions at the seed boundary on the image background. On the one hand, these features reflect the degeneracy of the evaluated traits which indicates their redundancy. On the other hand, the location of several loci related to color traits in the same region may indicate a more reliable identification of the association of the locus with a particular trait. Using many digital representations of the same trait looks redundant and confusing. The reasonable step would be to select a single or a few numerical characteristics that are most efficient in the identification of QTLs. We believe that various numerical representations of the same biological trait are useful and allow the evaluation of its subtle details. Many traits with QTL in the same locus can support its significance.

3.2. Identification of QTLs Associated with Seed Features

Our analysis allowed us to identify a number of QTLs associated with wheat seed characteristics, shape/size, and shell color. Such analyses have long been intensively conducted based on both QTL and GWAS investigations [8,11,41,46,47]. Reference [11] investigated three-dimensional characteristics of seed size and shape based on the analysis of images of seed obtained in two projections and the use of Fourier analysis-based descriptors using two populations, one of which was SynOpDH. They found a QTL that affects a number of shape characteristics. For seed length, eight QTLs were found for chromosomes 2A, 2D, 4B, 5A, 5B, 6A, 7A, and 7D. In our work, we detected a smaller number of QTLs for this trait located on the chromosomes 2A, 2D, 3B, 5B (2 QTLS) and 7D. For seed width, three QTLs on chromosomes 2A, 5A and 6A were detected by [11], whereas we detected four QTLs for this trait located on the chromosomes 1D, 2D, 3B, and 4D. This anomaly could be due to the use of data from several different environments in many years by reference [11].

Reference [47] previously analyzed 92 accessions from the ITMI population for a large number of traits, including such traits as TKW and kernel color (KC), in different locations and years. We did not find any overlap of QTL for TKW with the traits characterizing seed size and shape in our work. For seed color, 15 QTLs were reported by [47]. A comparison of our results with those from this work showed that of the 15 QTLs, one exactly matched the one found in our work. This is Q.KC_Pu07-3B [47] bounded by markers AX-94979462 and IAAV6088 and located in our work on chromosome 3B at position 306.179 (Table S1). In our work, several QTLs associated with seed shell color characteristics correspond to this locus (see marker HSV_dCH_1-3Bm (Table S1) and also listed in Table 2.) It is also interesting to note the QTLs Q.KC_Mo07-3D and Q. KC_Mo08-3D [47] bounded by markers D_GDS7LZN02IJRXZ_309 on the left and CAP12_c2615_128 on the right located on chromosome 3D at 76 cM. In our work, we found a series of color-related QTL localized on chromosome 3D at position 100–102, bounded by markers CAP12_c2615_128 on the left and BS00067163_51 on the right. Thus, the QTLs from our work and that of reference [47] are in the vicinity on the chromosome. For the other QTLs associated with color, we found no coincidence. For example, reference [47] identified five QTLs associated with shell color on chromosome 5A. However, in our work, only two loci at other positions on this chromosome were associated with color.

Among the loci associated with color, the site on chromosome 3D (ar 100 cM) bounded by the markers *CAP12_c2615_128* and *BS00067163_51* is perhaps the most interesting. As indicated above, it is located next to the color QTL identified in [47]. In our work, 34 different traits characterizing seed color are associated with it. All of them are color parameters in various color spaces. Our analysis allowed us to localize the physical coordinates of this site on chromosome 3D: 573.6–580.8 Mbp according to IWGS v2.0 genome annotation (Table S5). Interestingly, reference [48] recently performed the analysis of the *PHS-3D* QTL associated with seed resistance to pre-harvest sprouting for a population of synthetic hexaploid wheat [49]. It turned out that on the physical map this region is located on chromosome 3D at positions 571.9–574.3 Mbp which overlaps with the physical localization of the QTL we identified. Reference [48] also showed that plant genotypes susceptible to pre-harvest sprouting are characterized by a ~2.4 Mbp deletion involving 20 genes in this region of the genome. It turned out that the gene encoding the transcription factor Myb10-D, which confers resistance to pre-harvest sprouting by activation of flavonoid and abscisic acid biosynthesis pathways, was located in this region. Note that plants that do not contain deletions in this region and are resistant to pre-harvest sprouting have reddish/brown coloring of the seed shell.

3.3. Epistatic QTLs

In our work, we identified several QTLs whose contribution to the trait are non-additive. Currently, there are only a few examples of epistatic QTLs analysis in wheat [46,50–52]. We found 34 QTL pairs exhibiting epistatic interactions. Our results show a predominance of epistatic QTLs for color features (30 pairs of QTLs). Onc pair of QTL each was identified for the area, width, rugosity, and solidity of seed. These results demonstrate a possible interaction of genes located at different loci in the formation of color traits.

Epistatic QTLs for yield, flour color, and seed weight traits were investigated for the RIL population of durum wheat [51]. QTL epistatic interactions on chromosomes 1A and 1B and chromosomes 5B and 7B were determined for thousand-seed weight. Reference [46] analyzed yield traits including 1000-seed weight, seed length, and seed width of bread wheat in the RIL Chuannong18 × T1208 population. Epistatic QTLs were found for 114 QTL pairs, including 10 for seed length, 17 for seed width, and seven for 1000-seed weight. The authors noted the complex nature of the effect of epistatic interactions on the seed properties. Thus, more epistatic pairs for geometric seed traits were identified in these works compared to ours. However, in our work, the most intense epistatic interactions were shown for seed shell color, a trait not reported in [46]. Interestingly, QTL pairs that are localized on chromosomes 3B (position 306) and 3D (position 100), which are also characterized by a large number of additive QTLs, are often represented. This again indicates the importance of these regions for the formation of seed shell color in wheat.

3.4. Gene Prioritization

Our analysis allowed us to identify a number of candidate genes associated with seed size/shape and their color, based on bioinformatics analysis and annotation of genes according to data in the literature and the KEGG database. We identified eight loci associated in the genome with seed size/shape traits for which we found 28 orthologous genes involved in gene networks controlling these traits. Some of genes are transcription factors (EREBP, HD-ZIP, and MYB) that may be involved in the regulation of seed growth and development. In particular, transcription factors associated with the response to ethylene (EREBP) are known to be involved in the determination of seed size, seed weight, and accumulation of seed oil and protein in *A. thaliana* [53]. Reference [54] identified two transcription factors from the AP2/EREBP family, TaPARG, located on 2A and 2D chromosomes of wheat which regulate several yield-related traits, including seed weight.

Several genes represent families of enzymes related to ubiquitin modification (E3 ubiquitin–protein ligase, ubiquitin thioesterase protein OTUB1). Ubiquitins and related enzymes are known to play an important role in seed development by controlling cell proliferation [55]. For example, genes of the E3 ligase family are involved in amylose biosynthesis in wheat [56]. The *TaGW2-6A* gene from this family is shown to control seed size [57].

Another type of enzyme that was frequently found among the candidates we identified are endochitinases. These are enzymes involved in defense against pathogens such as bacteria or fungi in seeds [58,59]. However, this is not their only role in seed formation and function. It has been shown from proteomic data in rice that *chitinase 14* interacts with the *GW2* (*RING-type E3 ubiquitin ligase*) gene. It was also shown that *GW2* controls seed size through the regulation of chitinase 14 and phosphoglycerate kinase levels or activities [60]. Other genes that we detected (cytokinin dehydrogenase, aarF domain-containing kinase, and eukaryotic translation initiation factor 2C) may also be associated with seed development in wheat [20].

For QTLs associated with seed color, we also found a number of possible candidates among genes encoding enzymes of plant pigment biosynthesis pathways. On chromosome 2A, we found several genes that are involved in plant pigment biosynthesis. Among them, one gene, which we annotated as NCED, is involved in the carotenoid biosynthesis pathway. In rice, mutants of this gene lead to changes in pericarp seed coloration [61]. The expression of this enzyme is controlled by abscisic acid [62], and *NCED* is also involved in ABA biosynthesis [48]. The functions of this gene in seed development are well known [63]: it is an important regulator in seed development, in the zygotic embryogenesis, and dormancy. Another gene related to carotenoid biosynthesis that we found among the primate genes is *CYP707A ((+)-abscisic acid 8′-hydroxylase)* (Table 2). Its functions are closely related to the *NCED* gene, and its participation in the same processes related to seed development has been shown [63]. Interestingly, two of these genes are located near loci associated with seed coat redness (*Q.YCrCb_dCCr_1-2A.3* and *HSV_dCH_3-6At/Lab_ma-6At*).

We found two genes that are involved in the flavonoid biosynthesis pathway (Table 2) that provide different coloration of seeds in cereals [64,65]. These include the homologue FG2 (*flavonol-3-O-glucoside L-rhamnosyltransferase*), for which mutations result in a phenotype with seed color change in soybean [66]. *Shikimate O-hydroxycinnamoyltransferase* has been shown to be elevated in expression in wheat plants with high seed antioxidant activity [67].

In the QTL region of *Q.Lab_dCb_3-3Ai*, we found two genes involved in phenylpropanoid biosynthesis. They both encode *4-coumarate–CoA ligase*. This enzyme catalyzes the conversion of p-coumaric acid to p-coumaroyl CoA, which further serves as a source of biosynthesis of both lignin (a structural component of the seed shell) and flavonoids. In transcriptome-wide association studies in *Brassica napus*, 4CL expression during seed development was shown to positively correlate with seed coat content, i.e., the fraction of seed mass attributable to the coat [68]. Interestingly, in gene expression analysis in *B. napus* plants with brown seed coloration, the expression level of genes encoding this enzyme was higher than in plants with yellow coloration [69].

Modern genomics advances in wheat genome sequencing and genetic marker technologies allow QTLs to be linked to the physical coordinates of the wheat genome. Such analysis is now an important complement to QTL identification [70–72]. The genes we have identified as possible candidates associated with seed size/shape and color formation in the ITMI/MP can be further investigated in more detail using genetic and molecular methods to establish the mechanisms controlling these important traits.

4. Materials and Methods

4.1. Materials

We studied seeds from 114 accessions of the well-known ITMI/MP of bread wheat (*T. aestivum L.*). The ITMI mapping population was obtained by pollination of the *T. aestivum* (var. Opata 85) flower with the pollen of the synthetic hexaploid spring wheat W7984 [47]. Plants of each genotype were grown in season in 2014 on the experimental fields of IPK in Gatersleben, Germany.

4.2. Seed Imaging Protocol and Image Processing

Seeds were imaged in March 2020. We supposed that the storage time affected seed traits for different genotypes in the same manner. The imaging of seeds was performed according to the protocol described earlier [1]: seeds were scattered in an amount of up to 20 pieces on the table on a white sheet of A4 paper. A ColorChecker color calibration card (x-rite ColorChecker® Classic Mini, X-Rite, Grand Rapids, MI, USA, https://xritephoto.com/camera; accessed on 20 January 2022) was placed in the image area and used for color correction and obtaining image scale. The lighting was adjusted to avoid shadows. Images were taken with a digital camera Canon EOS 600D equipped with a Canon EF 100 mm f/2.8 Macro USM lens and saved in files in JPG or PNG format. Examples of images are shown in Figure S2. Digital image processing was performed using the SeedCounter application for desktop PC [73] with color analysis capabilities [1]. We used two images per genotype for our analysis: 15 and 5 seeds. Splitting was initially used to check for reproducibility of the evaluated trait values. No significant differences between mean values of the seed traits were observed between these replicates according to an *F*-test (results not shown). Therefore, we used the average values of the images of 20 seeds from two replicates as input for QTL analysis.

4.3. Quantitative Characteristics of Seed Shape, Size, and Color

The analysis of digital images for each seed yielded a set of 55 quantitative characteristics described earlier [1]. Size was defined by seed length (sL), width (sW), and projected area (sA). Seed shape characteristics included circularity (sCi), roundness (sRo), rugosity (sRg), and solidity (sSo).

The circularity and roundness indices reflect how close the shape of a contour is to a circle but are calculated differently. Circularity is a measure of the similarity of a 2D figure to a circle [43]. For objects with rugged contours, the closeness of the shape to a circle is more correctly described by the roundness parameter since this value does not depend on the roughness of the contour line. This index is calculated as the ratio of the area of the shape (area) to the square of the length of the major axis [40]. For a shape other than a circle, the index takes values less than unity. The rugosity index (sRg) is defined as the ratio of the contour perimeter to the convex perimeter [43]. The index of solidity (sSo) is the ratio of the contour area to the area of its convex hull [74].

To describe the color characteristics of the seeds, we used a color representation in the form of four-color spaces: RGB, HSV, Lab, and YCrCb [44,75,76]. Each of them represents color as three components. The component values of one space can be obtained by transforming the component values of the other. The color features included two types of descriptors, which were independently calculated for each color space.

The first type of descriptors: mean values of component intensities for seed pixels. To calculate them, the mean and standard deviations of intensities for each of the color component channels were first estimated, then the pixels whose intensities differ from the mean by more than three standard deviations were excluded from the analysis. The mean value was calculated for the remaining pixels and used further. The descriptors of the average component values are indicated by a small letter m. For example, for the RGB color space, these are the three parameters: RGB_mR, RGB_mG, and RGB_mB. For other spaces, the indications are similar.

The second type of descriptors are dominant seed colors. These descriptors provide an illustration of representative colors in an image or its region [77]. To determine dominant colors, all seed pixels were grouped by color similarity into three clusters. The clusters were ranked by the number of pixels they contained. In each of the three clusters, the values of the three color components for the centroid were determined. This procedure was performed for each color space and resulted, respectively, in nine color descriptors. For example, for the RGB space, these are RGB_dC*j*_*i* parameters, where $j = 1,2,3$ is the color component designation, and $i = 1,2,3$ is the number of the dominant cluster. For example, the RGB_dCR_1 parameter is the R component for the first dominant color in the RGB space. The use of three dominant colors allows for a more accurate estimation of the shades of seed coloration.

As a result, three size characteristics, four shape characteristics, and 48 color characteristics were determined for each seed. Characteristics were calculated for each seed of the 114 wheat genotypes. The mean values for the genotype were estimated and used in the QTL analysis.

4.4. Statistical Analysis

To get an idea of the similarity of genotypes in the space of seed traits, we used the principal component method implemented in the PAST program [78].

4.5. Genotyping and QTL Analysis

Fresh flag leaves were used for the DNA extraction for the purpose of genotyping which was performed using Illumina (San Diego, CA, USA) Infinium technology. An optimized array (wheat 20K Infinium SNP array) was used. This array is the refined version of the 15K chip [79] of 90K iSELECT SNP-chip as previously reported [80]. To make it more informative, 5385 markers from the 35K Wheat Breeders Array [81] were also added. All sequences of the markers, a complete genetic map and the list of 92 RILs with genotypic data are available in reference [47].

To capture the variance explained by the molecular markers such as SNPs mapped to any genome, different methods were proposed (such as single marker analyses, interval mapping, and composite interval mapping) and implemented in different computer programs (*Qgene*, *QTL Cartographer* and *PLABQTL*) [82] which have successfully been used to detected QTLs for various traits in wheat including seed-related traits such as seed longevity [37,83,84]. In light of the limitations of the above-mentioned methods, a more refined method known as "inclusive composite interval mapping" was proposed and implemented in the *QTLIciMapping 4.2.53* (http://www.isbreeding.net/ (latest released in September 2019, accessed on 2 February 2022) which is considered as the most modern method of QTL detection [47]. We have recently detected several QTLs for *Fusarium* head blight [85] and seed longevity [50] in wheat and germination-related traits in tobacco [86] by applying the *QTLIciMapping* tool. Therefore, we convened the *IciMapping 4.2.53* to detect the putative additive QTLs of the traits under consideration by applying the inclusive composite interval mapping (ICIM) command where 1.0 cM was the walking speed. An LOD score of >2.0 ≤3 was applied to detect QTLs as significant and >3.0 as highly significant [87].

In order to discover digenic epistasis QTLs to find clues for latent variation, the ICIM-EPI command was used where LOD was kept at 5.0 cM. Here, the epistasis QTLs with LOD ≥5 and explaining ≥5% phenotypic variance were reported. All QTLs were assigned names according to the rules set out in the Catalog of Gene Symbols [88]. All additive and epistasis QTLs were visualized using the "circlize" package in R [89].

4.6. Seed Traits Similarity by QTL Location

Preliminary analysis of QTLs demonstrated that loci for several traits often overlap. In particular, the same locus may associate with several characteristics of the seed shell color. In this regard, we decided to analyze the similarity of traits by their localization in the genome. To do this, we compiled a list of loci with which they were associated for each of the 55 features. Based on the overlap of the list of these loci, we calculated the Ochiai index [90] for each pair of features. This index was suggested for ecological studies to estimate associations between species and groups of sites representing habitat types. In our work, the index reflects the degree of overlap of the lists of loci between two traits. The greater the similarity between sets of loci for two traits, the greater the index value. It equals 1 when loci are identical and 0 when there is no loci overlap. Based on this measure, we performed clustering of features in the PAST [78] package and built a tree of the trait similarity.

4.7. QTL Gene Prioritization

In order to identify possible candidate genes associated with seed traits, we prioritized them based on several conditions and using the gene annotation provided in the KEGG database [91]. The analysis included loci for which the LOD value exceeded three and consisted of several steps.

In the first step of the analysis, we determined the physical localization of the markers by aligning their sequences to the IWGS 2.1 wheat genome assembly sequence [92]. Genome sequence and annotation data were obtained from URGI (https://urgi.versailles.inra.fr/download/iwgsc/IWGSC_RefSeq_Assemblies/v2.1; accessed on 10 January 2022). We considered "high confidence" gene annotations only. Marker sequences were obtained from reference [80] and the Gramene marker database (https://archive.gramene.org/markers/; accessed on 10 January 2022) [93]. Marker sequences were aligned using BLASTn of the BLAST+ package [94] using e-value = 1×10^{-17} (other parameters were set by default). This allowed us to search similar sequences with length above 50 nt and avoid noise. Thus, for each of our selected QTLs, we obtained a list of IWGS 2.1 wheat genome annotation genes. Note that it was not possible to determine the physical boundaries of the QTLs for several loci because, for one of the sequences, the alignment did not occur on the chromosome corresponding to the marker.

Since plant pigments can be synthesized in various tissues and organs, when prioritizing genes for seed color QTL, we additionally filtered genes by expression level in the seed. For this purpose, we used the expression data presented for wheat in the expVIP database [95]. Data in text format were downloaded from URGI (https://urgi.versailles.inra.fr/download/iwgsc/IWGSC_RefSeq_Annotations/v1.1/iwgsc_refseqv1.1_rnaseq_mapping_2017July20.zip; accessed on 10 January 2022). We used data from RNA-seq experiments in which the column "High level tissue" contains "seed". We selected genes as expressed if their TPM ≥ 1 in these experiments. To perform filtering, we developed scripts in Python, taking into consideration the Gene ID conversion between annotation ver. 2.1 (genome) and 1.2 (transcriptome).

In the second stage of analysis, we generated a list of orthologous protein groups from the KEGG database [91], which are associated with the formation of the traits of seed size and color. It is well known that the color of the seed shell is determined by the presence of specific plant pigments in it [22]. Therefore, we selected orthogroups involved in KEGG pathways of the biosynthesis of these pigments and a number of their precursors. The list of such pathways includes tryptophan metabolism (map00380), terpenoid backbone biosynthesis (map00900), carotenoid biosynthesis (map00906), phenylpropanoid biosynthesis (map00940), flavonoid biosynthesis (map00941), anthocyanin biosynthesis (map00942), isoflavonoid biosynthesis (map00943), and flavone and flavonol biosynthesis (map00944). We obtained 307 KEGG orthogroups for these pathways.

Seed size depends on a multitude of biological processes occurring at the molecular level, including protein ubiquitination, response to hormonal signals, protein biosynthesis and transport, etc. Therefore, it was not possible to isolate the pathways corresponding to these processes based only on their description in the KEGG database. However, the genes involved in seed development have been fairly well experimentally studied in *A. thaliana* and rice (*Oryza sativa*). Therefore, we used three recent literature reviews describing the molecular processes of seed development in *Arabidopsis* and rice [20,21,38]. We combined set of genes from three papers and removed duplicated IDs. During compilation, we converted gene IDs from reference [20] from RAP to MSU format using the "ID converter" tool at the website of the OryzaExpress database ([96]; http://bioinf.mind.meiji.ac.jp/OryzaExpress/ID_converter.php; accessed on 20 January 2022). We identified KEGG orthogroups for the selected genes and used them for our analysis.

The assignment of KEGG orthologous groups to wheat genes by their sequence was performed using BlastKOALA [97] and KofamKOALA [98] tools. The orthogroups were assigned to genes by at least one of the methods. We prioritized genes whose orthogroups were in the lists associated with traits of seed coat color and size.

5. Conclusions

A QTL search for seven traits of seed size/shape and 48 traits of coat color evaluated on the basis of digital image analysis of the ITMI/MP identified 212 additive and 34 pairs of epistatic QTLs on all the chromosomes of wheat genome except chromosomes 1A and 5D. The number of QTLs varied from one to ten for one single trait. The majority of the traits yielded three to four QTLs. We demonstrated that one locus could control dozens of seed characteristics. Analysis of the loci overlap showed that this is typical for color traits and rarely occurred for seed size/shape traits. For a number of highly significant QTLs, we identified the physical location of their markers on the wheat chromosomes. Additionally, we demonstrated that the overlap between QTL regions was low for seed size/shape traits and high for coat color traits. Using the literature and KEGG data, we identified sets of genes in *Arabidopsis* and rice from the networks controlling seed size and color. This information along with the coordinates of the markers in the wheat genome was used for the prioritization of wheat genes within QTL regions. We identified 29 candidate genes from the seed size-related loci and 14 for genes from the loci associated with seed coat color. The genes we have identified as possible candidates associated with seed size/shape and color formation in the ITMI/MP can be further investigated in more detail using genetic and molecular methods to establish the mechanisms controlling these important traits. Our results demonstrate the complex nature of the genetic control of the wheat seed traits and the efficiency of the image analysis methods for obtaining novel QTLs for seed characteristics.

Supplementary Materials: The following supporting information can be downloaded at https://www.mdpi.com/article/10.3390/plants11162105/s1: Figure S1: PCA biplot of seed size/shape and color traits of ITMI/MP performed using seed coat traits (mean seven values for seed size/shape and 12 values for color components of four-color spaces); Figure S2: Examples of seed images used for digital phenotyping; Table S1: Complete list of QTLs identified through composite interval mapping. Left and right flanking markers linked to QTL are also provided along with LOD, PVE%, and additive effect (+ = provided by Opata and -= provided by W7984 parent). Similar asterisks (also highlighted in similar color) indicate likely identical loci. Markers highlighted in red are involved in multiple QTLs. Table S2: Pairs of epistatic quantitative trait loci detected in the ITMI/MP. Table S3: List of 197 (42 from rice and 155 from Arabidopsis) genes involved in molecular processes of the grain development. Input Gene ID = Gene ID from review papers; KEGG Gene ID = GID for gene in KEGG database; NA: not found in KEGG; KEGG Gene Name = Gene name in KEGG database and KEGG Orthology ID = Orthogroup ID in KEGG database. Table S4: Pathway ID and KEGG orthologous groups. Pathway KEGG = pathway ID and KEGG orthology group = Orthogroup ID in KEGG database. Table S5: List of QTLs used in gene prioritization. Table S6: List of 823 genes associated with grain color.

Author Contributions: Conceptualization, M.A.R.A., A.B. and D.A.A.; methodology, M.A.R.A., V.S.K. and E.G.K.; software, E.G.K.; validation, M.A.R.A., E.G.K. and D.A.A.; formal analysis, M.A.R.A., E.G.K., M.A.G. and N.A.S.; investigation, M.A.R.A., E.G.K., N.A.S. and D.A.A.; resources, A.B.; data curation, M.A.R.A., E.G.K. and M.A.G.; writing—original draft preparation, M.A.R.A., E.G.K. and D.A.A.; writing—review and editing, M.A.R.A., A.B. and D.A.A.; visualization, M.A.R.A. and E.G.K.; supervision, A.B. and D.A.A.; project administration, A.B. and D.A.A.; funding acquisition, A.B. and D.A.A. All authors have read and agreed to the published version of the manuscript.

Funding: This research received no external funding, whereas the publication of this article was funded by the Open Access Fund of the Leibniz Association, Russian Science Foundation (21-76-30003) and Russian Ministry of Science and Higher Education (FWNR-2022-0020).

Institutional Review Board Statement: Not applicable.

Informed Consent Statement: Not applicable.

Data Availability Statement: Not applicable.

Acknowledgments: Seed phenotyping was supported by the RSF project 21-76-30003. The data analysis was performed using computational resources of the ICG SB RAS "Bioinformatics" Joint Computational Center, supported by the budget project № FWNR-2022-0020.

Conflicts of Interest: The authors declare no conflict of interest.

References

1. Afonnikov, D.A.; Komyshev, E.G.; Efimov, V.M.; Genaev, M.A.; Koval, V.S.; Gierke, P.U.; Börner, A. Relationship between the Characteristics of Bread Wheat Grains, Storage Time and Germination. *Plants* **2021**, *11*, 35. [CrossRef] [PubMed]
2. Brinton, J.; Uauy, C. A reductionist approach to dissecting grain weight and yield in wheat. *J. Integr. Plant Biol.* **2019**, *61*, 337–358. [CrossRef] [PubMed]
3. Huang, Y.; Kong, Z.; Wu, X.; Cheng, R.; Yu, D.; Ma, Z. Characterization of three wheat grain weight QTLs that differentially affect kernel dimensions. *Theor. Appl. Genet.* **2015**, *128*, 2437–2445. [CrossRef] [PubMed]
4. Zhang, X.; Deng, Z.; Wang, Y.; Li, J.; Tian, J. Unconditional and conditional QTL analysis of kernel weight related traits in wheat (*Triticum aestivum* L.) in multiple genetic backgrounds. *Genetica* **2014**, *142*, 371–379. [CrossRef] [PubMed]
5. Baasandorj, T.; Ohm, J.B.; Manthey, F.; Simsek, S. Effect of kernel size and mill type on protein, milling yield, and baking quality of hard red spring wheat. *Cereal Chem.* **2015**, *92*, 81–87. [CrossRef]
6. Li, S.; Wang, L.; Meng, Y.; Hao, Y.; Xu, H.; Hao, M.; Lan, S.; Zhang, Y.; Lv, L.; Zhang, K.; et al. Dissection of genetic basis underpinning kernel weight-related traits in common wheat. *Plants* **2021**, *10*, 713. [CrossRef]
7. Ali, A.; Ullah, Z.; Alam, N.; Naqvi, S.M.; Jamil, M.; Bux, H.; Sher, H. Genetic analysis of wheat grains using digital imaging and their relationship to enhance grain weight. *Sci. Agric.* **2020**, *77*, e20190069. [CrossRef]
8. Alemu, A.; Feyissa, T.; Tuberosa, R.; Maccaferri, M.; Sciara, G.; Letta, T.; Abeyo, B. Genome-wide association mapping for grain shape and color traits in Ethiopian durum wheat (*Triticum turgidum* ssdurum). *Crop J.* **2020**, *8*, 757–768. [CrossRef]

9. Kumari, S.; Jaiswal, V.; Mishra, V.K.; Paliwal, R.; Balyan, H.S.; Gupta, P.K. QTL mapping for some grain traits in bread wheat (*Triticum aestivum* L.). *Physiol. Mol. Biol. Plants* **2018**, *24*, 909–920. [CrossRef] [PubMed]
10. Kumar, A.; Mantovani, E.E.; Seetan, R.; Soltani, A.; Echeverry-Solarte, M.; Jain, S.; Simsek, S.; Doehlert, D.; Alamri, M.S.; Elias, E.M.; et al. Dissection of genetic factors underlying wheat kernel shape and size in an elite× nonadapted cross using a high density SNP linkage map. *Plant Genome* **2015**, *9*, 0081. [CrossRef]
11. Williams, K.; Sorrells, M.E. Three-dimensional seed size and shape QTL in hexaploid wheat (*Triticum aestivum* L.) populations. *Crop Sci.* **2014**, *54*, 98–110. [CrossRef]
12. Gegas, V.C.; Nazari, A.; Griffiths, S.; Simmonds, J.; Fish, L.; Orford, S.; Sayers, L.; Doonan, J.H.; Snape, J.W. A genetic framework for grain size and shape variation in wheat. *Plant Cell* **2010**, *22*, 1046–1056. [CrossRef] [PubMed]
13. Breseghello, F. MESorrells QTL analysis of kernel size and shape in two hexaploid wheat mapping populations. *Field Crops Res.* **2007**, *101*, 172–179. [CrossRef]
14. Huang, X.; Börner, A.; Röder, M.; Ganal, M. Assessing genetic diversity of wheat (*Triticum aestivum* L.) germplasm using microsatellite markers. *Theor. Appl. Genet.* **2002**, *105*, 699–707. [CrossRef] [PubMed]
15. Ma, L.; Li, T.; Hao, C.; Wang, Y.; Chen, X.; Zhang, X. Ta GS 5-3A, a grain size gene selected during wheat improvement for larger kernel and yield. *Plant Biotechnol. J.* **2016**, *14*, 1269–1280. [CrossRef]
16. Ma, M.; Wang, Q.; Li, Z.; Cheng, H.; Li, Z.; Liu, X.; Song, W.; Appels, R.; Zhao, H. Expression of Ta CYP 78A3, a gene encoding cytochrome P450 CYP 78A3 protein in wheat (*Triticum aestivum* L.), affects seed size. *Plant J.* **2015**, *83*, 312–325. [CrossRef]
17. Ma, D.; Yan, J.; He, Z.; Wu, L.; Xia, X. Characterization of a cell wall invertase gene TaCwi-A1 on common wheat chromosome 2A and development of functional markers. *Mol. Breed.* **2012**, *29*, 43–52. [CrossRef]
18. Su, Z.; Hao, C.; Wang, L.; Dong, Y.; Zhang, X. Identification and development of a functional marker of TaGW2 associated with grain weight in bread wheat (*Triticum aestivum* L.). *Theor. Appl. Genet.* **2011**, *122*, 211–223. [CrossRef]
19. Tikhenko, N.; Alqudah, A.M.; Borisjuk, L.; Ortleb, S.; Rutten, T.; Wu, D.; Nagel, M.; Himmelbach, A.; Mascher, M.; Röder, M.S.; et al. DEFECTIVE ENDOSPERM-D1 (Dee-D1) is crucial for endosperm development in hexaploid wheat. *Commun. Biol.* **2020**, *3*, 791. [CrossRef]
20. Chen, K.; Łyskowski, A.; Jaremko, Ł.; Jaremko, M. Genetic and molecular factors determining grain weight in rice. *Front. Plant Sci.* **2021**, *12*, 605799. [CrossRef]
21. Li, N.; Xu, R.; Li, Y. Molecular networks of seed size control in plants. *Annu. Rev. Plant Biol.* **2019**, *70*, 435–463. [CrossRef] [PubMed]
22. Lachman, J.; Martinek, P.; Kotíková, Z.; Orsák, M.; Šulc, M. Genetics and chemistry of pigments in wheat grain: A review. *J. Cereal Sci.* **2017**, *74*, 145–154. [CrossRef]
23. Khlestkina, E. Genes determining the coloration of different organs in wheat. *Russ. J. Genet. Appl. Res.* **2013**, *3*, 54–65. [CrossRef]
24. Glagoleva, A.Y.; Shoeva, O.Y.; Khlestkina, E.K. Melanin pigment in plants: Current knowledge and future perspectives. *Front. Plant Sci.* **2020**, *11*, 770. [CrossRef] [PubMed]
25. Ma, D.; Wang, C.; Feng, J.; Xu, B. Wheat grain phenolics: A review on composition, bioactivity, and influencing factors. *J. Sci. Food Agric.* **2021**, *101*, 6167–6185. [CrossRef]
26. Li, Y.; Ma, D.; Sun, D.; Wang, C.; Zhang, J.; Xie, Y.; Guo, T. Total phenolic, flavonoid content, and antioxidant activity of flour, noodles, and steamed bread made from different colored wheat grains by three milling methods. *Crop J.* **2015**, *3*, 328–334. [CrossRef]
27. Loskutov, I.G.; Khlestkina, E.K. Wheat, barley, and oat breeding for health benefit components in grain. *Plants* **2021**, *10*, 86. [CrossRef]
28. Khlestkina, E.K.; Pshenichnikova, T.A.; Usenko, N.I.; Otmakhova, Y.S. Prospects of molecular genetic approaches in controlling technological properties of wheat grain in the context of the" grain–flour–bread" chain. *Vavilov J. Genet. Breed.* **2016**, *20*, 511–527. [CrossRef]
29. Groos, C.; Gay, G.; Perretant, M.R.; Gervais, L.; Bernard, M.; Dedryver, F.; Charmet, G. Study of the relationship between pre-harvest sprouting and grain color by quantitative trait loci analysis in a white× red grain bread-wheat cross. *Theor. Appl. Genet.* **2002**, *104*, 39–47. [CrossRef]
30. Lin, M.; Zhang, D.; Liu, S.; Zhang, G.; Yu, J.; Fritz, A.K.; Bai, G. Genome-wide association analysis on pre-harvest sprouting resistance and grain color in US winter wheat. *BMC Genom.* **2016**, *17*, 794. [CrossRef]
31. Kumar, A.; Kumar, J.; Singh, R.; Garg, T.; Chhuneja, P.; Balyan, H.S.; Gupta, P.K. QTL analysis for grain colour and pre-harvest sprouting in bread wheat. *Plant Sci.* **2009**, *177*, 114–122. [CrossRef]
32. Pang, Y.; Liu, C.; Wang, D.; Amand, P.S.; Bernardo, A.; Li, W.; He, F.; Li, L.; Wang, L.; Yuan, X.; et al. High-resolution genome-wide association study identifies genomic regions and candidate genes for important agronomic traits in wheat. *Mol. Plant* **2020**, *13*, 1311–1327. [CrossRef] [PubMed]
33. Guo, Y.; Zhang, G.; Guo, B.; Qu, C.; Zhang, M.; Kong, F.; Zhao, Y.; Li, S. QTL mapping for quality traits using a high-density genetic map of wheat. *PLoS ONE* **2020**, *15*, e0230601. [CrossRef] [PubMed]
34. Chu, J.; Zhao, Y.; Beier, S.; Schulthess, A.W.; Stein, N.; Philipp, N.; Röder, M.S.; Reif, J.C. Suitability of single-nucleotide polymorphism arrays versus genotyping-by-sequencing for genebank genomics in wheat. *Front. Plant Sci.* **2020**, *11*, 42. [CrossRef]
35. Colasuonno, P.; Marcotuli, I.; Gadaleta, A.; Soriano, J.M. From genetic maps to QTL cloning: An overview for durum wheat. *Plants* **2021**, *10*, 315. [CrossRef]

36. Arif, M.A.R.; Börner, A. An SNP based GWAS analysis of seed longevity in wheat. *Cereal Res. Commun.* **2020**, *48*, 149–156. [CrossRef]
37. Arif, M.A.R.; Börner, A. Mapping of QTL associated with seed longevity in durum wheat (*Triticum durum* Desf.). *J. Appl. Genet.* **2019**, *60*, 33–36. [CrossRef] [PubMed]
38. Li, N.; Xu, R.; Duan, P.; Li, Y. Control of grain size in rice. *Plant Reprod.* **2018**, *31*, 237–251. [CrossRef] [PubMed]
39. Yang, W.; Feng, H.; Zhang, X.; Zhang, J.; Doonan, J.H.; Batchelor, W.D.; Xiong, L.; Yan, J. Crop phenomics and high-throughput phenotyping: Past decades, current challenges, and future perspectives. *Mol. Plant* **2020**, *13*, 187–214. [CrossRef] [PubMed]
40. Afonnikov, D.A.; Genaev, M.A.; Doroshkov, A.V.; Komyshev, E.G.; Pshenichnikova, T.A. Methods of high-throughput plant phenotyping for large-scale breeding and genetic experiments. *Russ. J. Genet.* **2016**, *52*, 688–701. [CrossRef]
41. Williams, K.; Munkvold, J.; Sorrells, M. Comparison of digital image analysis using elliptic Fourier descriptors and major dimensions to phenotype seed shape in hexaploid wheat (*Triticum aestivum* L.). *Euphytica* **2013**, *190*, 99–116. [CrossRef]
42. Martín-Gómez, J.J.; Rewicz, A.; Goriewa-Duba, K.; Wiwart, M.; Tocino, Á.; Cervantes, E. Morphological description and classification of wheat kernels based on geometric models. *Agronomy* **2019**, *9*, 399. [CrossRef]
43. Cervantes, E.; Martín, J.J.; Saadaoui, E. Updated methods for seed shape analysis. *Scientifica* **2016**, *2016*, 5691825. [CrossRef] [PubMed]
44. Komyshev, E.; Genaev, M.; Afonnikov, D. Analysis of color and texture characteristics of cereals on digital images. *Vavilov J. Genet. Breed.* **2020**, *24*, 340. [CrossRef]
45. Goriewa-Duba, K.; Duba, A.; Wachowska, U.; Wiwart, M. An evaluation of the variation in the morphometric parameters of grain of six Triticum species with the use of digital image analysis. *Agronomy* **2018**, *8*, 296. [CrossRef]
46. Ren, T.; Fan, T.; Chen, S.; Li, C.; Chen, Y.; Ou, X.; Jiang, Q.; Ren, Z.; Tan, F.; Luo, P.; et al. Utilization of a Wheat55K SNP array-derived high-density genetic map for high-resolution mapping of quantitative trait loci for important kernel-related traits in common wheat. *Theor. Appl. Genet.* **2021**, *134*, 807–821. [CrossRef] [PubMed]
47. Arif, M.A.R.; Shokat, S.; Plieske, J.; Ganal, M.; Lohwasser, U.; Chesnokov, Y.V.; Kocherina, N.V.; Kulwal, P.; Kumar, N.; McGuire, P.E.; et al. A SNP-based genetic dissection of versatile traits in bread wheat (*Triticum aestivum* L.). *Plant J.* **2021**, *108*, 960–976. [CrossRef] [PubMed]
48. Lang, J.; Fu, Y.; Zhou, Y.; Cheng, M.; Deng, M.; Li, M.; Zhu, T.; Yang, J.; Guo, X.; Gui, L.; et al. Myb10-D confers PHS-3D resistance to pre-harvest sprouting by regulating NCED in ABA biosynthesis pathway of wheat. *New Phytol.* **2021**, *230*, 1940–1952. [CrossRef]
49. Yang, J.; Tan, C.; Lang, J.; Tang, H.; Hao, M.; Tan, Z.; Yu, H.; Zhou, Y.; Liu, Z.; Li, M.; et al. Identification of qPHS. sicau-1B and qPHS. sicau-3D from synthetic wheat for pre-harvest sprouting resistance wheat improvement. *Mol. Breed.* **2019**, *39*, 132. [CrossRef]
50. Arif, M.A.R.; Agacka-Mołdoch, M.; Qualset, C.O.; Börner, A. Mapping of additive and epistatic QTLs linked to seed longevity in bread wheat (*Triticum aestivum* L.). *Cereal Res. Commun.* **2022**, 1–7. [CrossRef]
51. Roncallo, P.F.; Akkiraju, P.C.; Cervigni, G.L.; Echenique, V.C. QTL mapping and analysis of epistatic interactions for grain yield and yield-related traits in *Triticum turgidum* L. var. durum. *Euphytica* **2017**, *213*, 277. [CrossRef]
52. Roncallo, P.F.; Cervigni, G.L.; Jensen, C.; Miranda, R.; Carrera, A.D.; Helguera, M.; Echenique, V. QTL analysis of main and epistatic effects for flour color traits in durum wheat. *Euphytica* **2012**, *185*, 77–92. [CrossRef]
53. Jofuku, K.D.; Omidyar, P.K.; Gee, Z.; Okamuro, J.K. Control of seed mass and seed yield by the floral homeotic gene APETALA2. *Proc. Natl. Acad. Sci. USA* **2005**, *102*, 3117–3122. [CrossRef] [PubMed]
54. Li, B.; Li, Q.; Mao, X.; Li, A.; Wang, J.; Chang, X.; Hao, C.; Zhang, X.; Jing, R. Two novel AP2/EREBP transcription factor genes TaPARG have pleiotropic functions on plant architecture and yield-related traits in common wheat. *Front. Plant Sci.* **2016**, *7*, 1191. [CrossRef] [PubMed]
55. Li, N.; Li, Y. Ubiquitin-mediated control of seed size in plants. *Front. Plant Sci.* **2014**, *5*, 332. [CrossRef] [PubMed]
56. Parveen, A.; Rahim, M.S.; Sharma, A.; Mishra, A.; Kumar, P.; Fandade, V.; Kumar, P.; Bhandawat, A.; Verma, S.K.; Roy, J. Genome-wide analysis of RING-type E3 ligase family identifies potential candidates regulating high amylose starch biosynthesis in wheat (*Triticum aestivum* L.). *Sci. Rep.* **2021**, *11*, 11461. [CrossRef]
57. Lv, Q.; Li, L.; Meng, Y.; Sun, H.; Chen, L.; Wang, B.; Li, X. Wheat E3 ubiquitin ligase TaGW2-6A degrades TaAGPS to affect seed size. *Plant Sci.* **2022**, *320*, 111274. [CrossRef] [PubMed]
58. Pollard, A.T. Seeds vs fungi: An enzymatic battle in the soil seedbank. *Seed Sci. Res.* **2018**, *28*, 197–214. [CrossRef]
59. Gomez, L.; Allona, I.; Casado, R.; Aragoncillo, C. Seed chitinases. *Seed Sci. Res.* **2002**, *12*, 217–230. [CrossRef]
60. Lee, K.H.; Park, S.W.; Kim, Y.J.; Koo, Y.J.; Song, J.T.; Seo, H.S. Grain width 2 (GW2) and its interacting proteins regulate seed development in rice (*Oryza sativa* L.). *Bot. Stud.* **2018**, *59*, 23. [CrossRef] [PubMed]
61. Amer Hamzah, M.; Mohd Kasim, N.A.; Shamsuddin, A.; Mustafa, N.; Mohamad Rusli, N.I.; Teh, C.Y.; Ho, C.L. Nucleotide variations of 9-cis-epoxycarotenoid dioxygenase 2 (NCED2) and pericarp coloration genes (Rc and Rd) from upland rice varieties. *3 Biotech* **2020**, *10*, 105. [CrossRef] [PubMed]
62. Sano, N.; Marion-Poll, A. ABA metabolism and homeostasis in seed dormancy and germination. *Int. J. Mol. Sci.* **2021**, *22*, 5069. [CrossRef] [PubMed]
63. Matilla, A.J.; Carrillo-Barral, N.; Rodríguez-Gacio, M.d.C. An update on the role of NCED and CYP707A ABA metabolism genes in seed dormancy induction and the response to after-ripening and nitrate. *J. Plant Growth Regul.* **2015**, *34*, 274–293. [CrossRef]

64. Shoeva, O.Y.; Mock, H.P.; Kukoeva, T.V.; Börner, A.; Khlestkina, E.K. Regulation of the flavonoid biosynthesis pathway genes in purple and black grains of *Hordeum vulgare*. *PLoS ONE* **2016**, *11*, e0163782. [CrossRef]
65. Khlestkina, E. The adaptive role of flavonoids: Emphasis on cereals. *Cereal Res. Commun.* **2013**, *41*, 185–198. [CrossRef]
66. Rojas Rodas, F.; Rodriguez, T.O.; Murai, Y.; Iwashina, T.; Sugawara, S.; Suzuki, M.; Nakabayashi, R.; Yonekura-Sakakibara, K.; Saito, K.; Kitajima, J.; et al. Linkage mapping, molecular cloning and functional analysis of soybean gene Fg2 encoding flavonol 3-O-glucoside (1→6) rhamnosyltransferase. *Plant Mol. Biol.* **2014**, *84*, 287–300. [CrossRef]
67. Ma, D.; Xu, B.; Feng, J.; Hu, H.; Tang, J.; Yin, G.; Xie, Y.; Wang, C. Dynamic Metabolomics and Transcriptomics Analyses for Characterization of Phenolic Compounds and Their Biosynthetic Characteristics in Wheat Grain. *Front. Nutr.* **2022**, *9*, 844337. [CrossRef]
68. Zhang, Y.; Zhang, H.; Zhao, H.; Xia, Y.; Zheng, X.; Fan, R.; Tan, Z.; Duan, C.; Fu, Y.; Li, L.; et al. Multi-omics analysis dissects the genetic architecture of seed coat content in Brassica napus. *Genome Biol.* **2022**, *23*, 86. [CrossRef]
69. Hong, M.; Hu, K.; Tian, T.; Li, X.; Chen, L.; Zhang, Y.; Yi, B.; Wen, J.; Ma, C.; Shen, J.; et al. Transcriptomic analysis of seed coats in yellow-seeded Brassica napus reveals novel genes that influence proanthocyanidin biosynthesis. *Front. Plant Sci.* **2017**, *8*, 1674. [CrossRef]
70. Sun, L.; Huang, S.; Sun, G.; Zhang, Y.; Hu, X.; Nevo, E.; Peng, J.; Sun, D. SNP-based association study of kernel architecture in a worldwide collection of durum wheat germplasm. *PLoS ONE* **2020**, *15*, e0229159. [CrossRef]
71. Kiseleva, A.A.; Leonova, I.N.; Pshenichnikova, T.A.; Salina, E.A. Dissection of novel candidate genes for grain texture in Russian wheat varieties. *Plant Mol. Biol.* **2020**, *104*, 219–233. [CrossRef] [PubMed]
72. Goel, S.; Singh, K.; Singh, B.; Grewal, S.; Dwivedi, N.; Alqarawi, A.A.; Abd_Allah, E.F.; Ahmad, P.; Singh, N.K. Analysis of genetic control and QTL mapping of essential wheat grain quality traits in a recombinant inbred population. *PLoS ONE* **2019**, *14*, e0200669. [CrossRef] [PubMed]
73. Komyshev, E.; Genaev, M.; Afonnikov, D. Evaluation of the SeedCounter, a mobile application for grain phenotyping. *Front. Plant Sci.* **2017**, *7*, 1990. [CrossRef] [PubMed]
74. Zdilla, M.J.; Hatfield, S.A.; McLean, K.A.; Cyrus, L.M.; Laslo, J.M.; Lambert, H.W. Circularity, solidity, axes of a best fit ellipse, aspect ratio, and roundness of the foramen ovale: A morphometric analysis with neurosurgical considerations. *J. Craniofacial Surg.* **2016**, *27*, 222. [CrossRef] [PubMed]
75. Gowda, S.N.; Yuan, C. ColorNet: Investigating the importance of color spaces for image classification. In *Computer Vision—ACCV 2018, Proceedings of the Asian Conference on Computer Vision, Perth, Australia, 2–6 December 2018*; Springer: Berlin/Heidelberg, Germany, 2018.
76. Busin, L.; Vandenbroucke, N.; Macaire, L. *Volume 151 of Advances in Imaging and Electron Physics, Chapter Chapter 2: Color Spaces and Image Segmentation*; Elsevier Inc.: Amsterdam, The Netherlands, 2008.
77. Cieplinski, L. MPEG-7 color descriptors and their applications. In *Computer Analysis of Images and Patterns, Proceedings of the 9th International Conference, CAIP 2001, Warsaw, Poland, September 2021*; Springer: Berlin/Heidelberg, Germany, 2001.
78. Hammer, Ø.; Harper, D.A.; Ryan, P.D. PAST: Paleontological statistics software package for education and data analysis. *Palaeontol. Electron.* **2001**, *4*, 9.
79. Soleimani, B.; Lehnert, H.; Keilwagen, J.; Plieske, J.; Ordon, F.; Naseri Rad, S.; Ganal, M.; Beier, S.; Perovic, D. Comparison between core set selection methods using different Illumina marker platforms: A case study of assessment of diversity in wheat. *Front. Plant Sci.* **2020**, *11*, 1040. [CrossRef]
80. Wang, S.; Wong, D.; Forrest, K.; Allen, A.; Chao, S.; Huang, B.E.; Maccaferri, M.; Salvi, S.; Milner, S.G.; Cattivelli, L.; et al. Characterization of polyploid wheat genomic diversity using a high-density 90 000 single nucleotide polymorphism array. *Plant Biotechnol. J.* **2014**, *12*, 787–796. [CrossRef]
81. Allen, A.M.; Winfield, M.O.; Burridge, A.J.; Downie, R.C.; Benbow, H.R.; Barker, G.L.; Wilkinson, P.A.; Coghill, J.; Waterfall, C.; Davassi, A.; et al. Characterization of a Wheat Breeders' Array suitable for high-throughput SNP genotyping of global accessions of hexaploid bread wheat (*Triticum aestivum*). *Plant Biotechnol. J.* **2017**, *15*, 390–401. [CrossRef]
82. Arif, M.A.R.; Afzal, I.; Börner, A. Genetic Aspects and Molecular Causes of Seed Longevity in Plants—A Review. *Plants* **2022**, *11*, 598. [CrossRef]
83. Agacka-Mołdoch, M.; Arif, M.A.R.; Lohwasser, U.; Doroszewska, T.; Qualset, C.O.; Börner, A. The inheritance of wheat grain longevity: A comparison between induced and natural ageing. *J. Appl. Genet.* **2016**, *57*, 477–481. [CrossRef]
84. Arif, R.; Nagel, M.; Neumann, K.; Kobiljski, B.; Lohwasser, U.; Börner, A. Genetic studies of seed longevity in hexaploid wheat using segregation and association mapping approaches. *Euphytica* **2012**, *186*, 1–13. [CrossRef]
85. Sgarbi, C.; Malbrán, I.; Saldúa, L.; Lori, G.A.; Lohwasser, U.; Arif, M.A.R.; Börner, A.; Yanniccari, M.; Castro, A.M. Mapping Resistance to Argentinean Fusarium (Graminearum) Head Blight Isolates in Wheat. *Int. J. Mol. Sci.* **2021**, *22*, 13653. [CrossRef] [PubMed]
86. Agacka-Mołdoch, M.; Rehman Arif, M.A.; Lohwasser, U.; Doroszewska, T.; Lewis, R.S.; Börner, A. QTL analysis of seed germination traits in tobacco (*Nicotiana tabacum* L.). *J. Appl. Genet.* **2021**, *62*, 441–444. [CrossRef] [PubMed]
87. Meng, L.; Li, H.; Zhang, L.; Wang, J. QTL IciMapping: Integrated software for genetic linkage map construction and quantitative trait locus mapping in biparental populations. *Crop J.* **2015**, *3*, 269–283. [CrossRef]

88. McIntosh, R.A.; Yamazaki, Y.; Dubcovsky, J.; Rogers, W.J.; Morris, C.F.; Sommers, D.J. Catalogue of gene symbols for wheat: 2008. In Proceedings of the 11th International Wheat Genetics, Brisbane, Australia, 24–29 August 2008; Appels, R., Eastwood, R., Lagudah, E., Langridge, Mackay, M., McIntyre, L., Eds.; Sydney University Press: Sydney, Australia, 2008.
89. Gu, Z.; Gu, L.; Eils, R.; Schlesner, M.; Brors, B. Circlize Implements and Enhances Circular Visualization in R. *Bioinformatics* **2014**, *30*, 2811–2812. [CrossRef]
90. Ochiai, A. Zoogeographic studies on the soleoid fishes found in Japan and its neighbouring regions. *Bull. Jpn. Soc. Sci. Fish.* **1957**, *22*, 526–530. [CrossRef]
91. Kanehisa, M.; Furumichi, M.; Tanabe, M.; Sato, Y.; Morishima, K. KEGG: New perspectives on genomes, pathways, diseases and drugs. *Nucleic Acids Res.* **2017**, *45*, D353–D361. [CrossRef]
92. Zhu, T.; Wang, L.; Rimbert, H.; Rodriguez, J.C.; Deal, K.R.; De Oliveira, R.; Choulet, F.; Keeble-Gagnère, G.; Tibbits, J.; Rogers, J.; et al. Optical maps refine the bread wheat *Triticum aestivum* cv. *Chin. Spring Genome Assembly. Plant J.* **2021**, *107*, 303–314.
93. Tello-Ruiz, M.K.; Naithani, S.; Gupta, P.; Olson, A.; Wei, S.; Preece, J.; Jiao, Y.; Wang, B.; Chougule, K.; Garg, P.; et al. Gramene 2021: Harnessing the power of comparative genomics and pathways for plant research. *Nucleic Acids Res.* **2021**, *49*, D1452–D1463. [CrossRef]
94. Camacho, C.; Coulouris, G.; Avagyan, V.; Ma, N. BLAST+: Architecture and applications. *BMC Bioinform.* **2009**, *10*, 421. [CrossRef]
95. Borrill, P.; Ramirez-Gonzalez, R.; Uauy, C. expVIP: A customizable RNA-seq data analysis and visualization platform. *Plant Physiol.* **2016**, *170*, 2172–2186. [CrossRef] [PubMed]
96. Hamada, K.; Hongo, K.; Suwabe, K.; Shimizu, A.; Nagayama, T.; Abe, R.; Kikuchi, S.; Yamamoto, N.; Fujii, T.; Yokoyama, K.; et al. OryzaExpress: An integrated database of gene expression networks and omics annotations in rice. *Plant Cell Physiol.* **2011**, *52*, 220–229. [CrossRef] [PubMed]
97. Kanehisa, M.; Sato, Y.; Morishima, K. BlastKOALA and GhostKOALA: KEGG tools for functional characterization of genome and metagenome sequences. *J. Mol. Biol.* **2016**, *428*, 726–731. [CrossRef] [PubMed]
98. Aramaki, T.; Blanc-Mathieu, R.; Endo, H.; Ohkubo, K.; Kanehisa, M.; Goto, S.; Ogata, H. KofamKOALA: KEGG Ortholog assignment based on profile HMM and adaptive score threshold. *Bioinformatics* **2020**, *36*, 2251–2252. [CrossRef] [PubMed]

plants

Article

QTL and Candidate Genes for Seed Tocopherol Content in 'Forrest' by 'Williams 82' Recombinant Inbred Line (RIL) Population of Soybean

Dounya Knizia [1,2], Jiazheng Yuan [3], Naoufal Lakhssassi [1], Abdelhalim El Baze [1], Mallory Cullen [1], Tri Vuong [4], Hamid Mazouz [2], Henry T. Nguyen [4], My Abdelmajid Kassem [3] and Khalid Meksem [1,*]

1 School of Agricultural Sciences, Southern Illinois University, Carbondale, IL 62901, USA; dounya.knizia@siu.edu (D.K.); naoufal.lakhssassi@siu.edu (N.L.); abdelhalim.elbaze@siu.edu (A.E.B.); cullenmallory@gmail.com (M.C.)
2 Laboratoire de Biotechnologies & Valorisation des Bio-Ressources (BioVar), Department de Biologie, Faculté des Sciences, Université Moulay Ismail, Meknes 50000, Morocco; h.mazouz@fs-umi.ac.ma
3 Plant Genomics and Biotechnology Laboratory, Department of Biological and Forensic Sciences, Fayetteville State University, Fayetteville, NC 28301, USA; jyuan@uncfsu.edu (J.Y.); mkassem@uncfsu.edu (M.A.K.)
4 Division of Plant Science and Technology, University of Missouri, Columbia, MO 65211, USA; vuongt@missouri.edu (T.V.); nguyenhenry@missouri.edu (H.T.N.)
* Correspondence: meksem@siu.edu

Abstract: Soybean seeds are rich in secondary metabolites which are beneficial for human health, including tocopherols. Tocopherols play an important role in human and animal nutrition thanks to their antioxidant activity. In this study, the 'Forrest' by 'Williams 82' (F×W82) recombinant inbred line (RIL) population (n = 306) was used to map quantitative trait loci (QTL) for seed α-tocopherol, β-tocopherol, δ -tocopherol, γ-tocopherol, and total tocopherol contents in Carbondale, IL over two years. Also, the identification of the candidate genes involved in soybean tocopherols biosynthetic pathway was performed. A total of 32 QTL controlling various seed tocopherol contents have been identified and mapped on Chrs. 1, 2, 5, 6, 7, 8, 9, 10, 12, 13, 16, 17, and 20. One major and novel QTL was identified on Chr. 6 with an R^2 of 27.8, 9.9, and 6.9 for δ-tocopherol, α-tocopherol, and total tocopherol content, respectively. Reverse BLAST analysis of the genes that were identified in *Arabidopsis* allowed the identification of 37 genes involved in soybean tocopherol pathway, among which 11 were located close to the identified QTLs. The tocopherol cyclase gene (*TC*) *Glyma.06G084100* is located close to the QTLs controlling δ-tocopherol (R^2 = 27.8), α-tocopherol (R^2 = 9.96), and total-tocopherol (R^2 = 6.95). The geranylgeranyl diphosphate reductase (*GGDR*) *Glyma.05G026200* gene is located close to a QTL controlling total tocopherol content in soybean (R^2 = 4.42). The two methylphytylbenzoquinol methyltransferase (*MPBQ-MT*) candidate genes *Glyma.02G002000* and *Glyma.02G143700* are located close to a QTL controlling δ-tocopherol content (R^2 = 3.57). The two γ-tocopherol methyltransferase (*γ-TMT*) genes, *Glyma.12G014200* and *Glyma.12G014300*, are located close to QTLs controlling (γ+ß) tocopherol content (R^2 = 8.86) and total tocopherol (R^2 = 5.94). The identified tocopherol seed QTLs and candidate genes will be beneficial in breeding programs to develop soybean cultivars with high tocopherol contents.

Keywords: soybean; RIL; forrest; Williams 82; linkage map; tocopherol; SNP

check for updates

Citation: Knizia, D.; Yuan, J.; Lakhssassi, N.; El Baze, A.; Cullen, M.; Vuong, T.; Mazouz, H.; T. Nguyen, H.; Kassem, M.A.; Meksem, K. QTL and Candidate Genes for Seed Tocopherol Content in 'Forrest' by 'Williams 82' Recombinant Inbred Line (RIL) Population of Soybean. *Plants* **2022**, *11*, 1258. https://doi.org/10.3390/plants11091258

Academic Editor: Matthew Nelson

Received: 4 April 2022
Accepted: 3 May 2022
Published: 6 May 2022

1. Introduction

Tocopherols and tocotrienols collectively constitute the tocochromanols family, known as Vitamin E. Tocochromanols are fat-soluble phenolic compounds, synthesized by photosynthetic organisms. In soybean, vitamin E is present almost exclusively as tocopherols. Tocopherols exist in four isoforms, α-tocopherol, γ-tocopherol, β-tocopherol, and δ-tocopherol which differ from each other by the number and the location of the methyl groups. α-tocopherol possesses three methyl groups, followed by γ-tocopherol and β-tocopherol that have two methyl groups, and finally δ-tocopherol with only one methyl group [1].

Tocopherols have an important role in human and animal nutrition thanks to their vitamin E activity. However, from a nutritional perspective among the four tocopherol isoforms, α-tocopherol is the most important due to the high vitamin E activity [2]. It also has been reported to play a role in the prevention of cardiovascular diseases and cancer [3,4]. In the human body, α-tocopherol is preferentially accumulated due to its affinity with the liver α-tocopherol transfer protein (α-TTP), which enriches plasma with α-tocopherol [5].

Soybean (*Glycine max* Merr.) is not only one of the main sources of vegetable oil and animal feed worldwide, but also used for production of biofuel, aquaculture feed, and as source of protein for the human diet due to a high protein (40–42%) and oil contents (18–22%) [6], which make it an important crop worldwide.

Soybean seeds are rich in secondary metabolites beneficial for human health including tocopherols. Total tocopherol content is relatively high in soybean seeds compared to other oilseeds crops, and γ-tocopherol is the predominant form, while α-tocopherol content is less than 10% of the total tocopherol content [3,4,7].

Since soybean oil provides 30% of the total worldwide oil consumption and ~70% of the vitamin E in the American diet comes from soybean oil, developing soybean cultivars with high seed α-tocopherol contents could have tremendous positive effects on the health benefits associated with eating soybeans and their market value.

In soybean, tocopherol seed content and composition vary from one cultivar to another, being controlled by several genetic and environmental factors. These factor make it one of the most complex quantitative traits [8] and many studies have focused on investigating the genetic and molecular factors underlying this trait [9–12].

A 'TK780' by 'B04009' RIL population ($n = 94$) was used to identified six QTLs involved in α-tocopherol biosynthesis on Chr. 9, 11 and 12 [7]. Also, using a 'OAC Bayfield' × 'OAC Shire' RIL population across three locations over 2 years, and 151 SSR markers, 26 SSR markers linked to QTLs with individual and total tocopherol content across 17 chromosomes were identified [13]. Likewise, using a 'Beifeng 9' by 'Freeborn' RIL population in six environments, 18, 13, 11, and 13 QTLs associated with seed α-tocopherol, γ-tocopherol, δ-tocopherol, and total tocopherol contents were identified respectively [14].

Tocopherol biosynthesis takes place at the plastid's envelope, where a combination of two precursors derived from different pathways occurs. The homogentisic acid (HGA), a product of the cytosolic shikimate pathway, is used to form the aromatic ring of tocopherols, while phytyl diphosphate (PDP), a product of either the methylerytrithol phosphate (MEP) pathway or the phytol recycling pathway [15], forms the prenyl tail. The condensation of these two precursors is catalyzed by the homogentisate phytyl transferase (HPT) and creates 2-methyl-6- phytyl-1,4-benzoquinol (MPBQ), which can be further methylated by MPBQ methyltransferase (MPBQ-MT) to 2,3-dimethyl-6-phytyl-1,4-benzoquinone (DMPBQ). The cyclization of the MPBQ and DMPBQ by the tocopherol cyclase produces γ-tocopherol and δ-tocopherol, respectively. The conversion of γ-tocopherol and δ-tocopherol to α-tocopherol and β-tocopherol is catalyzed by the γ-tocopherol methyltransferase (γ-TMT) and it represents the last step of the tocopherol biosynthesis pathway.

Many studies have elucidated the genes involved in tocopherol pathway in *Arabidopsis*. The tyrosine produced by the shikimate pathway is catalyzed by Tyrosine Aminotransferase (TAT) resulting in the formation of P-hydroxy Phenyl Pyruvate (HPP). The HPP will be catalyzed by p-hydroxyphenylpyruvate dioxygenase (HPPD) to produce the homogentisic acid, this enzyme is controlled by PDS1. In *A. thaliana,* mutants of *pds1* have shown a lack of tocopherols and plastoquinone with a lethal photobleached phenotype, this result showed the importance of PDS1 in the tocopherol biosynthesis pathway [16]. The overexpression of the *PDS1* gene in tobacco leaves or in *A. thaliana* seeds only gave moderately increased tocopherol concentrations [17,18]. The phytyl diphosphate (PDP) can be derived, either from the MEP pathway after reduction of geranylgeranyl diphosphate (GGDP) by the

Geranylgeranyl Diphosphate Reductase (GGDR) enzyme, or from the phytol recycling pathway. Many studies have investigated the phytol recycling pathway and have shown that mutants of *vte5-1* are devoid of phytol kinase. Also, *vte5-1* mutants have shown a reduction in total tocopherol content in seeds and leaves with 80% and 65% respectively, compared to the wild type [19]. The *VTE5* gene controls the phytol kinase that catalyzes the phytol phosphorylation producing Phytolmonophosphate which is catalyzed by Phytolphosphate kinase VTE6 leading to phytyl diphosphate (PDP) formation. Arabidopsis *vte6* mutants have shown tocopherol deficiency in leaves and a reduction in plant growth and seed longevity. The overexpression of the *VTE6* gene resulted in a two-fold increase in PDP that resulted in higher γ-tocopherol accumulation in seeds [20]. Homogentisate phytyl transferase (HPT) catalyzes the condensation of HGA and PDP to produce 2-methyl-6-phytyl-1,4- benzoquinone (MPBQ). In *Arabidopsis*, the HPT enzyme is encoded by the *VTE2* gene [21,22].

The *Arabidopsis* devoid of VTE2 have shown a complete deficiency in all tocopherol derivatives and all pathway precursors, which means that this is a crucial step in the tocopherol biosynthetic pathway [23]. The MPBQ-MT enzyme is encoded by the *VTE3* gene, which is a limiting step in producing α- and γ-tocopherol. In *Arabidopsis*, *vte3-2* mutants were lacking in α- and γ-tocopherol and exhibited a pale green phenotype, abnormal chloroplasts and did not survive beyond the seedling stage [24,25].

The *VTE1* gene catalyzes the conversion of 2-methyl-6-phytyl-1,4- benzoquinone (MPBQ) and 2,3-dimethyl-5-phytyl-1,4-benzoquinone (DMPBQ) to δ-tocopherol and γ-tocopherol, respectively. *vte1* mutants have a nonfunctional tocopherol cyclase enzyme (TC) and are totally devoid of all tocopherol forms, and accumulate DMPBQ, the γ-tocopherol precursor [26]. The overexpression of the *VTE1* gene in *Arabidopsis* plants showed an improvement in total tocopherol by 7-fold in leaves, as well as a major shift from α- to γ-tocopherol [27].

The VTE4 gene encodes the γ-tocopherol methyltransferase (γ-TMT) that catalyzes the methylation of the γ-tocopherol and δ-tocopherol to produce α-tocopherol and β-tocopherol, respectively. The co-expression of both *At-VTE3* and *At-VTE4* in soybean showed an accumulation of >95% of α-tocopherol, in addition to a 5-fold increase of seed vitamin E activity [28].

In soybean, few genes have been reported to play a role in seed tocopherol content. These genes are *γ-TMT1*, *γ-TMT2*, and *γ-TMT3* mainly [29]. Also, the Di-glucose binding protein with Leucine-rich repeat domain gene (*Glyma.02G099800*); Eukaryotic aspartyl protease family protein gene (*Glyma.02G100800*); Cytochrome b561/ferric reductase transmembrane protein family gene (*Glyma.02G101300*); Transmembrane amino acid transporter family protein gene (*Glyma.02G098200*); and plant U-box 26 gene (*Glyma.02G102900*) were identified and determined to be significantly associated with α-tocopherol, γ-tocopherol, δ-tocopherol and total tocopherol in soybean seeds [30].

In this study, the genetic factors associated with tocopherol content in soybean were investigated, QTL for seed α-tocopherol, β-tocopherol, δ-tocopherol, γ-tocopherol, and total tocopherol contents were mapped, the link between the biosynthesis genes for tocopherol and soybean seed tocopherol content was studied, and the *in-Silico* tocopherol biosynthetic pathway in soybean was reconstructed.

2. Results

2.1. The SNP-Based Genetic Map

The SNP-Based genetic map used in this study was described previously and identified QTLs that control seed isoflavone contents [31]. The map which covered 4029.9 cM, was composed of 2075 SNP markers, and was based on 306 RILs of F × W82 [31].

2.2. Tocopherol Contents Frequency Distribution, Heritability, and Correlation

The frequency distributions among different tocopherol contents were not always normal in the FxW82 RIL population based on Shapiro–Wilk's method for normality test.

Only total-tocopherol 2017 (T-Toc-2017) and δ-tocopherol 2020 (δ-Toc-2020) were normally distributed. The positive or negative skewness and kurtosis value (>3) were also identified in the RIL population (Table 1; Figure 1).

Table 1. Mean, range, CV (%), skewness, kurtosis, and value of Shapiro-Wilk normality test (W value) for seed tocopherol content of the RILs in Carbondale, IL. SE: Standard error.

Year	Trait	Mean (μg/g)	Range (μg/g)	CV (%)	SE	Skewness	Kurtosis	W Value ($p < 0.05$)
2017	δ-tocopherol17	95.86	94	19.73	1.09	0.01	2.21	0.98 **
	γ+β-tocopherol17	172.44	97.1	9.78	0.97	0.48	3.18	0.98 **
	α-tocopherol17	4.94	40.7	83.64	0.24	4.46	32.89	0.52 ***
	Total-tocopherol17	271.5	150.7	8.97	1.41	0.25	2.9	0.99
2020	δ-tocopherol20	94.5	76	14.67	0.81	0.2	3.1	0.99
	γ+β-tocopherol20	180.67	149.2	10.06	1.06	−0.36	5.59	0.97 ***
	α-tocopherol20	5.13	25.7	55.49	0.16	3.13	21.28	0.71 ***
	Total-tocopherol20	279.44	206.4	9.13	1.48	0.16	4.32	0.98 **

** $p < 0.01$, *** $p < 0.001$.

Each tocopherol component also showed a different degree of variation in the parameters of traits, and the variability appeared to not be greatly impacted by different environments. α-tocopherol 2017 (α-Toc-2017), displayed the highest coefficient of variation (CV) value (83.64%); however, the CV of (α-Toc-2020) was 55.49% indicating that phenotypic variability among tocopherol contents was constant year over these 2 years.

The broad sense heritability (h^2) of (μg/g of dry seed weight) for seed α-tocopherol (α-Toc), δ-tocopherol (δ-Toc), γ+β-tocopherol ((γ+β)-Toc), and total tocopherol (T-Toc) contents (in μg/g of dry seed weight) across two different years appeared to be quite diverse. δ-Toc had the highest heritability (71%) and the h^2 for T-Toc was 41% (Table 2). However, the h^2 values for (γ+β)-Toc and α-Toc were negative (−41% and −61%, respectively) implying that there was biologically meaningful phenotypic repulsion among these traits. The high heritability of seed δ-Toc contents suggested that a large portion of phenotypic variation could be detected in the mapped QTL. The RILs-Year interactions still played a significant role in the molecular formation among tocopherols in soybean seeds based on our two-way ANOVA analysis because the σGE^2 is relatively high (data not shown). It should be used as a parameter for trait improvement.

Due to cost effect of this undergraduate student-centered project, only technical replicates could be applied, and F value and probability could not be generated from the dataset (Table 2). Hence, we only calculated the Sum Sq and Mean Sq to determine σG^2 and σGE^2 for each trait (Table 2) using type I sum of squares (ANOVA (model)) function in R program but not σe^2 due to limited replicates.

2.3. Seed Tocopherol Contents QTL

We used both the interval mapping (IM) and composite interval mapping (CIM) methods of WinQTL Cartographer 2.5 [32] to identify QTLs that control seed α-Toc, δ-Toc, (γ+β)-Toc, and T-Toc contents; however, only QTLs identified by CIM method with LOD scores >2.5 are reported here. A total of 32 QTL that control these seed tocopherols contents have been identified and mapped on Chr. 1, 2, 5, 6, 7, 8, 9, 10, 12, 13, 16, 17, and 20 in this RIL population grown in both years (2017 and 2020) (Table 3, Figures S1 and S2).

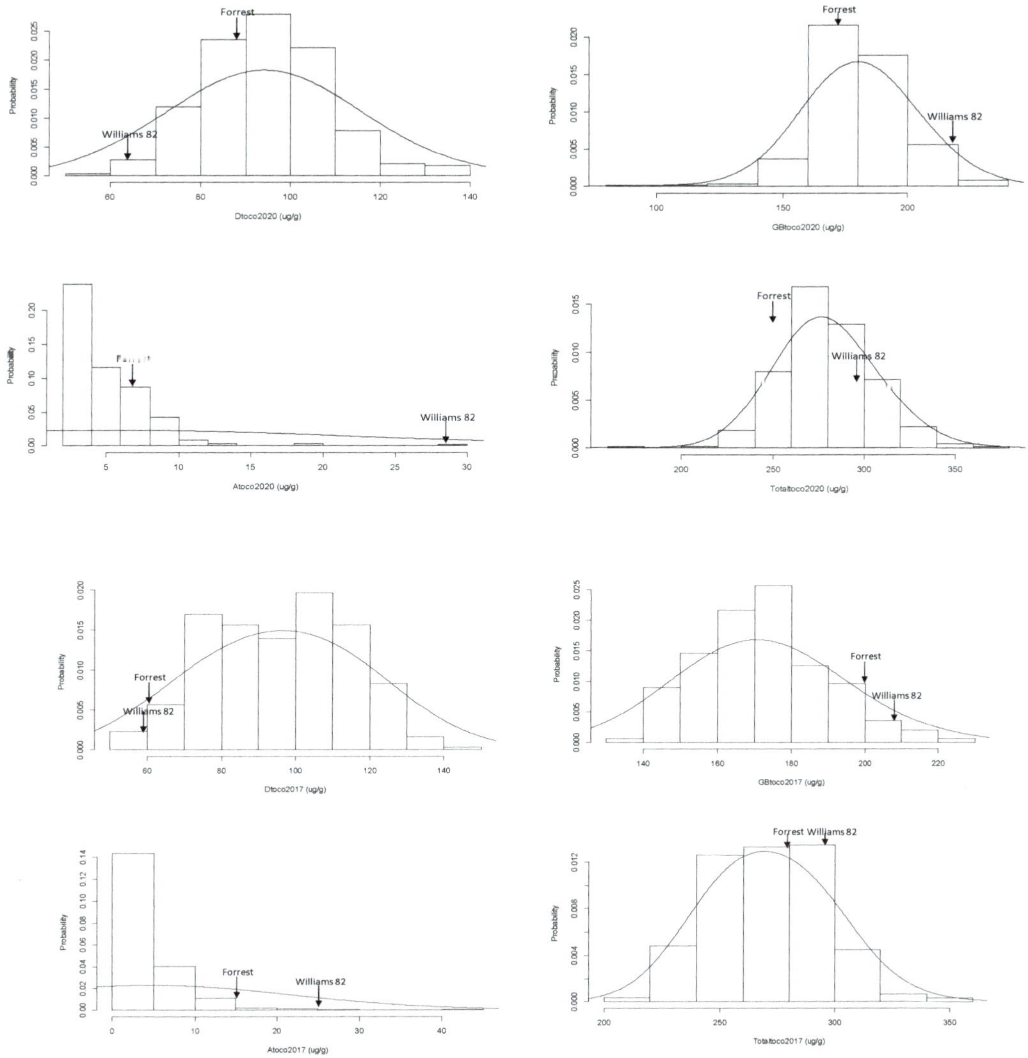

Figure 1. The distribution of seed tocopherol contents (µg/g of seed weight) in the FxW82 RIL population. The seed α-Tocopherol (α-Toc), δ-Tocopherol (δ-Toc), (γ+β)-Tocopherol ((γ+β)-Toc), and Total-Tocopherols (T-Toc) contents were tested in the RILs harvested in Carbondale, IL 2017 and 2020, respectively.

Table 2. The broad sense heritability (h^2) of tocopherol traits (δ-tocopherol, (γ+β)-tocopherol, α-tocopherol, and total-tocopherol) from the seeds harvested at Carbondale, IL in 2017 and 2020.

Response: δ-tocopherol				
	Df	Sum Sq	Mean Seq	H^2
Line	592	163,429	276.06	0.71
Year	1	391	391.28	
Line:Year	1	81	80.65	
Residuals	0	0	NA	

Table 2. *Cont.*

Response: γ+β-tocopherol				
	Df	Sum Sq	Mean Seq	H^2
Line	592	191,945	324.23	-0.41
Year	1	6	6.13	
Line:Year	1	457	456	
Residuals	0	0	NA	
Response: α-tocopherol				
	Df	Sum Sq	Mean Seq	H^2
Line	592	7479.9	12.635	−0.61
Year	1	20	20.041	
Line:Year	1	20	20.41	
Residuals	0	0	NA	
Response: Total-tocopherol				
	Df	Sum Sq	Mean Seq	H^2
Line	592	377,872	638.3	0.47
Year	1	205	205.28	
Line:Year	1	338	337.55	
Residuals	0	0	NA	

Table 3. QTLs that control seed α-Tocopherol (α-Toc), δ-Tocopherol (δ-Toc), (α+ß)-Tocopherol ((α+ß)-Toc), and Total-Tocopherols (T-Toc) contents in two environments over two years (A. 2017 and B. 2020). The two environments are in Carbondale, IL (2017) (A) and (2020) (B). Only solid QTL with LOD scores >2.5 and identified by CIM are reported.

A. QTL that Control Seed Tocopherols Contents in Carbondale, IL (2017)								
Trait	**QTL**	**Chr.**	**Marker/Interval**	**Position (cM)**	**LOD**	**R2**	**Additive**	**Environment**
α-Tocopherol	qα-Toc-1	6	Gm06_1537675-Gm06_1570293	173.7–178.7	6.1	9.96	1.648195	Carbondale, IL
	qα-Toc-2	6	Gm06_1858327-Gm06_2048675	192.6–197.6	6.77	9.95	1.477826	Carbondale, IL
δ-Tocopherol	qδ-Toc-1	1	Gm01_1887205-Gm01_1653315	174.2–179.2	3.06	3.1	−3.30536	Carbondale, IL
	qδ-Toc-2	2	Gm02_1481798-Gm02_9925870	133.5–140.2	3.4	3.57	5.481172	Carbondale, IL
	qδ-Toc-3	6	Gm06_1674534-Gm06_4447485	183.8–207	23.01	27.9	10.14229	Carbondale, IL
γ+ß-Tocopherol	qγ+ß-Toc-1	6	Gm06_1674534-Gm06_4368839	185.8–203.2	5.13	6.16	4.161084	Carbondale, IL
	qγ+ß-Toc-2	8	Gm08_3018731-Gm08_4266625	17.8–31.2	3.02	3.78	−3.23578	Carbondale, IL
	qγ+ß-Toc-3	12	Gm12_3820261-Gm12_3818392	0.5–1	4.14	5.23	3.852091	Carbondale, IL
	qγ+ß-Toc-4	12	Gm12_3805393-Gm12_3696093	2.5–18.5	7.18	8.86	5.050705	Carbondale, IL
	qγ+ß-Toc-5	13	Gm13_2587196-Gm13_2048499	189.1–210.7	3.79	5.43	3.906039	Carbondale, IL
Total-Tocopherols	qTotal-Toc-1	5	Gm05_3674925-Gm05_3256515	29.4–32.2	3.63	4.42	8.695197	Carbondale, IL
	qTotal-Toc-2	6	Gm06_1739930-Gm06_2073990	188.9–197.6	4.07	5.05	−5.69939	Carbondale, IL
	qTotal-Toc-3	6	Gm06_3849946-Gm06_4447485	200.1–207	5.67	6.95	−6.57105	Carbondale, IL
	qTotal-Toc-4	7	Gm07_3635708-Gm07_1829304	81.4–88.9	2.88	3.46	−5.31938	Carbondale, IL
	qTotal-Toc-5	9	Gm09_3483063-Gm09_3544488	74.8–78	3.13	3.78	−4.56753	Carbondale, IL
	qTotal-Toc-6	12	Gm12_3820261-Gm12_3696093	2.5–18.5	4.84	5.94	4.753021	Carbondale, IL

Table 3. *Cont.*

B. QTLs that control seed tocopherols contents in Carbondale, IL (2020)								
Trait	**QTL**	**Chr.**	**Marker**	**Position (cM)**	**LOD**	**R2**	**Additive**	**Envt.**
α-Tocopherol	qα-Toc-1	1	Gm01_3466825-Gm01_5255151	4.1–10.1	5.81	0.35	2.35	Carbondale, IL
	qα-Toc-2	2	Gm02_5141136-Gm02_1020061	137.1–139.8	2.9	0.04	0.82	Carbondale, IL
	qα-Toc-3	6	Gm06_1954068-Gm06_2015292	195–197	2.01	0.03	0.5	Carbondale, IL
δ-Tocopherol	qδ-Toc-1	1	Gm01_4912170-Gm01_4852475	91.6–93.6	2.57	0.04	−2.66	Carbondale, IL
	qδ-Toc-2	8	Gm08_1810148-Gm08_2201336	125.7–130.8	2.42	0.04	2.86	Carbondale, IL
	qδ-Toc-3	10	Gm10_3943637-Gm10_3935014	79.2–81.4	2.4	0.03	−2.53	Carbondale, IL
	qδ-Toc-4	16	Gm16_1079308-Gm16_3673245	0.5–12.5	3.71	0.05	−6.65	Carbondale, IL
	qδ-Toc-5	20	Gm20_3665142-Gm20_1046460	174.9–176.9	2.68	0.04	−5.95	Carbondale, IL
(γ+ß)-Tocopherol	q(γ+ß)-Toc-2	2	Gm02_5155733-Gm02_4311734	130.5–132.5	2.15	0.04	−9.75	Carbondale, IL
	q(γ+ß)-Toc-1	16	Gm16_1079308-Gm16_3673245	2.5–18.5	2.87	0.23	−10.56	Carbondale, IL
Total-Toopherol	qT-Toc-1	1	Gm01_3504836-Gm01_5566630	0.1–1.7	4.52	0.08	−14.62	Carbondale, IL
	qT-Toc-2	8	Gm08_2622664-Gm08_2852874	12.9–13.3	2.11	0.03	−4.44	Carbondale, IL
	qT-Toc-3	10	Gm10_3935014-Gm10_3890052	79.4–84.4	2	0.03	−4.27	Carbondale, IL
	qT-Toc-4	16	Gm16_1079308-Gm16_3673245	0.5–18.5	3.1	0.18	−13.92	Carbondale, IL
	qT-Toc-5	17	Gm17_3916734-Gm17_3929518	6.2–48.7	3.07	0.18	−11.81	Carbondale, IL
	qT-Toc-6	20	Gm20_3665142-Gm20_1046460	174.9–176.9	2.51	0.03	−10.65	Carbondale, IL

Five QTLs controlling α-tocopherol content in soybean were identified on Chrs. 6, 1 and 2 (Table 3, Figures S1 and S2). The qα-Toc-2-IL-2017 (192.6–197.6 cM) and qα-Toc-3-IL-2020 (195–197 cM) were collocated on Chr. 06. Additionolly, eight QTLs underlying δ-tocopherol content were identified on Chrs. 1,2,6,8,16,20. The qδ-Toc-3-IL-2017 located on Chr.6 explains 27.9% of the phenotype (Table 3, Figures S1 and S2). For the γ+β tocopherol content, ten QTLs were identified on Chrs. 2,6,8,12,13, and 16 (Table 3, Figures S1 and S2). Twelve QTLs controlling total tocopherol content were identified and mapped on Chrs. 1, 5, 6, 7, 8, 9, 10, 12, 16, 17, 20 (Table 3, Figures S1 and S2).

2.4. In Silico Reconstruction of the Tocopherol Biosynthetic Pathway in Soybean

The tocopherol biosynthetic pathway has been investigated in the model plant Arabidopsis thaliana. The genes and compounds involved in that pathway were previously reported [33]. To reconstruct the tocopherol biosynthesis pathway in soybean, the reverse BLAST of these genes was conducted using SoyBase.

Thirty-seven candidate genes underlying the soybean's tocopherol pathway were identified (Figure 2). In the Shikimat pathway five candidate genes were identified for Tyrosine Aminotransferase (TAT) including *Glyma.06g235500*, *Glyma.06g235900*, *Glyma.12g161500*, *Glyma.12g205900*, and *Glyma.13g295000*. Two candidate genes were identified for hydroxyphenylpyruvate dioxygenase (HPPD) (PDS1), *Glyma.14G030400*, and *Glyma.02G284600*. In the last step of the MEP pathway three candidate genes underlying the geranylgeranyl diphosphate reductase (GGDR) that catalyzes the production of phytyl diphosphate (PDP) from geranylgeranyl diphosphate (GGDP), were identified, *Glyma.02G273800*, *Glyma.05G026200*, and *Glyma.17G100700* (Figure 2).

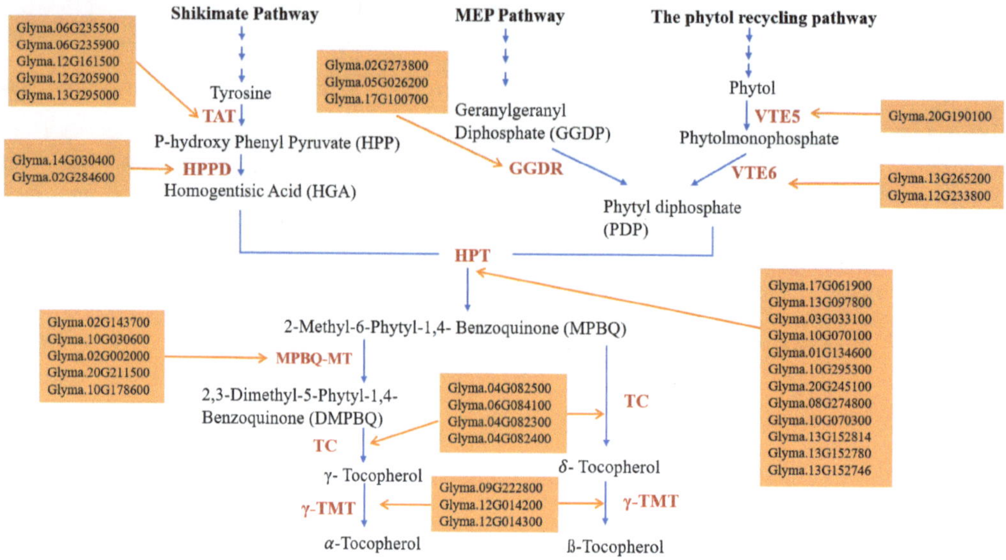

Figure 2. Tocopherol metabolic pathway [15] with identified candidate genes in soybean. TAT: Tyrosine Aminotransferase; HPPD: Hydroxyphenylpyruvate Dioxygenase; GGDR: Geranylgeranyl Diphosphate Reductase; HPT: Homogentisate Phytyl Transferase; MPBQ-MT: methylphytylbenzoquinol methyltransferase; TC: tocopherol cyclase; γ-TMT: gamma tocopherol methyltransferase. MEP Pathway: methylerythritol 4-phosphate pathway.

For the phytol recycling pathway, one candidate gene was identified to be the phytol kinase (VTE5) *Glyma.20G190100*, and two candidate genes were identified to be the phytol-phosphate kinase (VTE6) *Glyma.13G265200*, and *Glyma.12G233800* (Figure 2).

In the core tocopherol pathway, twelve candidate genes were identified for the HPT (VTE2), *Glyma.17G061900, Glyma.13G097800, Glyma.03G033100, Glyma.10G070100, Glyma.01G134600, Glyma.10G295300, Glyma.20G245100, Glyma.08G274800, Glyma.10G070300, Glyma.13G152814, Glyma.13G152780*, and *Glyma.13G152746*, four were identified for the TC (VTE1), *Glyma.04G082500, Glyma.06G084100, Glyma.04G082300*, and *Glyma.04G082400*. Five candidate genes were identified for the MPBQ-MT (VTE3), *Glyma.02G143700, Glyma.10G030600, Glyma.02G002000, Glyma.20G211500*, and *Glyma.10G178600*, in addition to three for the γ-TMT (VTE4), *Glyma.09G222800, Glyma.12G014200*, and *Glyma.12G014300* (Figure 2).

2.5. The Association between the Identified Tocopherol Pathway Candidate Genes and the Identified Tocopherol QTL

Among the identified candidate genes, 11 were located close to the identified QTLs on Chrs. 2, 5, 6, 10, 12, and 17 (Table 4, Figure 2). These candidate genes include the tocopherol cyclase candidate (TC) gene *Glyma.06G084100* that is located close to seven seed tocopherol QTLs controlling δ-tocopherol, α-tocopherol, and total tocopherol on Chr. 6 (Table 4).

On Chr. 2, the MPBQ-MT candidate genes *Glyma.02G002000* and *Glyma.02G143700* are located close to qδ-Toc-2-(2017). The *Glyma.02G002000* candidate gene is also located close to qα-Toc-2-(2020) and q(γ+ß)-Toc-1-(2020) (Table 4, Figure 2). The γ-TMT candidate genes *Glyma.12G014200* and *Glyma.12G014300* are positioned near to QTLs controlling γ+ß tocopherol and total tocopherol (Table 4, Figure 2). The HPT candidate gene *Glyma.17G061900*, and the GGDR candidate gene *Glyma.17G100700* are located close to QTLs controlling total tocopherol on Chr. 17 (Table 4, Figure 2). *Glyma.05G026200* is a GGDR candidate gene that is positioned near to a QTL underlying total tocopherol on Chr. 5 (Table 4, Figure 2).

Table 4. Tocopherol candidate genes located within or close to the tocopherol QTL identified in the FxW82 RIL population grown in Carbondale, IL over two years **A.** 2017 and **B.** 2020.

A. QTLs that Control Seed Tocopherols Contents in Carbondale, IL (2017)								
Trait	**QTL**	**Wm82.a4. v1 Gene Models**			**Glyma1.0 Gene Models**			
		Gene ID	**Start**	**End**	**Gene ID**	**Start**	**End**	**Dist. (Mbp)**
α-Tocopherol	qα-Toc-1	*Glyma.06G084100*	6,435,516	6,441,328	*Glyma06g08850*	6,460,802	6,466,636	4.8
	qα-Toc-2	*Glyma.06G084100*	6,435,516	6,441,328	*Glyma06g08850*	6,460,802	6,466,636	4.4
δ-Tocopherol	qδ-Toc-1							
	qδ-Toc-2	*Glyma.02G002000*	237,750	243,006	*Glyma02g00440*	237,612	245,017	1.2
		Glyma.02G143700	15,253,811	15,256,708	*Glyma02g16210*	14,623,815	14,626,862	4.6
	qδ-Toc-3	*Glyma.06G084100*	6,435,516	6,441,328	*Glyma06g08850*	6,460,802	6,466,636	2.01
(γ+β)-Tocopherol	q(γ+β)-Toc-1	*Glyma.06G084100*	6,435,516	6,441,328	*Glyma06g08850*	6,460,802	6,466,636	2.09
	q(γ+β)-Toc-2							
	q(γ+β)-Toc-3	*Glyma.12G014200*	1,020,484	1,023,995	*Glyma12g01680*	1,020,554	1,024,132	2.7
		Glyma.12G014300	1,028,051	1,031,954	*Glyma12g01690*	1,028,132	1,032,092	2.7
	q(γ+β)-Toc-4	*Glyma.12G014200*	1,020,484	1,023,995	*Glyma12g01680*	1,020,554	1,024,132	2.6
		Glyma.12G014300	1,028,051	1,031,954	*Glyma12g01690*	1,028,132	1,032,092	2.6
	q(γ+β)-Toc-5							
Total-Tocopherols	qT-Toc-1	*Glyma.05G026200*	2,284,067	2,286,242	*Glyma05g01000*	606,481	608,812	2.6
	qT-Toc-2	*Glyma.06G084100*	6,435,516	6,441,328	*Glyma06g08850*	6,460,802	6,466,636	4.3
	qT-Toc-3	*Glyma.06G084100*	6,435,516	6,441,328	*Glyma06g08850*	6,460,802	6,466,636	2.3
	qT-Toc-4							
	qT-Toc-5							
	qT-Toc-6	*Glyma.12G014200*	1,020,484	1,023,995	*Glyma12g01680*	1,020,554	1,024,132	2.6
		Glyma.12G014300	1,028,051	1,031,954	*Glyma12g01690*	1,028,132	1,032,092	2.6

B. QTLs that control seed tocopherols contents in Carbondale, IL (2020)								
Trait	**QTL**	**Wm82.a4. v1 Gene Models**			**Glyma1.0 Gene Models**			
		Gene ID	**Start**	**End**	**Gene ID**	**Start**	**End**	**Dist. (Mbp)**
α-Tocopherol	*qα-Toc-1*							
	qα-Toc-2	*Glyma.02G002000*	237,689	243,112	*Glyma02g00440*	237,612	245,017	0.7
	qα-Toc-3	*Glyma.06G084100*	6,466,090	6,471,839	*Glyma06g08850*	6,460,802	6,466,636	4.5
δ-Tocopherol	*qδ-Toc-1*							
	qδ-Toc-2							
	qδ-Toc-3	*Glyma.10G030600*	2,658,064	2,661,302	*Glyma10g03590*	2,650,012	2,653,309	1.28
		Glyma.10G070100	6,923,409	6,931,780	*Glyma10g08080*	6,888,551	6,893,731	2.95
		Glyma.10G070300	7,023,173	7,029,710	*Glyma10g08150*	6,986,426	6,992,505	3.04
	qδ-Toc-4							
	qδ-Toc-5							
(γ+β)-Tocopherol	*q(γ+β)-Toc-2*	*Glyma.02G002000*	237,689	243,112	*Glyma02g00440*	237,612	245,017	4.06
	q(γ+β)-Toc-1							
Total-Tocopherol	*qT-Toc-1*							
	qT-Toc-2							
	qT-Toc-3	*Glyma.10G030600*	2,658,064	2,661,302	*Glyma10g03590*	2,650,012	2,653,309	1.23
		Glyma.10G070100	6,923,409	6,931,780	*Glyma10g08080*	6,888,551	6,893,731	2.95
		Glyma.10G070300	7,023,173	7,029,710	*Glyma10g08150*	6,986,426	6,992,505	3.05
	qT-Toc-4							
	qT-Toc-5	*Glyma.17G061900*	4,728,685	4,734,790	*Glyma17g06940*	4,998,801	5,004,742	1.06
		Glyma.17G100700	7,920,291	7,923,450	*Glyma17g10890*	8,190,830	8,194,219	4.2
	qT-Toc-6							

2.6. *Association between the Identified Candidate Genes and the Previously Reported QTL*

Mapping the identified genes to the previously reported QTL regions associated with soybean seeds tocopherols was done using data from SoyBase and previous studies describing the QTL underlying tocopherol contents in soybean [7,13,34,35]. Six candidate genes were located within the identified seed tocopherol QTLs and ten were very close to some of these regions (Table 5).

Among these QTLs, qαTC-9 QTL was collocated with the γ-TMT3 (*Glyma.09G222800*) [7]. Also, γ-TMT2 (*Glyma.12G014300*) and γ -TMT1 (*Glyma.12G014200*) are located 703 kb and 711 kb, respectively, apart from qαTC-12 QTL associated with α-tocopherol content [7]. Whereas the fourth tocopherol candidate gene (*Glyma.06G084100*) was located 9.7 Mbp from the QTVEC2_2 QTL underlying total seed tocopherol content identified earlier [34]. The HPT gene (*Glyma.17G061900*) is located 4 Mbp apart from qαγR-17 QTL associated with seed α-tocopherol content [7]. Moreover, *Glyma.04G082500, Glyma.04G082300,* and *Glyma.04G082400*, the tocophetol cyclase candidate genes, were located within qδTC-4 QTL associated with δ-tocopherol [7]. The MPBQ-MT genes *Glyma.02G002000* and *Glyma.02G143700* were located within two QTL controlling the seed γ-tocopherol content. The first one is the seed tocopherol, γ-1-5 QTL [34] (https://soybase.org/; accessed on 3 April 2022) and the second one is the seed tocopherol, γ-2-5 [35] (https://soybase.org/; accessed on 3 April 2022). The two HPT candidate genes *Glyma.10G070100* and *Glyma.10G070300* and the MPBQ-MT candidate gene *Glyma.10G030600* are located 3.3, 3.18, and 7.6 Mbp, respectively from the seed total tocopherol, T-Toc 2-2 QTL [35] (https://soybase.org/; accessed on 3 April 2022). The MPBQ-MT candidate gene *Glyma.20G211500* and the HPT candidate gene (*Glyma.20G245100*) were located 0.22 and 0.9 Mbp apart from the seed tocopherol, QTVEC2_2 QTL controlling total tocopherol content [35] (Table 5).

Table 5. Tocopherol candidate genes associated to the previously reported QTLs.

Gene ID	Start	End	QTL	QTL Start	QTL End	Parents	Number Loci Tested	Lod Score	Interval Length	Reference
Glyma.09G222800	44,341,974	44,346,311	qαTC-9	43,927,286	44,366,371	TK780 X B04009	ND	13.1	ND	[7]
Glyma.12G014200	1,026,615	1,029,095	qαTC-12	1,507,927	1,790,872	TK780 X B04009	ND	7.8	ND	[7]
Glyma.12G014300	1,033,151	1,037,054	qαTC-12	1,507,927	1,790,872	TK780 X B04009	ND	7.8	ND	[7]
Glyma.04G082500	6,948,445	6,954,177	qδTC-4	6,780,105	7,188,146	TK780 X B04009	ND	5.5	ND	[7]
Glyma.04G082400	6,946,447	6,947,480	qδTC-4	6,780,105	7,188,146	TK780 X B04009	ND	5.5	ND	[7]
Glyma.04G082300	6,945,685	6,946,469	qδTC-4	6,780,105	7,188,146	TK780 X B04009	ND	5.5	ND	[7]
Glyma.06G084100	6,466,090	6,471,839	Seed tocopherol, alpha 1-2	16,106,296	16,256,544	OAC Bayfield X Hefeng 25	107	ND	ND	[35]
Glyma.14G030400	2,204,142	2,206,424	Seed tocopherol, alpha 2-1	675,214	2204,996	Hefeng 25 X OAC Bayfield	606	ND	ND	[34]
Glyma.02G143700	14,826,295	14,829,286	Seed tocopherol, gamma 1-5	13,316,369	37,285,448	OAC Bayfield X Hefeng 25	107	ND	ND	[35]
			Seed tocopherol, gamma 2-5	14,288,241	45,267,040	Hefeng 25 X OAC Bayfield	606	ND	56.73	[34]
Glyma.02G002000	237,689	243,112	Seed tocopherol, gamma 1-5	13,316,369	37,285,448	OAC Bayfield X Hefeng 25	107	ND	ND	[35]
			Seed tocopherol, gamma 2-5	14,288,241	45,267,040	Hefeng 25 X OAC Bayfield	606	ND	56.73	[34]
Glyma.13G097800	21,299,008	21,305,797	Seed tocopherol, delta 1-3	15,248,933	15,306,234	OAC Bayfield X Hefeng 25	107	ND	ND	[35]
Glyma.17G061900	4,728,685	4,734,790	qαγR-17	8,786,113	9,025,866	TK780 X B04009	ND	4.1	ND	[7]
Glyma.17G100700	792,0291	7,923,450	Seed tocopherol, gamma 3-6	5,891,979	36,718,722	OAC Bayfield X OAC Shire	550	2.6	67.66	[13]
Glyma.12g161500	30,805,424	30,815,155	Seed tocopherol, total 3-5	24,129,662	37,556,592	OAC Bayfield X OAC Shire	550	3.4	29.62	[13]
Glyma.12g205900	38,082,220	38,086,113	Seed tocopherol, alpha 3-3	24,129,662	37,556,592	OAC Bayfield X OAC Shire	550	3.5	29.62	[13]
			Seed tocopherol, total 3-5	24,129,662	37,556,592	OAC Bayfield X OAC Shire	550	3.4	29.62	[13]
Glyma.13g295000	38,800,738	38,805,839	Seed tocopherol, delta 3-2	37,603,911	40,131,770	OAC Bayfield X OAC Shire	550	2.6	17.11	[13]
			Seed tocopherol, delta 3-3	31,449,060	43,325,731	OAC Bayfield X OAC Shire	550	3.8	52.94	[13]

2.7. Organ-Specific Expression of the Identified Candidate Genes

To investigate the role of the identified 37 candidate genes, the expression analysis of these genes was performed in Williams 82 cv. using the publicly available RNA-seq database at SoyBase (https://soybase.org/; accessed on 3 April 2022). The tissues that were included in this dataset were leaves, nodules, roots, pods, and seeds. Amongst the 37 candidate genes, no RNAseq data was available for the TC candidate gene *Glyma.04G082400*, the HPT candidate gene *Glyma.08G274800*, and the MPBQ-MT candidate gene *Glyma.10G178600*. The rest of the tocopherol candidate genes presented different gene expression patterns. Most of the identified candidate genes were expressed in all the analyzed tissues except for the HTP candidate gene, *Glyma.03G033100*, that was not expressed in any of the tissues. While the two GGDR candidate genes, *Glyma.05G026200* and *Glyma.17G100700*, the HPT candidate gene, *Glyma.13G097800*, and the MPBQ-MT candidate gene, *Glyma.02G143700*, were highly expressed in flowers. The two GGDR candidate genes *Glyma.05G026200* and *Glyma.17G100700*, the MPBQ-MT candidate genes *Glyma.02G143700*, *Glyma.02G002000*, and *Glyma.10G030600*, the TC candidate gene *Glyma.06G084100*, and the γ-TMT candidate gene *Glyma.12G014300* were abundantly expressed in leaves. The GGDR candidate genes *Glyma.05G026200*, *Glyma.02G273800* and *Glyma.17G100700*, and the TAT candidate gene *Glyma.12G161500* were highly expressed in seeds. The TAT candidate genes *Glyma.06G235900* and *Glyma.12G205900*, the two GGDR candidate genes *Glyma.05G026200* and *Glyma.17G100700*, and the MPBQ-MT candidate genes *Glyma.02G143700* and *Glyma.10G030600* were highly expressed in pods. The γ-TMT candidate gene *Glyma.09G222800*, the GGDR candidate genes *Glyma.05G026200* and *Glyma.02G273800* were highly expressed in roots. Whereas the TAT candidate genes *Glyma.12G161500* and *Glyma.13G295000*, the GGDR candidate genes *Glyma.05G026200* and *Glyma.02G273800*, and the MPBQ-MT candidate gene *Glyma.10G030600* were highly expressed in the nodules (Figure 3A, Table S1).

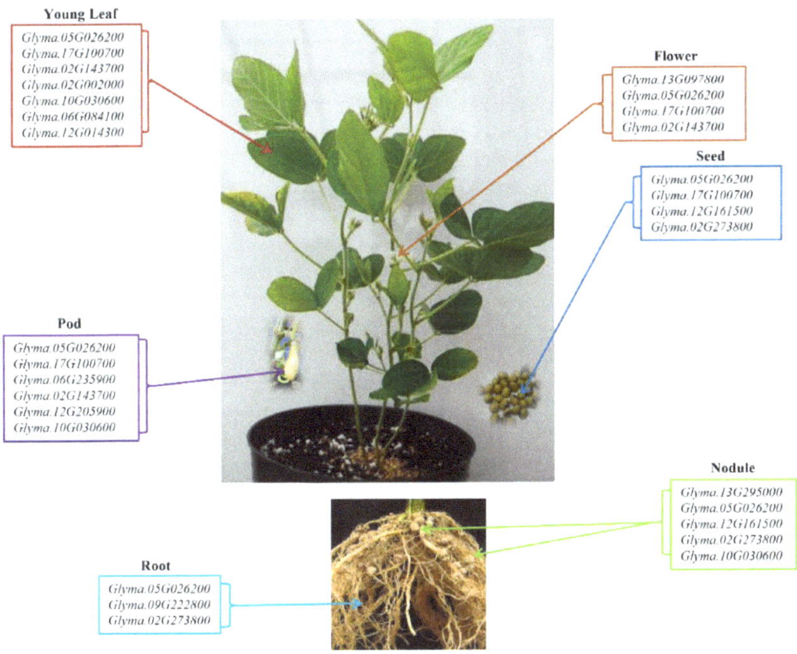

(**A**)

Figure 3. *Cont.*

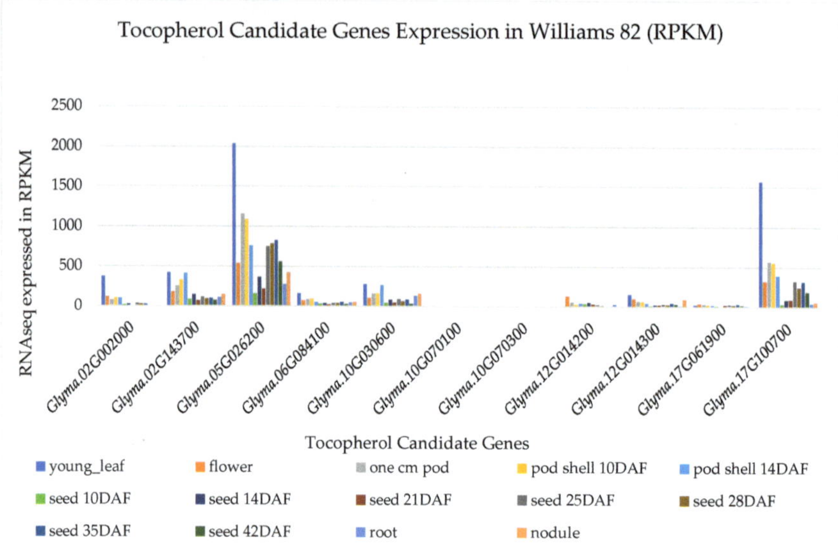

(**B**)

Figure 3. (**A**). Tissue specific expression of the identified tocopherol candidate genes. (**B**). Expression pattern of the 11 tocopherol candidate genes located within tocopherol QTL in Williams 82 (RPKM) were retrieved from publicly available RNA-seq data from Soybase database (http://www.soybase.org/soyseq; accessed on 3 April 2022).

Amongst the identified candidate genes, eleven were located close to the tocopherol seed QTLs identified in FxW82 RIL population, in tocopherol seed content in soybean. *Glyma.05G026200* and *Glyma.17G100700* are highly expressed in seeds in Williams 82 cv., followed by *Glyma.02G002000*, *Glyma.02G143700*, *Glyma.10G030600*, *Glyma.12G014300*, and *Glyma.06G084100* that have moderate expression profiles in seeds. The rest of the genes have a low expression profile in seeds except for *Glyma.10G070100* and *Glyma.10G070300* that have a limited expression profile, with very low to no expression in seeds (Figure 3B).

3. Discussion

Tocopherols are lipophilic antioxidants that are important for human health due to their ability to prevent the oxidation of unsaturated fatty acids by scavenging the free radicals and prevent cell membrane damage [13]. Soybean seeds contain the highest tocopherol concentrations among all legume species [36]. The dominant tocopherol isoform in soybean seeds is γ-tocopherol with amounts reaching up to 70% of the total tocopherol content, while α-tocopherol isoform has a lower concentration of about 10% of the total tocopherol content. The α-tocopherol isoform has the highest vitamin E activity [4] and has the highest affinity with the hepatic tocopherol transfer protein. Therefore, improving soybean seed tocopherol composition and content is crucial.

Several studies have revealed the genetic and molecular bases underlying tocopherol content in soybean [7,13,14,29,30,34,35] as summarized recently in [37].

Among the QTLs identified in these studies, qαTC-9 QTL was collocated with the *γ-TMT3 Glyma.09G222800*; [7], however, the QTL identified in this study on Chr. 9 was more than 40 Mbp apart from this gene. Also, *γ-TMT2* (*Glyma.12G014300*) and *γ -TMT1* (*Glyma.12G014200*) are located 703 kb and 711 kb, respectively, apart from the qαTC-12 QTL associated with α-tocopherol content [7]. Similarly, in this study, these candidate genes are located 2.7, 2.6 and 2.6 Mbp apart from *q(γ+ß)-Toc-3-(2017)*, *q(γ+ß)-Toc-4-(2017)*, and *qT-Toc-6-(2017)*, respectively, on Chr. 12 (Table 4, Figures S1 and S2). Also, the TC candidate gene (*Glyma.06G084100*) was located 9.7 Mbp from the QTVEC2_2 QTL underlying the total tocopherol identified earlier [34]. Likewise, this gene is located close to seven seed tocopherol QTL, 2 Mbp from the *qδ-Toc-3-(2017)* (R^2 = 27.8), 4.8 and 4.4 Mbp from *qα-Toc-1-(2017)* and *qα-Toc-2-(2017)*, respectively, 4.3 and 2.3 Mbp from *qT-Toc-2-(2017)* and *qT-Toc-3-(2017)*, respectively, 2.09 Mbp from *q(γ+ß)-Toc-1-(2017)*, and 4.5 Mbp from *qα-Toc-2-(2020)* on Chr. 6 (Table 4, Figures S1 and S2). On Chr. 2, the *MPBQ-MT* candidate genes *Glyma.02G002000* and *Glyma.02G143700* are located 1.2 and 4.6 Mbp, respectively from *qδ-Toc-2-(2017)*. Also, the *Glyma.02G002000* candidate gene is located 0.7 and 4.06 Mbp apart from *qα-Toc-2-(2020)* and *q(γ+ß)-Toc-1-(2020)*, respectively (Table 4, Figures S1 and S2). This is coherent with previous studies [34,35], where these two candidate genes were located within seed tocopherol, gamma 1-5 [35] (https://soybase.org/; accessed on 3 April 2022) and seed tocopherol, gamma 2-5 [34] (https://soybase.org/; accessed on 3 April 2022).

The QTL associate with δ-tocopherol explains 27.87% of the phenotype, and the one associated with α-tocopherol explains only 9.96% of the phenotype. A TC (*Glyma.06G084100*) gene was identified close to these QTLs, the TC enzyme is involved directly in the conversion of MPBQ to δ-tocopherol, and indirectly in the conversion to α-tocopherol (Table 4, Figure 2). α-tocopherol is the most known potent fat-soluble antioxidant, it is preferentially absorbed and accumulated in humans [38], its activity has been demonstrated in the prevention and treatment of heart disease, cancer and Alzheimer's disease [39]. Alpha-tocopherol has been designated as the most beneficial tocopherol compound among health professionals. Unfortunately, this compound is present in small amount in soybean oil when compared to sunflower, canola or corn oil [40]. Therefore, improving α-tocopherol content in soybean is a priority for the soy-industry, the identification in the two years data of the qα-Toc-2-IL-2017 (192.6–197.6 cM) and qα-Toc-3-IL-2020 (195–197 cM) that were collocated on Chr. 06 will provide an opportunity for breeding lines with high α-tocopherol.

Interestingly, the *HPT* candidate gene (*Glyma.17G061900*) is located 4 Mbp apart from qαγR-17 QTL associated with α-tocopherol [7]. Similarly, this candidate gene is located 1.06 Mbp apart from *qT-Toc-5-(2020)* identified here on Chr. 17 (Table 4, Figures S1 and S2). The two *HPT* candidate genes *Glyma.10G070100* and *Glyma.10G070300* and the *MPBQ-MT* candidate gene *Glyma.10G030600* are located 3.3, 3.18, and 7.6 Mbp, respectively from the seed tocopherol, total 2-2 QTL [34,35] (https://soybase.org/; accessed on 3 April 2022). Likewise, in this study these genes are located 2.95, 3.04 and 1.28, respectively from *qδ-Toc-2-(2020)*, and *qT-Toc-3-(2020)* on Chr. 10 (Table 4, Figures S1 and S2). The *MPBQ-MT* candidate gene *Glyma.20G211500* and the *HPT* candidate gene (*Glyma.20G245100*) were located 0.22 and 0.9 Mbp apart from seed tocopherol, QTVEC2_2 QTL controlling total tocopherol content identified earlier [34], however, the QTL identified in this study on Chr. 20 was located more than 40 Mbp apart from these genes. Moreover, *Glyma.04G082500*, *Glyma.04G082300*, and *Glyma.04G082400* tocopherol cyclase candidate genes were located within qδTC-4 QTL associated with δ-tocopherol and identified earlier [7].

Although previous studies have reported some soybean genes as candidates for the tocopherol biosynthesis pathway [41], the present study shows the most comprehensive analysis of the whole soybean genome, showing the potential candidate genes for the tocopherol biosynthetic pathway in soybean.

Most QTL regions that were identified in 2017 were not found in 2020 except the qα-Toc-2-IL-2017 (192.6–197.6 cM) and qα-Toc-3-IL-2020 (195–197 cM) that were collocated on Chr. 06. This could be explained by the difference in weather conditions between 2017 and 2020. In August 2017 the temperature ranged between 8 and 33.3 °C, while in August 2020 the temperature ranged between 13.3 and 32.8 °C (https://www.extremeweatherwatch.com/; accessed on 3 April 2022). It has been proven that temperature stress during all stages of development affect soybean seed tocopherol concentrations [42].

The QTL region identified on Chr.7, qTotal-Toc-4-IL-2017 (81.4–88.9 cM), is 7.6 cM from a QTL region previously identified [7]. Likewise, the QTL region identified on Chr. 20 is 61.21 cM away from a QTL region identified in previous studies [34,35]. Interestingly, the QTL region identified on Chr.12 qγ+ß-Toc-3-IL-2017, qγ+ß-Toc-4-IL-2017, qTotal-Toc-6-IL-

2017 (0.5–18.5 cM) and Chr.8 qγ+ß-Toc-2-IL-2017 and qT-Toc-2-IL-2020 (12.9–31.2 cM) were reported in previous studies [7,34]. Which make them important regions to investigate further for candidate genes. The rest of the QTLs are novel (Table 4, Figures S1 and S2).

Although previous studies identified QTL regions for soybean seed tocopherol content on Chr.6, all the identified QTLs map to the region between 74.5 and 118.5 cM (Table S2) [7,34,35]. The QTL regions identified in this study on Chr.6 clusters between 173.7 and 207 cM, which is the region that encompass an important gene in the tocopherol biosynthesis pathway, namely the tocopherol cyclase candidate gene, *Glyma.06G084100*. This QTL on Chr.6 is responsible for 27.8% of δ-tocopherol, 9.96% of α-tocopherol, 6.16% of γ+ß-tocopherol, and 6.95% of total tocopherol content

4. Materials and Methods

4.1. Plant Materials

The $F_{6:13}$ 'Forrest' × 'Williams 82' RIL population (n = 306) described previously was used in this study [30,43]. The parents and RILs were grown in Carbondale, southern Illinois in 2017 and 2020, and seeds were harvested at maturity of all RILs and parents.

4.2. Tocopherols Quantification

At maturity, seeds of the parents and RILs were harvested and analyzed for α-Tocopherol (α-Toc), δ-Tocopherol (δ-Toc), α+ß-Tocopherol ((γ+ß)-Toc), and total-Tocopherols (T-Toc) using a protocol developed and validated in the Nguyen Lab, the University of Missouri. Briefly, approx. 1gr. of soybean seeds were ground to fine powders with a Thomas Wiley Mini-Mill, followed by lyophilizing for 48 hrs. Approx. 200mg of powder were mixed with 2mL 200-proof ethanol and vortexed, followed by an incubation with agitation at 75 °C for 2 hrs. The products were then filtered into HPLC vials for analysis along with standard solutions of tocopherols. Quantification of tocopherols was performed by employing an external calibration curve method, in which each curve was created with the six standard solutions of 0.62, 1.25, 2.5, 5, 10, and 20 μg/mL.

4.3. DNA Isolation, SNP Genotyping, and Genetic Map Construction

DNA Isolation, SNP Genotyping, and the construction of the F×W82 genetic linkage map have been described earlier [30]. Briefly, SNP genotyping was performed with Illumina Infinium SoySNP6K BeadChips (Illumina, Inc. San Diego, CA, USA) and the genetic map was constructed with JoinMap 4.0 software with a LOD score of 3.0 and maximum distance of 50 cM as described earlier [30].

4.4. Seed Tocopherols QTL Detection

We used WinQTL Cartographer 2.5 [31] and both interval mapping (IM) and composite interval mapping (CIM) methods to identify QTL that control seed α-Toc, δ-Toc, γ+ß-Toc, and T-Toc in this RIL population; however, only QTL detected with CIM are reported here. QTL identified via IM are reported in the supplementary data section (Table S4A,B). MapChart 2.2 [32] was used to draw chromosomes with CIM tocopherols QTL locations.

4.5. Tocopherols Candidate Genes Identification

The reverse blast of the genes underlying the tocopherol pathway in *Arabidopsis* was conducted using the available data at SoyBase (https://soybase.org/; accessed on 3 April 2022). The sequences of the *Arabidopsis* genes were obtained from the Phytozome database (https://phytozome-next.jgi.doe.gov; accessed on 3 April 2022), these sequences were used for Blast in SoyBase. The obtained genes that control the tocopherol biosynthetic pathway were mapped to the identified tocopherol QTL.

4.6. Expression Analysis

The expression analysis of the identified tocopherol candidate genes that are located within or close the identified seed tocopherol QTLs was performed using the publicly available data from SoyBase (https://soybase.org/; accessed on 3 April 2022) to produce the expression profiles of these candidate genes in the soybean reference genome Williams 82 in Glyma1.0 Gene Models version.

5. Conclusions

In conclusion, 32 QTL controlling seed tocopherol contents on Chr. 1, 2, 5, 6, 7, 8, 9, 10, 12, 13, 16, 17, and 20 were identified. 37 candidate genes involved in soybean tocopherol biosynthetic pathway have also been identified among which 11 were located close to the QTL regions identified in this study. Two of these candidate genes were highly expressed in seeds *Glyma.05G026200* and *Glyma.17G100700*, followed by *Glyma.06G084100*, *Glyma.02G002000*, *Glyma.02G143700*, *Glyma.10G030600*, and *Glyma.12G014300* with moderate expression profiles in seeds (Figure 3B).

Forrest and Williams 82 sequences of the eleven candidate genes located close to the identified QTLs were compared, and the results have shown that three of them have SNPs between the Forrest and Williams 82 sequences, *Glyma.06G084100, Glyma.17G061900* and *Glyma.17G100700* (Figure 4). The *TC* candidate gene *Glyma.06G084100* has 5 SNPs in the coding sequence, one of them caused a missense mutation (T379A) (Figure 4) in addition to 12 SNPs and 2 InDels in the 5′UTR region (Table S3). The *HPT* candidate gene, *Glyma.17G061900*, has only one SNP located in the coding sequence that caused a missense mutation (G326A) (Figure 4). For the *GGDR* candidate gene, *Glyma.17G100700*, there is also only one SNP that caused a silent mutation (Figure 4). These SNPs could play a role in the difference of tocopherol content between Forrest and Williams 82 cultivars. *Glyma.06G084100* is associated with the qδ-Toc-3-(2017) (R2 = 27.8), qα-Toc-1-(2017), qα-Toc-2-(2017), qT-Toc-2-(2017), qT-Toc-3-(2017), q(γ+ß)-Toc-1-(2017), and qα-Toc-2-(2020) on Chr. 6 (Table 3, Figures S1 and S2). While *Glyma.17G061900* and *Glyma.17G100700* are associated to qT-Toc-5-(2020) on Chr. 17 (Table 2, Figures S1 and S2). These genes could be used in breeding programs or gene editing technology to develop soybean lines and cultivars that produce high amounts of the beneficial tocopherols (vitamin E) for human consumption.

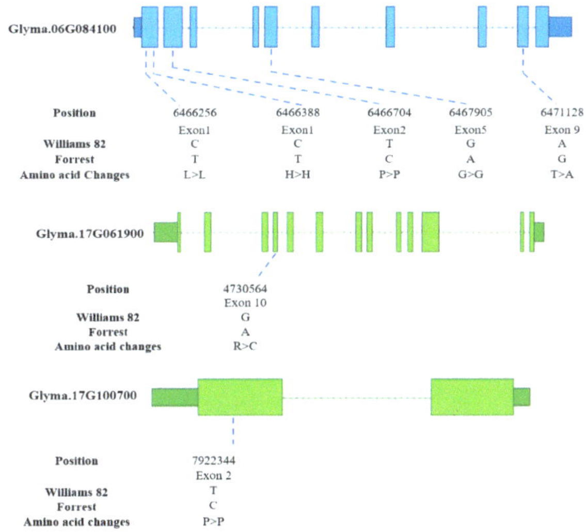

Figure 4. Positions of SNPs between Forrest and Williams 82 cultivars in *Glyma.06G084100*, *Glyma.17G061900* and *Glyma.17G100700* coding sequences. In the gene model diagram, the light blue/light green boxes represent exons, blue/green bars represent introns, dark blue/dark green boxes represent 3′UTR or 5′UTR. SNPs were positioned relative to the genomic position in the genome version W82.a2.

6. Patents

Patent resulting from this work is under submission.

Supplementary Materials: The following supporting information can be downloaded at: https://www.mdpi.com/article/10.3390/plants11091258/s1, Table S1. Expression profiles of the tocopherol biosynthesis candidate genes in soybean based on RNAseq data available from RNAsequencing data (http://www.soybase.org/soyseq; accessed on 3 April 2022). Table S2. QTLs identified in previous studies on Chr. 6. Table S3. Positions of SNPs between Forrest and Williams 82 cultivars in the promoter of the tocopherol cyclase candidate gene (*GmTC06, Glyma.06G084100*). Table S4. A. QTLs controlling seed tocopherols contents in Carbondale, IL (2017)—Identified by Interval Mapping Method. B. QTLs controlling seed tocopherols contents in Carbondale, IL (2020)—Identified by Interval Mapping Method. Figure S1. Positions of QTL that control seed α-Tocopherol (α-Toc), δ-Tocopherol (δ-Toc), (γ+β)-Tocopherol ((γ+β)-Toc), and Total-Tocopherols (T-Toc) contents on chromosomes 1, 2, 5, 6, 7, 8, 9, 10, 12, 13, 16, 17, and 20. The QTL have been identified in F×W82 grown in two environments in Carbondale, IL over two years (2017 and 2020). Legend: (γ + β) = (G + β) and (Carb-IL) = Carbondale, IL. Figure S2. QTL that control seed α-Tocopherol (α-Toc), δ-Tocopherol (δ-Toc), (γ+β)-Tocopherol ((γ+β)-Toc), and Total-Tocopherols (T-Toc) contents identified by IM and CIM methods in the F×W82 RIL population grown in two environments in Carbondale, IL over two years (2017 and 2020).

Author Contributions: Conceptualization, K.M., H.T.N. and M.A.K.; D.K. conceived and wrote the manuscript, methodology, D.K., J.Y., T.V. and N.L.; validation, K.M., M.A.K. and H.T.N.; formal analysis, D.K. and J.Y.; investigation, K.M. and D.K.; resources and data curation, K.M., M.A.K. and H.T.N.; review and editing, D.K., J.Y., N.L., M.C., A.E.B., T.V., H.M., M.A.K., K.M. and H.T.N.; visualization, J.Y.; supervision, K.M., M.A.K., J.Y., H.M. and H.T.N.; project administration, K.M., M.A.K., H.T.N. and K.M. supervised and conceived the work, performed data interpretation, designed the experiment, and edited the manuscript. All authors have read and agreed to the published version of the manuscript.

Funding: This research was funded by the United Soybean Board: Development and Releases of High Tocopherol Soybean Germplasm Project USB # 2220-162-0116.

Institutional Review Board Statement: Not applicable.

Informed Consent Statement: Not applicable.

Data Availability Statement: Data supporting reported results are available on request from the corresponding author.

Acknowledgments: We thank all the student workers at Southern Illinois University Carbondale, who assisted in planting the recombinants inbred lines, DNA extraction, and greenhouse and field work. Technical assistance in the HPLC analysis by Haiying Shi at the University of Missouri is greatly appreciated.

Conflicts of Interest: The authors declare no conflict of interest.

References

1. Rani, A.; Kumar, V.; Verma, S.K.; Shakya, A.K.; Chauhan, G.S. Tocopherol Content and Profile of Soybean: Genotypic Variability and Correlation Studies. *J. Am. Oil Chem. Soc.* **2007**, *84*, 377–383. [CrossRef]
2. Kamal-Eldin, A.; Appelqvist, L.-Å. The chemistry and antioxidant properties of tocopherols and tocotrienols. *Lipids* **1996**, *31*, 671–701. [CrossRef] [PubMed]
3. Grusak, M.A.; DellaPenna, D. Improving the nutrient composition of plants to enhance human nutrition and health. *Annu. Rev. Plant Physiol. Plant Mol. Biol.* **1999**, *50*, 133–161. [CrossRef] [PubMed]
4. Bramley, P.M.; Elmadfa, I.; Kafatos, A.; Kelly, F.J.; Manios, Y.; Roxborough, H.; Schuch, W.; Sheehy, P.J.A.; Wagner, K.H. Vitamin E: A critical review. *J. Sci. Food Agric.* **2000**, *80*, 913–938. [CrossRef]
5. Kono, N.; Ohto, U.; Hiramatsu, T.; Urabe, M.; Uchida, Y.; Satow, Y.; Arai, H. Impaired α-TTP-PIPs interaction underlies familial vitamin E deficiency. *Science* **2013**, *340*, 1106–1110. [CrossRef]
6. Pagano, M.C.; Miransari, M. 1—The importance of soybean production worldwide. In *Abiotic and Biotic Stresses in Soybean Production*; Miransari, M., Ed.; Academic Press: San Diego, CA, USA, 2016; pp. 1–26.
7. Park, C.; Dwiyanti, M.S.; Nagano, A.J.; Liu, B.; Yamada, T.; Abe, J. Identification of quantitative trait loci for increased α-tocopherol biosynthesis in wild soybean using a high-density genetic map. *BMC Plant Biol.* **2019**, *19*, 510. [CrossRef]

8. Seguin, P.; Tremblay, G.; Pageau, D.; Liu, W. Soybean Tocopherol Concentrations Are Affected by Crop Management. *J. Agric. Food Chem.* **2010**, *58*, 5495–5501. [CrossRef]
9. Marwede, V.; Schierholt, A.; Möllers, C.; Becker, H.C. Genotype × Environment Interactions and Heritability of Tocopherol Contents in Canola. *Crop Sci.* **2004**, *44*, 728–731. [CrossRef]
10. Ujiie, A.; Yamada, T.; Fujimoto, K.; Endo, Y.; Kitamura, K. Identification of Soybean Varieties with High α-Tocopherol Content. *Breed. Sci.* **2005**, *55*, 123–125. [CrossRef]
11. Carrera, C.S.; Seguin, P. Factors Affecting Tocopherol Concentrations in Soybean Seeds. *J. Agric. Food Chem.* **2016**, *64*, 9465–9474. [CrossRef]
12. Dwiyanti, M.S.; Maruyama, S.; Hirono, M.; Sato, M.; Park, E.; Yoon, S.H.; Yamada, T.; Abe, J. Natural diversity of seed α-tocopherol ratio in wild soybean (*Glycine soja*) germplasm collection. *Breed. Sci.* **2016**, *66*, 653–657. [CrossRef] [PubMed]
13. Shaw, E.J.; Rajcan, I. Molecular mapping of soybean seed tocopherols in the cross 'OAC Bayfield' × 'OAC Shire'. *Plant Breed.* **2017**, *136*, 83–93. [CrossRef]
14. Liu, H.; Cao, G.; Wu, D.; Jiang, Z.; Han, Y.; Li, W. Quantitative trait loci underlying soybean seed tocopherol content with main additive, epistatic and QTL × environment effects. *Plant Breed.* **2017**, *136*, 924–938. [CrossRef]
15. Muñoz, P.; Munné Bosch, S. Vitamin E in Plants: Biosynthesis, Transport, and Function. *Trends Plant Sci.* **2019**, *24*, 1040–1051. [CrossRef]
16. Norris, S.R.; Shen, X.; Della Penna, D. Complementation of the Arabidopsis pds1 Mutation with the Gene Encoding p-Hydroxyphenylpyruvate Dioxygenase. *Plant Physiol.* **1998**, *117*, 1317–1323. [CrossRef]
17. Falk, J.; Andersen, G.; Kernebeck, B.; Krupinska, K. Constitutive overexpression of barley 4-hydroxyphenylpyruvate dioxygenase in tobacco results in elevation of the vitamin E content in seeds but not in leaves1. *FEBS Lett.* **2003**, *540*, 35–40. [CrossRef]
18. Li, Y.; Wang, Z.; Sun, X.; Tang, K. Current Opinions on the Functions of Tocopherol Based on the Genetic Manipulation of Tocopherol Biosynthesis in Plants. *J. Integrat. Plant Biol.* **2008**, *50*, 1057–1069. [CrossRef]
19. Valentin, H.E.; Lincoln, K.; Moshiri, F.; Jensen, P.K.; Qi, Q.; Venkatesh, T.V.; Karunanandaa, B.; Baszis, S.R.; Norris, S.R.; Savidge, B.; et al. The Arabidopsis vitamin E pathway gene5-1 Mutant Reveals a Critical Role for Phytol Kinase in Seed Tocopherol Biosynthesis. *Plant Cell* **2005**, *18*, 212–224. [CrossRef]
20. Vom Dorp, K.; Hölzl, G.; Plohmann, C.; Eisenhut, M.; Abraham, M.; Weber, A.P.M.; Hanson, A.D.; Dörmann, P. Remobilization of Phytol from Chlorophyll Degradation Is Essential for Tocopherol Synthesis and Growth of Arabidopsis. *Plant Cell* **2015**, *27*, 2846–2859. [CrossRef]
21. Collakova, E.; DellaPenna, D. Isolation and functional analysis of homogentisate phytyltransferase from Synechocystis sp. PCC 6803 and Arabidopsis. *Plant Physiol.* **2001**, *127*, 1113–1124. [CrossRef]
22. Schledz, M.; Seidler, A.; Beyer, P.; Neuhaus, G. A novel phytyltransferase from Synechocystis sp. PCC 6803 involved in tocopherol biosynthesis. *FEBS Lett.* **2001**, *499*, 15–20. [CrossRef]
23. Sattler, S.E.; Cheng, Z.; DellaPenna, D. From Arabidopsis to agriculture: Engineering improved Vitamin E content in soybean. *Trends Plant Sci.* **2004**, *9*, 365–367. [CrossRef] [PubMed]
24. Motohashi, R.; Ito, T.; Kobayashi, M.; Taji, T.; Nagata, N.; Asami, T.; Yoshida, S.; Yamaguchi-Shinozaki, K.; Shinozaki, K. Functional analysis of the 37 kDa inner envelope membrane polypeptide in chloroplast biogenesis using a Ds-tagged Arabidopsis pale-green mutant. *Plant J.* **2003**, *34*, 719–731. [CrossRef]
25. Cheng, Z.; Sattler, S.; Maeda, H.; Sakuragi, Y.; Bryant, D.A.; DellaPenna, D. Highly Divergent Methyltransferases Catalyze a Conserved Reaction in Tocopherol and Plastoquinone Synthesis in Cyanobacteria and Photosynthetic Eukaryotes. *Plant Cell* **2003**, *15*, 2343–2356. [CrossRef] [PubMed]
26. Porfirova, S.; Bergmuller, E.; Tropf, S.; Lemke, R.; Dormann, P. Isolation of an Arabidopsis mutant lacking vitamin E and identification of a cyclase essential for all tocopherol biosynthesis. *Proc. Natl. Acad. Sci. USA* **2002**, *99*, 12495–12500. [CrossRef]
27. Kanwischer, M.; Porfirova, S.; BergmüLler, E.; DöRmann, P. Alterations in Tocopherol Cyclase Activity in Transgenic and Mutant Plants of Arabidopsis Affect Tocopherol Content, Tocopherol Composition, and Oxidative Stress. *Plant Physiol.* **2005**, *137*, 713–723. [CrossRef] [PubMed]
28. Van Eenennaam, A.L.; Lincoln, K.; Durrett, T.P.; Valentin, H.E.; Shewmaker, C.K.; Thorne, G.M.; Jiang, J.; Baszis, S.R.; Levering, C.K.; Aasen, E.D.; et al. Engineering vitamin E content: From Arabidopsis mutant to soy oil. *Plant Cell* **2003**, *15*, 3007–3019. [CrossRef] [PubMed]
29. Dwiyanti, M.S.; Yamada, T.; Sato, M.; Abe, J.; Kitamura, K. Genetic variation of γ-tocopherol methyltransferase gene contributes to elevated α-tocopherol content in soybean seeds. *BMC Plant Biol.* **2011**, *11*, 152. [CrossRef]
30. Sui, M.; Jing, Y.; Li, H.; Zhan, Y.; Luo, J.; Teng, W.; Qiu, L.; Zheng, H.; Li, W.; Zhao, X.; et al. Identification of Loci and Candidate Genes Analyses for Tocopherol Concentration of Soybean Seed. *Front. Plant Sci.* **2020**, *11*, 539460. [CrossRef]
31. Knizia, D.; Yuan, J.; Bellaloui, N.; Vuong, T.; Usovsky, M.; Song, Q.; Betts, F.; Register, T.; Williams, E.; Lakhssassi, N.; et al. The Soybean High Density 'Forrest' by 'Williams 82' SNP-Based Genetic Linkage Map Identifies QTL and Candidate Genes for Seed Isoflavone Content. *Plants* **2021**, *10*, 2029. [CrossRef]
32. Wang, Z.; Chen, Z.; Cheng, J.; Lai, Y.; Wang, J.; Bao, Y.; Huang, J.; Zhang, H. QTL Analysis of Na^+ and K^+ Concentrations in Roots and Shoots under Different Levels of NaCl Stress in Rice (*Oryza sativa* L.). *PLoS ONE* **2012**, *7*, e51202. [CrossRef] [PubMed]
33. Fritsche, S.; Wang, X.; Jung, C. Recent Advances in our Understanding of Tocopherol Biosynthesis in Plants: An Overview of Key Genes, Functions, and Breeding of Vitamin E Improved Crops. *Antioxidants* **2017**, *6*, 99. [CrossRef] [PubMed]

34. Li, H.; Wang, Y.; Han, Y.; Teng, W.; Zhao, X.; Li, Y.; Li, W. Mapping quantitative trait loci (QTLs) underlying seed vitamin E content in soybean with main, epistatic and QTL × environment effects. *Plant Breed.* **2016**, *135*, 208–214. [CrossRef]
35. Li, H.; Liu, H.; Han, Y.; Wu, X.; Teng, W.; Liu, G.; Li, W. Identification of QTL underlying vitamin E contents in soybean seed among multiple environments. *Theor. Appl. Genet.* **2010**, *120*, 1405–1413. [CrossRef] [PubMed]
36. Grela, E.R.; Günter, K.D. Fatty acid composition and tocopherol content of some legume seeds. *Anim. Feed Sci. Technol.* **1995**, *52*, 325–331. [CrossRef]
37. Kassem, M.A. *Soybean Seed Composition: Protein, Oil, Fatty Acids, Amino Acids, Sugars, Mineral Nutrients, Tocopherols, and Isoflavones*; Springer International Publishing AG: Cham, Switzerland, 2021.
38. Rigotti, A. Absorption, transport, and tissue delivery of vitamin E. *Mol. Aspects Med.* **2007**, *28*, 423–436. [CrossRef]
39. Talegawkar, S.A.; Johnson, E.J.; Carithers, T.; Taylor, H.A.; Bogle, M.L.; Tucker, K.L. Total α-Tocopherol Intakes Are Associated with Serum α-Tocopherol Concentrations in African American Adults. *J. Nutr.* **2007**, *137*, 2297–2303. [CrossRef]
40. Grilo, E.C.; Costa, P.N.; Gurgel, C.S.S.; Beserra, A.F.D.L.; Almeida, F.N.D.S.; Dimenstein, R. Alpha-tocopherol and gamma-tocopherol concentration in vegetable oils. *Food Sci. Technol.* **2014**, *34*, 379–385. [CrossRef]
41. Vinutha, T.V.; Bansal, N.; Kumari, K.; Prashat, G.R.; Sreevathsa, R.; Krishnan, V.; Kumari, S.; Dahuja, A.; Lal, S.K.; Sachdev, A.; et al. Comparative Analysis of Tocopherol Biosynthesis Genes and Its Transcriptional Regulation in Soybean Seeds. *J. Agric. Food Chem.* **2017**, *65*, 11054–11064. [CrossRef]
42. Chennupati, P.; Seguin, P.; Liu, W. Effects of High Temperature Stress at Different Development Stages on Soybean Isoflavone and Tocopherol Concentrations. *J. Agric. Food Chem.* **2011**, *59*, 13081–13088. [CrossRef]
43. Wu, X.; Vuong, T.D.; Leroy, J.A.; Grover Shannon, J.; Sleper, D.A.; Nguyen, H.T. Selection of a core set of RILs from Forrest × Williams 82 to develop a framework map in soybean. *Theor. Appl. Genet.* **2011**, *122*, 1179–1187. [CrossRef] [PubMed]

plants

Article

Genome-Wide Association Reveals Trait Loci for Seed Glucosinolate Accumulation in Indian Mustard (*Brassica juncea* L.)

Erwin Tandayu, Priyakshee Borpatragohain, Ramil Mauleon and Tobias Kretzschmar *

Faculty of Science and Engineering, Lismore Campus, Southern Cross University, East Lismore, NSW 2480, Australia; e.tandayu.10@student.scu.edu.au (E.T.); priyakshee.borpatra.gohain@scu.edu.au (P.B.); ramil.mauleon@scu.edu.au (R.M.)
* Correspondence: Tobias.Kretzschmar@scu.edu.au

Abstract: Glucosinolates (GSLs) are sulphur- and nitrogen-containing secondary metabolites implicated in the fitness of Brassicaceae and appreciated for their pungency and health-conferring properties. In Indian mustard (*Brassica juncea* L.), GSL content and composition are seed-quality-determining traits affecting its economic value. Depending on the end use, i.e., condiment or oil, different GSL levels constitute breeding targets. The genetic control of GSL accumulation in Indian mustard, however, is poorly understood, and current knowledge of GSL biosynthesis and regulation is largely based on *Arabidopsis thaliana*. A genome-wide association study was carried out to dissect the genetic architecture of total GSL content and the content of two major GSLs, sinigrin and gluconapin, in a diverse panel of 158 Indian mustard lines, which broadly grouped into a South Asia cluster and outside-South-Asia cluster. Using 14,125 single-nucleotide polymorphisms (SNPs) as genotyping input, seven distinct significant associations were discovered for total GSL content, eight associations for sinigrin content and 19 for gluconapin. Close homologues of known GSL structural and regulatory genes were identified as candidate genes in proximity to peak SNPs. Our results provide a comprehensive map of the genetic control of GLS biosynthesis in Indian mustard, including priority targets for further investigation and molecular marker development.

Keywords: *Brassica juncea*; genome-wide association studies; glucosinolates (GSL); seed quality

Citation: Tandayu, E.; Borpatragohain, P.; Mauleon, R.; Kretzschmar, T. Genome-Wide Association Reveals Trait Loci for Seed Glucosinolate Accumulation in Indian Mustard (*Brassica juncea* L.). *Plants* **2022**, *11*, 364. https://doi.org/10.3390/plants11030364

Academic Editor: Abdelmajid Kassem

Received: 14 December 2021
Accepted: 26 January 2022
Published: 28 January 2022

1. Introduction

Glucosinolates (GSLs) are a class of well-studied sulphur (S)- and nitrogen (N)- containing secondary metabolites almost exclusively found in Brassicaceae, which include the economically and nutritionally important crops *B. napus* (canola and rapeseed), *B. juncea* (Indian mustard), *B. oleracea* (cabbage) and *B. rapa* (Chinese cabbage, turnip) [1–3]. Most of our knowledge on GSL biosynthesis, its regulation and its links to other metabolic pathways is based on the closely related model plant, *Arabidopsis thaliana* [1,4,5]. GSLs are categorised into three major classes, depending on the amino acid they are derived from: (i) aliphatic GSLs, predominantly derived from methionine and, to a lesser extent, from leucine, isoleucine and valine; (ii) aromatic GSLs, mostly derived from phenylalanine or tyrosine and (iii) indolic GSLs, derived from tryptophan. The synthesis of GSLs proceeds in three major steps: (i) chain elongation of precursor amino acids (only for methionine and phenylalanine), (ii) GSL core structure formation and (iii) GSL side chain modification. A recent comprehensive inventory from the literature and pathway databases (KNApSAcK, KEGG and AraCyc) listed as many as 113 genes associated with GSLs in *Arabidopsis* that were identified and characterised over the last two decades [4]. This includes 53 biosynthetic genes found in the KEGG or AraCyc databases, 32 experimentally confirmed biosynthetic genes, 23 transcriptional components and five transporters. While the GSL biosynthetic pathways are well understood in *Arabidopsis*, the respective regulatory and metabolic networks in the allotetraploid Brassica crops (*B. napus* and *B. juncea*) are suggested to be much more complex due to their intricate evolutionary history [6].

Indian mustard is an economically important *Brassica*, cultivated for two distinct markets. In India, Bangladesh, China and the Ukraine, and more recently in Canada and Australia, it is grown as an oilseed crop [7]. On the other hand, in Europe, North America, Argentina and China, it is primarily grown for condiment production (e.g., mustard and "wasabi" paste). Both end uses rely on GSL content as a trait to be selected either for or against during varietal improvement. "Canola" is a trademark term of the Canadian Canola Association used to describe rape or oilseed cultivars with "double low" GSLs (<30 μmol/g in defatted seed meal) and less than 2% erucic acid [8]. In *B. juncea* grown as a canola-type oilseed crop, GSLs have largely been selected against, which enables seed meal to be used for animal feed after oil extraction. Breeding for low-GSL *B. juncea* was spearheaded by Canadian breeders through the introgression of low-GSL "Bronowski" alleles from canola *B. napus* into an Indian high-GSL *B. juncea* line [9]. The resulting donor genotype for the low-GSL trait has been extensively used in breeding for low GSLs in Canadian and Australian germplasm [10]. As such, canola-quality *B. juncea* has become a viable alternative oil crop [11–13]. For the condiment market, high GSL levels, high sinigrin in particular, are desirable [14]. Sinigrin, when hydrolysed, produces allyl-isothiocyante (AITCs), including sulphoraphane, responsible for the pungency of mustard and demonstrated to possess tumour suppression properties [15,16]. Notably, Indian mustard predominantly accumulates the aliphatic GSLs 2-propenyl-GSL (sinigrin) and 3-butenyl-GSL (gluconapin), and, to a lesser extent, 2-hydroxy-3-butenyl-GSL (progoitrin) [14,17,18].

Enhancing the health-beneficial GSL levels in varieties aimed for vegetable or condiment use and reducing the overall GSL and erucic acid levels, while increasing desirable fatty acids in oilseed cultivars, remain among the key seed quality traits for *B. juncea* variety improvement [17]. A better understanding of the genetic bases of trait variation and corresponding beneficial alleles would aid in the development of molecular markers for varietal improvement and an accelerated rate of genetic gain [19,20]. Earlier, classical QTL mapping deciphered beneficial allelic variations for seed quality traits in *B. juncea* [21–23]. Recently, genome-wide association study (GWAS) has become the more popular choice to dissect the genetic basis of these complex traits. Compared with classical quantitative trait locus (QTL) mapping, which is generally confined to alleles and novel recombination within a bi-parental population, GWAS is able to tap into the allelic pool of broader populations that have undergone natural and artifical selection throughout domestication history. Since GWAS utilises a broader allelic pool, more variation is investigated. Furthermore, actual causal variants tend to be much closer to detected associated markers in GWAS, owing to the longer recombination history than in the case of a bi-parental population. As a result, GWAS offers a higher mapping resolution of the underlying genomic regions associated with the trait of interest. In *Brassica* crops, GWAS has been successfully employed for dissecting the genetic architecture of seed quality traits such as GSL accumulation, fatty acid composition and shattering resistance [24–28]. In *B. juncea*, high-density single-nucleotide polymorphisms (SNPs) were identified through different strategies, including double-digest restriction-associated DNA (dd-RAD) [29], RNA sequencing [30], specific-locus amplified fragment sequencing (SLAF-seq) [31], genotyping-by-sequencing (GBS) [29] and resequencing [17]. With these, GWAS has been utilised to investigate seed GSL content using high-density SNPs [17,32]. Akhatar et al., 2020, employed GWAS for seed quality traits including GSL content at varying nitrogen levels under field conditions, while Yang et al., 2021, performed GWAS on a set of vegetable and oilseed *B. juncea*, in conjunction with deploying two new genome sequences representing vegetable and oilseed varieties. Among the candidate genes proposed in these studies, only a *MYB28*, a major regulatory gene for aliphatic GSL biosynthesis, could be linked to the current inventory [4] of GSL genes in *Arabidopsis*. This suggests that a large number of possible genetic mechanisms may yet be uncovered through GWAS. Thus, the aim of this study was to perform

GWAS on a set of oilseed Indian mustard to further elucidate the genetic basis and add to the current understanding of seed GSL accumulation in Indian mustard.

2. Results

2.1. Genotype Data

A total of 69,594 SNP sites, with 61,931 (89%) anchored onto chromosomes, was obtained from the variant calling. An initial filtering for SNPs anchored onto chromosomes for 60% call rate, non-maf (minor allele frequency) filtered and 10% maximum marker heterozygosity resulted in 15,263 SNPs (26% overall with missing SNP calls), and missing states were imputed. Following imputation, a final set of 14,125 SNPs resulted from filtering for 5% minor allele frequency and 20% maximum heterozygosity and was used for GWAS.

2.2. Cluster, Population Structure and Principal Component Analyses of B. juncea Diverse Panel

The diversity panel consisted of 158 accessions from 28 countries, representing South Asia (53%, mostly from India and Pakistan), Asia (13%, other than South Asia), Europe (11%), North America (6%), Australia (6%), Africa (6%) and unknown origin (8%) (Table S1). Three approaches—(i) hierarchical clustering, (ii) population structure and (iii) PCA—revealed a genetic structure composed of two population clusters broadly reflecting geographical origin. UPGMA-based hierarchical clustering revealed one major cluster comprising accessions from the South Asian countries of India and Pakistan (blue-coloured branches), while the other major cluster contained accessions from outside of South Asia (green-coloured branches) (Figure 1a). Not all lines, however, matched this trend, including a few accessions from India, Nepal, Afghanistan and Bangladesh that located within the outside-South-Asia cluster and a few entries from Europe, Zimbabwe and China that fell within the South Asia cluster. A third minor cluster was largely composed of accessions from China and a few from Bhutan. ADMIXTURE suggested a similar structure as UPGMA (Figure 1b). At K = 2, cluster 1 was composed of accessions from India and Pakistan, while cluster 2 was mostly composed of accessions from outside India and Pakistan, a trend consistent with a previous report [20]. Using a 70% membership probability cut-off at K = 2, 46% of accessions fell into cluster 1 while 37% of accessions fell into cluster 2, and the remaining 17% were classified as admixed samples. The admixed samples comprised 13 South Asian (India, Bangladesh, Afghanistan, Nepal and Bhutan) accessions and 14 accessions from outside South Asia. With increasing K until K = 4, geographical origin was still traceable to clustering. At K = 3, accessions from India and Pakistan were dispersed into clusters 1 and 2, while accessions from outside India and Pakistan mostly constituted cluster 3. This was similar at K = 4, with further sub-structuring of accessions from outside India and Pakistan comprising clusters 3 and 4. A ten-fold cross validation error plot of ADMIXTURE runs using K = 1 to 12 (Figure 1c) showed that the error started to plateau at K = 4, suggesting this as a sensible K choice, while the lowest error was observed at K = 8. A PC plot reflecting the K = 2 assignment of ADMIXTURE clearly separated the two clusters at PC1 with admix samples interspersed between the clusters (Figure 1d). Further, only 18.7% of variation was explained by PC1, with succeeding PCs explaining less than 5% of variation.

2.3. Variance Components, Basic Descriptive Statistics and Correlations between Total GSLs, Sinigrin and Gluconapin

Residual distribution showed an approximately normal distribution with a mean of zero for total GSLs, sinigrin and gluconapin (Figure S1). Sinigrin and gluconapin combined accounted for ~95–99% of the total GSLs for nearly all samples in the diverse panel (Supplemental File S1). Nearly the entire proportion of variation for total GSLs, sinigrin and gluconapin concentrations was accounted for by the samples, based on variance components analysis using Restricted Maximum Likelihood (REML) (Table S2). This was further reflected by high broad heritability values of ~98% for the single major GLSs sinigrin and gluconapin, and a slightly lower value of ~88% for the total GSLs. Sinigrin had a higher range of concentrations (1.61–225.09 μmol/g^{-1}) compared to gluconapin

(0.01–174.57 µmol/g^{-1}). However, the gluconapin concentration was more variable, with a coefficient of variation (CV) of 108% compared to sinigrin concentrations with a CV of 79.61%. Figure 2 reflects the distribution of raw values of total GSLs (Figure 2a), gluconapin (Figure 2b) and sinigrin (Figure 2c) concentrations matched with the cluster assignment from ADMIXTURE. Notably, accessions in different clusters accumulated different single major GSLs (Figure 2b,c). As reflected in the distributions, cluster 1 and a few admixed samples predominantly accumulated gluconapin, while the majority of cluster 2 and admixed samples lacked gluconapin. Contrastingly, cluster 2 and the majority of admixed samples predominantly accumulated sinigrin, while most of cluster 1 still accumulated sinigrin at the lower ranges (Figure 2c). Given this finding, we compared the correlations of sinigrin and gluconapin with total GSLs in the full panel and within the clusters in which it predominantly accumulated (Figure 2d–g). Gluconapin had a weak correlation (r = 0.08, non-significant) with total GSLs in the full panel (Figure 2d) and a moderately positive correlation (r = 0.56) in cluster 1 (Figure 2f). A few outlier points in cluster 1 (Figure 2f) accumulated high sinigrin as their major GSL. While there was only a moderate correlation (r = 0.51) between sinigrin and total GSLs in the full panel (Figure 2e), a near perfect positive correlation (r = 0.99) was observed in cluster 2 (Figure 2g). The five outlier samples in cluster 2 (Figure 2g) comprised four accessions accumulating lower sinigrin concentrations, although it was still their major GSL, and one accession that accumulated high gluconapin as its major GSL. There were non-significant weak correlations of sinigrin and gluconapin with total GSLs within the clusters where they were not predominantly accumulated (Figure S2a,b). Sinigrin and gluconapin had significant negative correlations, having the strongest negative value (r = −0.64) in the full panel and a weak (r = −0.37) to moderate (r = −0.37) value within clusters 1 and 2, respectively (Figure S2c–e).

2.4. GWAS Using Multiple Models

Four GWAS models were tested and resulting q-q plots in each traits (Figure S3) were compared to assess which models best limited spurious associations, due to structure and relatedness. BLINK and FarmCPU returned a better correlation between observed and expected −log10 *p*-values in the lower range and returned a limited number of deviations at high log10 *p*-values. SUPER returned highly inflated −log10 *p*-values even in the lower ranges, suggestive that many detected loci were from spurious associations, which might also explain the exceptionally high number of significant SNPs detected under this model (Supplemental File S2). MLMM, despite detecting the lowest number of associations, also showed *p*-value inflation to some degree. The Manhattan plots from BLINK and FarmCPU (Figure 3) displayed a number of single SNPs associated above the Bonferroni threshold for total GSLs (Figure 3a), sinigrin (Figure 3b) and gluconapin (Figure 3c). BLINK detected four associated SNPs with the total GSL concentration, two on A02 and one each on A10 and B06, while FarmCPU detected four, one each on chromosomes A02, B01 and B02 and two on B08 (Figure 3a). One association at SNP A02_11235033 was detected in all four models (−log10 (*p*) = 6.09–10.39). The association at B02_725738 from FarmCPU was the strongest association (−log10 (*p*) = 9.38) for total GSLs considering only the BLINK and FarmCPU models. For sinigrin concentration, five SNPs were associated in BLINK, one SNP each on chromosomes A01, B01 and B08 and two on B04 (Figure 3b). FarmCPU also detected five associated SNPs, one each on chromosomes A03, B01 and B06 and two on B08 (Figure 3b). Two associations, at SNP B01_43311767 (−log10 (*p*) = 7.51–10.34) and at SNP B08_24075810 (−log10 (*p*) = 6.02–7.68), were commonly detected by both models. The strongest association (−log10 (*p*) = 10.34) was at SNP B01_43311767 in BLINK. Compared to total GSLs and sinigrin, more SNPs were found to be significantly associated with gluconapin concentration. BLINK detected a total of 14 associated SNPs, comprising one SNP each on chromosomes A03, A06, A08, A10, B02, B03, B04 and B05, two on B01 and four on A02. FarmCPU returned six associated SNPs on chromosomes A02, A06, A08, B02, B03 and B05. These associations were distinct, although the association at SNP B02_48309648 in BLINK and SNP B02_48309753 in FarmCPU were only 105 bp apart. SNP B02_48309648 in BLINK represented the strongest association (−log10 (p) = 17.47). FarmCPU appeared the most suitable model for total GSLs and sinigrin with respect to the control of spurious associations as most observed p-values correlated with expected p-values, with only a few p-values deviating sharply at the tail end (Figure S2a,b). For gluconapin, BLINK associated a higher number of SNPs, while controlling best for spurious associations (Figure S2c).

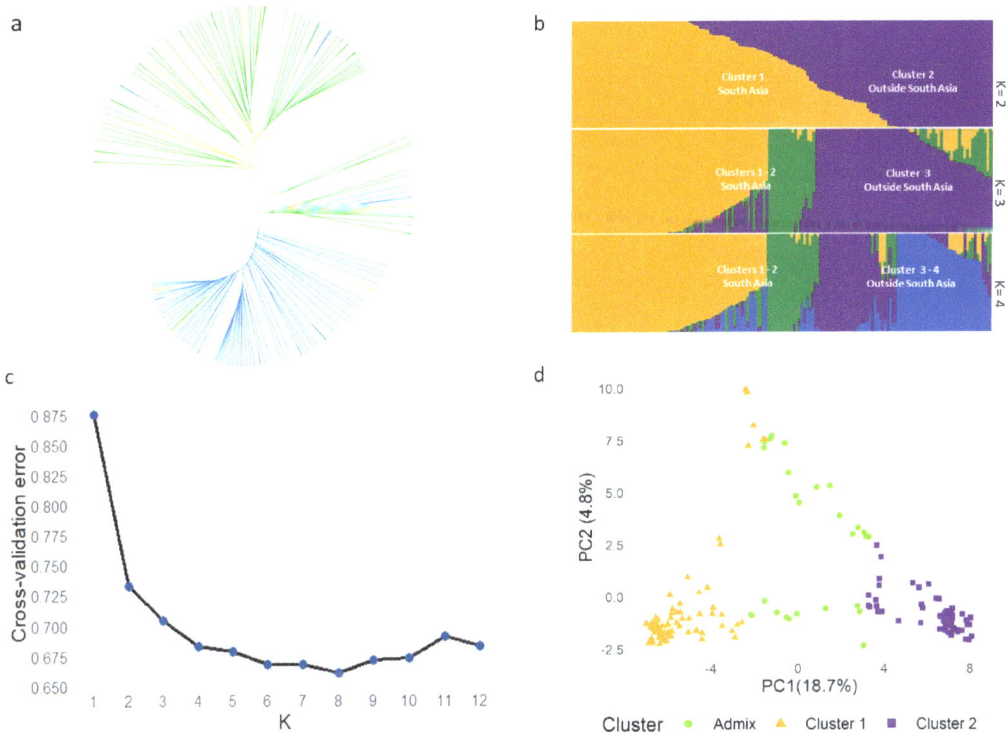

Figure 1. (**a**) Cluster analysis based on genetic distance using an UPGMA tree with branches coloured based on geographical origin: India and Pakistan and rest of South Asia (blue), rest of Asia, Europe, North America, Africa and Australia (green) and unknown origin (yellow). (**b**) Population structure as depicted by a sorted bar plot of ancestry proportions for K = 2–4, inferred with ADMIXTURE. (**c**) Ten-fold cross-validation error of ADMIXTURE analyses of K = 1 to 12. (**d**) Principal component analysis (PCA) coloured based on cluster assignment (threshold of 70% membership probability) at K = 2 in ADMIXTURE. Orange triangles used for cluster 1, purple squares for cluster 2 and green dots for admixture cluster.

2.5. Significant GWAS Hits Had Known and Potential GSL Genes in Their Vicinity

The LD decay plot based on 14,125 SNPs suggested no effective LD (threshold of r^2 = 0.1) at distances above 500 kb (Figure S4); hence, the search for potential candidate genes (using the *B. juncea* var. *tumida* V1.5 annotation) proximal to the trait-associated SNPs was limited to 250 kb upstream and downstream of the SNP position. Based on their homology with *Arabidopsis* genes and respective annotation, candidate genes were classified as known or potential GSL genes (Table 1).

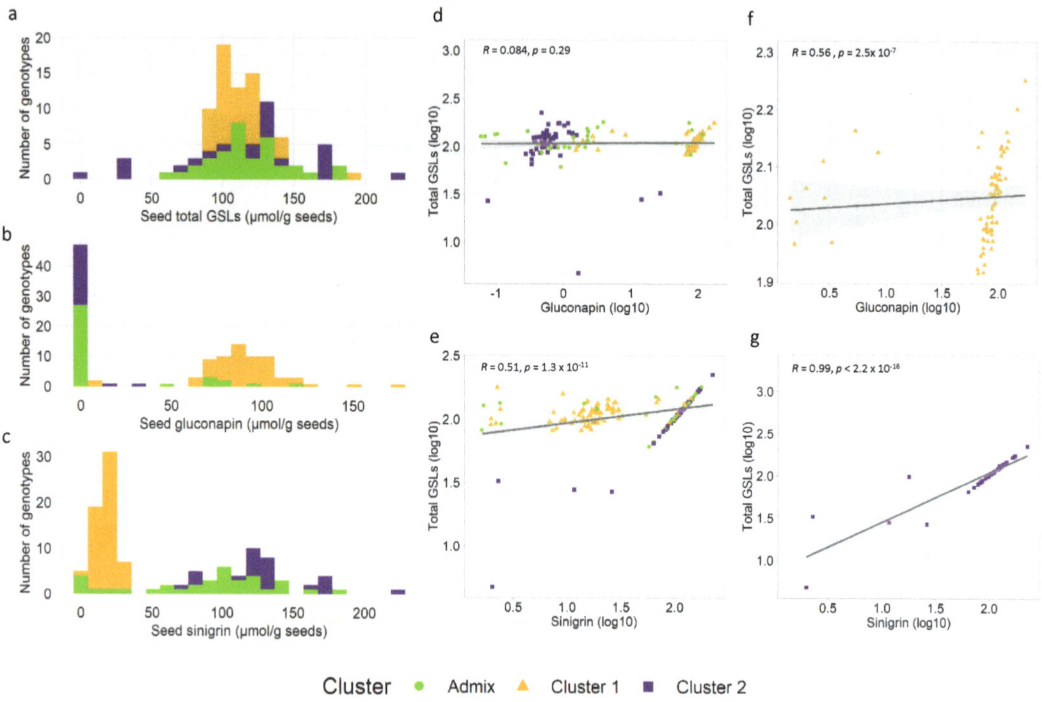

Figure 2. Distribution of raw mean values of (**a**) total GSLs, (**b**) gluconapin and (**c**) sinigrin, reflecting the ADMIXTURE cluster assignment at K = 2 of each accession (orange for cluster 1, purple for cluster 2 and green for admixture cluster). Correlations using log-transformed values of (**d**) gluconapin and total GSLs and (**e**) sinigrin and total GSLs in the full diversity panel. Correlation using log-transformed values of (**f**) gluconapin and total GSLs in cluster 1 and (**g**) sinigrin and total GSLs in cluster 2. Orange used triangles for cluster 1, purple squares for cluster 2 and green dots for admixture cluster.

For total GSLs, homologues of two known GSL genes were identified near SNP A02_3567961, a significant SNP detected in BLINK and explaining around 7% of the observed trait variation (phenotypic variation explained—PVE). These were *GSTF11* [33–35] at 39.61 kb upstream and *SCPL17* [36] at 68.54 kb downstream. SNP A02_11235033, the most reliable association detected in all four models and accounting for 6% PVE, was located 128.81 kb upstream of a homologue of *OBP2*, encoding a known regulator of GSL biosynthesis [37]. SNP B02_7295738, which was detected in both FarmCPU and SUPER with 11% PVE, was found located near two potential GSL genes. Homologues to the potential GSL gene *amino acid permease 4* (*AAP4*) at 213.38 kb upstream and *SAL1* at 246.34 kb were found. SNP B08_66155255, detected only in FarmCPU, albeit at 37% PVE, was a genic SNP within a potential GSL gene, a putative *CYP18-3*. Moreover, at 17.56 kb, another potential GSL gene, a *putative 2-oxoglutarate-dependent dioxygenase* gene was found.

For sinigrin, SNP A03_27702263 with 4% PVE, detected by FarmCPU and SUPER, had homologues of several known GSL regulatory genes in proximity. These included a putative *MYB28* [38,39] at 118.32 kb upstream, as well as a putative *MYB34* [40,41] and a *MAM1* [42,43] homologue at 115.48 kb and 160.65 kb downstream. SNP B04_9016612 with 7% PVE, significant in BLINK, was found close to a homologue of the known GSL gene $FMO_{GS\text{-}OX5}$ [44,45] at 1.51 kb upstream. B04_17138489 with 12% PVE, which was significantly associated only in BLINK, was flanked by a potential GSL gene homologous to *phosphoserine aminotransferase 1* (*PSAT1*) at 12.75 kb.

For gluconapin, SNP A02_34185026, detected by BLINK at 11% PVE, was found to be flanking an *LSU2* homologue, a potential GSL gene, at 5.75 kb downstream. SNP A02_34995417 with 1% PVE, detected in BLINK, was found to be located near additional homologues of *MYB28* and *MYB34* at 81.62 kb and 96.36 kb downstream, respectively. SNP A10_999168, solely detected by BLINK, was flanked by potential GSL genes monothiol glutaredoxin S11 (GRXS11) at 105.45 kb upstream and a UDP-glycosyltransferase 71C3 (UGT71C3) at 115.13 kb downstream. Variation within these two potential GSL genes may have contributed to the 11% PVE of this SNP. With 7% PVE, SNP B01_44925254, detected in BLINK and SUPER, was located near potential GSL genes *RETICULATA-RELATED 3* (*RER3*) at 105.45 kb upstream and a *Cysteine Synthase D1* (*CYSD1*) at 213.34 kb upstream. The strongest association from both BLINK and FarmCPU was only 105 bp apart and was considered the same association, SNP B02_48309648-753 with 3% PVE. This association was 180 kb upstream of a *HY5* homologue, encoding a known regulator of GSL biosynthesis [46]. SNP B03_474869, detected in FarmCPU with 6% PVE, was located near a potential GSL gene, *SULPHUR DEFICIENCY-INDUCED 2* (*SDI2*), at 23.76 kb. On the other hand, SNP B03_7408562, detected in BLINK and MLMM, was found to be near a potential GSL gene, *aldehyde dehydrogenase family 2 member B7* (*ALDH2B7*), located at 135.05 kb downstream.

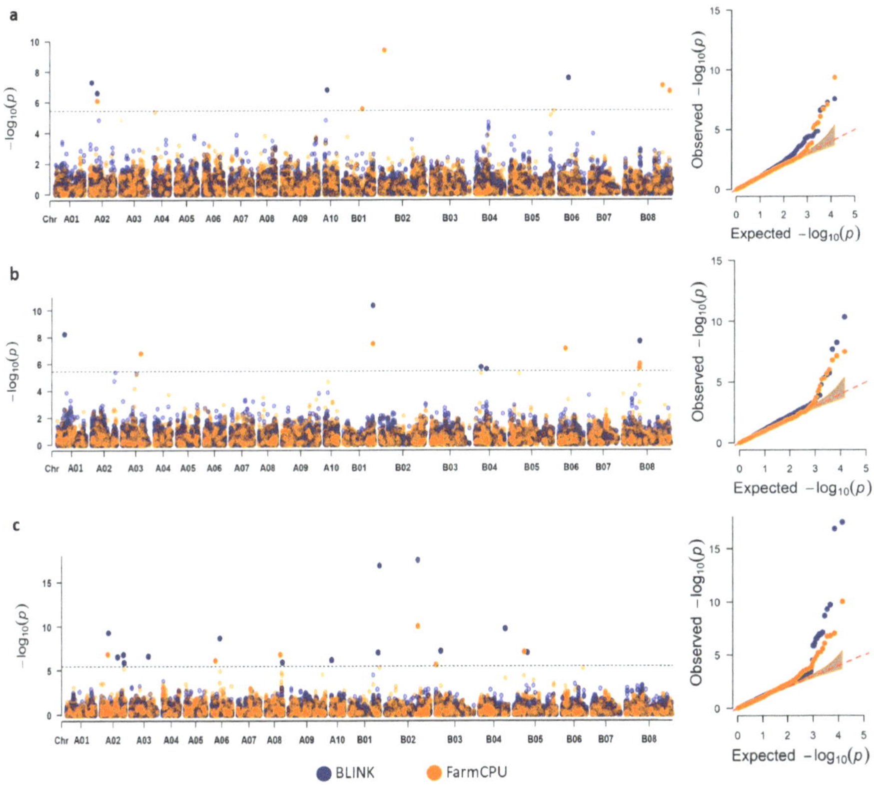

Figure 3. Manhattan and q-q plots for the GWAS of (**a**) total GSLs, (**b**) sinigrin and (**c**) gluconapin using BLINK (purple dots) and FarmCPU (orange dots) models. The horizontal line represents significance threshold at 5% after Bonferroni multiple test correction (−log10 (p) = 5.45).

Table 1. Peak SNP characteristics and respective candidate genes homologous to known and potential *Arabidopsis* genes involved in GSL metabolism.

Trait [a]	Peak SNP	*p*-Value	PVE [b]	Model [c]	Candidate Gene	Homologous Gene in *Arabidopsis*	% Amino Acid Identity	Distance to Peak SNP [kb]	*Arabidopsis* ID	Gene Description
TGSL	A02_3567961	5.08×10^{-8}	6.63	B	BjuA041358	*GSTF11* [33–35]	69.16	−39.61	AT3G03190.1	GSL core structure synthesis
					BjuA041338	*SCPL17* [36]	61.43	68.54	AT3G12203.3	GSL side-chain modification
	A02_11235033	2.54×10^{-7}	5.83	B, F, M, S	BjuA045411	*OBP2* [37]	84.92	−128.81	AT1G07640.1	GSL regulation
	B02_7295738	4.09×10^{-10}	11.41	F, S	BjuB047551	*AAP4*	94.21	213.38	AT5G63850.1	potential GSL gene
					BjuB047557	*SAL1*	64.76	246.34	AT5G63980.1	potential GSL gene
	B08_66155255	1.90×10^{-7}	37.03	F	BjuB019211	*CYP18-3*	65.48	−0.7	AT4G38740.1	potential GSL gene
					BjuB019215	*Probable 2-ODD* [d]	61.77	−17.56	AT5G05600.1	potential GSL gene
SIN	A03_27702263	1.50×10^{-7}	3.76	F, S	BjuA042263	*MYB28* [38,39]	79.95	−118.32	AT5G61420.2	GSL regulation
					BjuA042229	*MYB34* [40,41]	71.32	115.48	AT5G60890.1	GSL regulation
					BjuA042223	*MAM1* [42,43]	82.72	160.65	AT5G23010.3	GSL side-chain elongation
	B04_9016612	5.04×10^{-6}	6.84	B	BjuB028146	$FMO_{GS\text{-}OX5}$ [44,45]	69.41	−1.51	AT1G12140.3	GSL side-chain modification
	B04_17138489	2.51×10^{-6}	11.71	B	BjuB028703	*PSAT1*	83.57	12.75	AT4G35630.1	potential GSL gene
GNP	A02_34185026	1.64×10^{-7}	11.24	B	BjuA033112	*LSU2*	86.022	5.75	AT5G24660.1	potential GSL gene
	A02_34995417	1.29×10^{-6}	0.72	B	BjuA002140	*MYB28* [38,39]	67.46	81.62	AT5G61420.1	GSL regulation
					BjuA001524	*MYB34* [40,41]	72.00	96.36	AT5G60890.1	GSL regulation
	A10_999168	6.85×10^{-7}	10.72	B	BjuA037371	*GRXS11*	96.97	−105.45	AT1G06830.1	potential GSL gene
					BjuA037341	*UGT71C3*	79.19	115.13	AT1G07260.1	potential GSL gene
	B01_44925254	1.38×10^{-17}	7.15	B, S	BjuB006588	*RER3*	78.44	−105.02	AT3G08640.1	potential GSL gene
					BjuB006607	*CYSD1*	73.27	−213.34	AT3G04940.2	potential GSL gene
	B02_48309648-753	3.35×10^{-18}	2.80	B, F	BjuB009816	*HY5* [46]	89.94	180.71	AT5G11260.1	GSL regulation
	B03_474869	2.49×10^{-6}	6.03	F	BjuB005751	*SDI2*	82.50	23.76	AT1G04770.1	potential GSL gene
	B03_7408562	7.07×10^{-8}	4.78	B, M	BjuB003011	*ALDH2B7*	91.01	135.05	AT1G23800.1	potential GSL gene

[a] TGSL (Total GSLs), SIN (sinigrin), GNP (gluconapin). [b] Phenotypic variance explained by marker calculated in GAPIT. In case of co-detection with BLINK, PVEs obtained were from BLINK; if otherwise, PVEs were from FarmCPU. [c] BLINK(B), FarmCPU (F), MLMM (M), SUPER (S). [d] Probable 2-oxoglutarate-dependent dioxygenase.

3. Discussion

3.1. Population Structure

The population structure of a diversity panel can confound GWA analysis through spurious associations [47]. Population clustering based on the UPGMA tree and ADMIXTURE (at K = 2) reflected a broad grouping based on geographical origin, with one group composed mainly of genotypes from South Asia (India and Pakistan) and another group from outside of South Asia (Figure 1a,b). Recent studies on population structure in Indian mustard reported a similar trend, with optimal Ks in the range of K = 2–3 [20,48]. In our study, ADMIXTURE K = 2 split the panel based on geographical origin (Figure 1b), with further sub-structuring of the two broad clusters until K = 4, a sensible K choice based on the cross-validation error (Figure 1c). The admixed samples may have resulted from interbreeding of the two population groups in variety improvement efforts. PCA was also concordant with the other methods (Figure 1d). The distribution of total GSLs, sinigrin and gluconapin in our diversity panel resembled that of a different panel of 190 accessions of diverse geographical origin, quantified for the same chemical traits [14] (Figure 2a–c). ADMIXTURE clustering reflected in the distribution of sinigrin and gluconapin confirmed previous reports on the correlations of GSL profiles with origin. Accessions from South Asian countries India and Pakistan (cluster 1) contained mostly gluconapin and lower levels of sinigrin, while accessions from outside of South Asia (cluster 2) mostly contained sinigrin in seeds [49–52]. This structure depicts crop divergence leading to these two varietal subgroups based on different end uses [18,53]. A strong selection for the health-beneficial GSL in East-European-type mustard for leafy vegetable and condiment cultivation was attributed to a predominant accumulation of sinigrin in samples originating from outside of South Asia. On the other hand, in India, cultivation was geared towards edible oil use, with yield and increasing the oil content as the main focus of selection for varietal improvement and not for a specific GSL type [18,51]. Thus, accessions from the Indian subcontinent, though predominantly accumulating gluconapin, also accumulated a lower proportion of sinigrin in our panel, consistent with earlier reports [9,50,51]. As such, the individual correlations of sinigrin and gluconapin with total GSLs were reflected more accurately at subgroup level than in the full panel (Figure 2f,g). In cluster 1, a weaker correlation between gluconapin and total GSLs reflected the presence of other GSLs in the total GSLs in these accessions. Conversely, in cluster 2, an almost perfect correlation was observed between sinigrin and total GSLs, attributed to higher homogeneity of the GSL profile. While this structure and the inter-trait correlations might have confounding effects on the GWAS, the resulting q-q plots for the two selected models suggested that these covariates were well accounted for and corrected in our analysis.

3.2. Candidate Genes Identified in the Vicinity of Associated SNPs

With the development of newer models with improved statistical power, GWAS recently incorporated multiple model approaches to maximise the power of QTL detection [25,54–56]. FarmCPU and BLINK are two of the newest models, with demonstrated superiority in statistical power compared to earlier GWAS methods [57,58]. The single SNP peaks observed from our GWAS using BLINK and FarmCPU were characteristic results for these models. Compared with other earlier models that display large peaks with multiple SNPs characteristic of "Manhattan" plots, these models highlight only the most significant marker in each association [54,57,58]. We located several strong homologues of known *Arabidopsis* GSL biosynthetic and regulatory genes, as well as potential GSL genes, in the vicinity of most of the significantly associated SNPs (Table 1). The majority of SNPs showed minor effects of around 10% PVE or less, as expected for a complex quantitative trait. An exception was SNP B08_66155255, with 37.03% PVE for total GSLs. These known and potential GSL genes were annotated as such in SuCCombase (https://plant-scc.org, accessed on 7 September 2021) [59], a curated repository of genes involved in the metabolism of sulphur-containing compounds including GSLs. While the "known" genes were listed in the inventory of GSL biosynthetic pathways in *Arabidopsis* [4,5], the "potential GSL genes" were identified from published co-expression data, which pinpoint genes that might be involved in GSL biosynthesis, yet lack experimental support.

Of the seven listed candidate genes for total GSL concentration, three were found on chromosome A02, two on B02 and two on B08 (Table 1). *BjuA041358* and *BjuA041338* were homologues of two GSL structural genes, *GSTF11* and *SCPL17*, respectively, and were linked to SNP A02_3567961. *GSTF11* encodes glutathione S-transferase F11, responsible for converting the intermediate derivative aci-nitro compounds to reduced glutathione (GSH) conjugates during aliphatic GSL core structure synthesis [33–35], making *BjuA041358* the stronger candidate. *SCPL17*, on the other hand, is involved in the production of benzoyloxy GSLs in *Arabidopsis* [36], making *BjuA041338* a less likely candidate. SNP A02_11235033 was a high-confidence association, considering that it was detected in all four models. The only candidate gene in this region was *BjuA045411*, a homologue of *OBP2* encoding a DNA-binding-with-one-finger (DOF) transcription factor [60], demonstrated to regulate indolic GSL in *Arabidopsis* [37]. Since nearly all GSLs in *B. juncea* are aliphatic, however, this *OBP2* homologue would need to have a divergent role to account for the total GSL variation. The association at SNP B02_7295738, the SNP with the second highest PVE (11%) for total GSLs, was linked to two potential GSL genes: *BjuB047551*, a homologue of *AAP4* encoding an amino acid permease 4, and *BjuB047557*, a homologue of *SAL1* encoding an inositol polyphosphate 1-phosphatase [59]. Given the high predicted peptide sequence similarity (94%), the *AAP4* homologue was likely the better candidate gene compared to the *SAL1* homologue at 65% similarity. Despite an exceptionally high PVE of 37.03% for total GSLs, no homologues to known GSL genes were found in the vicinity of SNP B08_66155255. However, SNP B08_66155255 was located within the gene model of *BjuB019211*, a homologue of *CYP18-3*, a putative peptidyl-prolyl cis-trans isomerase potentially involved in GSL metabolism, as suggested by co-expression with known GSL genes [59]. Furthermore, around 18 kb upstream, a probable 2-oxoglutarate-dependent dioxygenase encoding gene was located. Known GSL genes *AOP2* and *AOP3* similarly encode 2-oxoglutarate-dependent dioxygenases, which catalyse the side-chain oxygenation in the aliphatic GSL core synthesis [61,62]. The high PVE of SNP B08_66155255 merits further investigation.

Of the five candidate genes associated with sinigrin, homologues of three known GSL regulatory genes were found in the vicinity of SNP A03_27702263. *BjuA042263*, *BjuA042229* and *BjuA042223* were homologues of *MYB28*, *MYB34* and *MAM1*, respectively. *MYB28*, also known as *HAG1* (*HIGH ALIPHATIC GLUCOSINOLATE 1*), positively regulates aliphatic GSLs [38,39], with gain-of-function and knock-down mutants showing contrasting levels of aliphatic GSLs and transcript levels of corresponding biosynthetic genes [38]. *MYB28* was further identified and validated through combined multi-omics approaches, including GWAS, as the major gene controlling leaf and seed GSL content in *B. napus* [25], suggesting that natural variation at this locus drives phenotypic variation. In oilseed *B. juncea*, targeted silencing of a *MYB28* orthologue led to the down-regulation of GSL biosynthesis [6], making *BjuA042263* a very strong candidate for this QTL region and a high priority for our further validation efforts. On the other hand, *MYB34* mainly exerts its role in the roots to regulate indolic GSL synthesis [40,41] and *MAM1* is a methylthioalkylmalate synthase involved in the GSL side-chain elongation of short-chained aliphatic GSLs [42,43], suggesting their respective *B. juncea* homologues to be less likely causal for the effects associated with SNP A03_27702263. SNP B04_9016612, with 7% PVE, was a genic SNP within *BjuB028146*, a homologue of $FMO_{GS\text{-}OX5}$ encoding a flavin-containing monooxygenase. $FMO_{GS\text{-}OX5}$ functions in aliphatic GSL side-chain modification by S-oxygenation of the basic aliphatic GSL derivatives [44,45], making *BjuB028146* a high-priority candidate gene. *BjuB028703*, homologous to the potential GSL gene *PSAT1*, was located near SNP B04_17138489, with 12% PVE. *PSAT* encodes a putative phosphoserine aminotransferase in the serine biosynthetic pathway [63]. Although this locus had a high PVE, *PSAT1* has not been directly associated with aliphatic GSL metabolism. However, serine is a substrate for tryptophan biosynthesis, which in turn is a precursor for the production of indolic

GSLs [64]. Furthermore, in *Arabidopsis*, it is regulated by *MYB34* and *MYB51*, two activators of indolic GSL biosynthesis [63].

Ten candidate genes, three known and seven potential GSL genes, can be speculated to contribute to gluconapin variation. Among these, *BjuA033112* a homologue of *LSU2 (RESPONSE TO LOW SULPHUR 2)*, was found less than 6 kb from SNP A02_34185026. While LSU proteins are of unknown function, they were demonstrated to be important stress-related hubs [65] and considered marker genes of sulphur metabolism [66], making *BjuA033112* a good candidate to account for the considerable 11% PVE of this locus. Interestingly, *MYB28* and *MYB34* homologues, additional copies of which were already implicated in the variation in sinigrin concentration on chromosome A03, were found in the vicinity of SNP A02_34995417, although this SNP contributed little to the observed gluconapin variation. *BjuA002140* was a homologue of *MYB28*, while *BjuA001524* was a homologue of *MYB34*. Copy number variation (CNV) of *MYB28* homologues on different chromosomes might have led to the divergence that specifically accounts for sinigrin and gluconapin accumulation in different genetic backgrounds. Recently, CNV was uncovered on *MYB28* loci through pairwise sequencing of a vegetable variety, T84-66, and an Australian oilseed variety, AU213 [17]. Among the associations with high PVE (11%) was SNP A10_999168, located near homologues of two potential GSL genes. *BjuA037341* was a homologue of *UGT71C3* encoding an UDP-glycosyltransferase, and *BjuA037371* a homologue of *GRXS11* encoding monothiol glutaredoxin, implicated in nitrogen signalling [67]. The direct involvement of UDP-glycosyltransferase *UGT74B1* [68] and of *UGT74C1* in aliphatic GSL core synthesis [69] suggests that *BjuA037341* is the higher-confidence candidate for this association. Having been detected in BLINK and SUPER, SNP B01_44925254 was a reliable and strong association ($-\log10$ (p) = 16.86) for gluconapin. However, homologues of only two potential GSL genes were found in proximity. These were *BjuB006588*, homologous to *RER3* encoding *RETICULATA-RELATED 3*, and *BjuB006607*, homologous to *CYSD1*, a cysteine synthase and a member of the O-acetylserine(thiol)lyase (*OASTL*) gene family. *OASTLs* include *OASA1*, an S assimilation pathway gene that catalyses the biosynthesis of cysteine and a precursor for GSL formation [70].

A *LONG HYPOCOTYL 5* (*HY5*) homologue, *BjuB009816*, was located near the high-confidence gluconapin associations SNP B02_48309748-53 at a PVE of 3%. *HY5*, a transcription regulator, was shown to partly control the light regulation of GSL biosynthetic genes, as well as many genes in the sulphate assimilation pathway [46]. Additionally, *hy5 Arabidopsis* mutants showed altered expression of GSL biosynthetic genes and MYB TFs associated with aliphatic GSL regulation [46]. *BjuB005751*, a homologue of another potential GSL gene, *SDI2* encoding SULPHUR DEFICIENCY-INDUCED 2, was located near SNP B03_474869. Under sulphur-limiting conditions in *Arabidopsis*, SDI2 acts as a repressor of aliphatic GSL biosynthesis at transcript and metabolite levels [71]. Despite being detected under non-limiting sulphur conditions, this *B. juncea SDI2* homologue could affect GSL composition. Lastly, *BjuB003011* a homologue of a potential GSL gene *ALDH2B7* encoding an aldehyde dehydrogenase family 2 protein, was located near SNP B03_7408562. While two models detected this association for gluconapin, no literature support was found for the involvement of *ALDH2B7* in GSL biosynthesis, aside from it being listed as a potential GSL gene in SuCCombase [4,59].

We found no overlap in proposed candidate genes with the GWAS study by Akhatar et al., 2020, probably owing to different aims, translating to differences in panel composition, different methods of GSL quantification and differences in cultivation. Furthermore, they limited the candidate gene search to a narrow window of 25 kb upstream and downstream of peak SNPs. The Akhatar et al., 2020, study was conducted under field conditions, with the aim to study the effects of various nitrogen levels. They used only 92 accessions, which were phenotyped for GSL content using Near-Infrared Reflectance Spectroscopy (NIRS) on intact seeds to predict total GSLs. In contrast, we phenotyped a larger, more diverse panel grown under controlled conditions, using quantitative approaches for several specific GSLs. Their study detected associations using a relaxed $-\log10$ (p) $\geq$ 3 threshold and proposed proximate candidate genes encoding for shikimate kinases (chromosome A04), chorismate mutase (chromosomes A06 and B04), jasmonate O-methyltransferase (chromosome B03), branched-chain-amino-acid transaminase (chromosome B06), cytochrome P450 enzyme CYP81G1 (chromosome B06) and MYB44 transcription factor (chromosome B06). Of these candidates, only the *CYP81G1* was listed as a potential GSL potential gene in SuCCombase and no genes had homologues of known and validated function in GSL biosynthesis or regulation. In contrast, our GWAS study used a controlled-environment growing condition, coupled with HPLC-MS-based analysis for the accurate quantification of individual GSLs, and applied a stringent Bonferroni threshold for the detection of associations. Yang et al., 2021, identified only two major control loci in a panel of 183 mixed vegetable and oilseed accessions phenotyped for individual GSLs using HPLC and genotyped at a density of 689,411 SNPs. *MYB28* (chromosome A02 and A09) was highlighted as a priority candidate gene, supporting the role of *MYB28* as a key regulator of GSL accumulation in *B. juncea*. Thus, our findings add value to previous studies and provide an exceptional resource of novel candidate gene homologues to known structural and regulatory genes of GSL metabolism. Further validation through allele mining and gene expression profiling is warranted, especially for associations explaining high levels of phenotypic variation and detected in multiple models.

4. Materials and Methods

4.1. Plant Materials and Growing Conditions

A diversity panel of 158 Indian mustard accessions from 28 countries, which had undergone two rounds of single seed descent (SSD) (Table S1), were grown in a CONVIRON® plant growth chamber (model: PGCFLEX, Winnipeg, MB, Canada) at Southern Cross University Lismore, New South Wales (28.8° S, 153.3° E), from March to mid-May 2020. Several seeds per accession were sown at 5 mm depth in a 10-cm-diameter free-draining plastic pot filled with commercial potting soil and thinned to one plant per pot two weeks after emergence. Each accession was grown in triplicate in a complete randomised block design. Three-week-old seedlings were supplied with 25 mL of diluted to half strength liquid fertiliser Canna A + B (CANNA Australasia, Subiaco Western Australia, delivered through syringe plunger, per pot. The growing conditions were set at 16 h of artificial lighting at 22 °C and eight hours of dark at 16 °C. Harvesting was done when all siliques were dried, and harvested siliques were further air-dried at 40 °C for 72 h before threshing.

4.2. Glucosinolate Analysis

In total, three biological replicates per accession (consisting of two individual seeds each) were used for quantifying GSL concentrations, following the method by Borpatragohain et al., 2019 [72]. In brief, two seeds per sample were placed in an Eppendorf safe-lock tube, to which 1.5 mL of 70% methanol and a 5 mm stainless-steel bead was added. The samples were then homogenised using a Qiagen Retsch MM 301 TissueLyser II (Qiagen Retsch, Hilden, Germany)) at 30 Hz for 45 s. Next, the samples were centrifuged for 15 min at 15,000 rpm at 7 °C using a Sigma laboratory tabletop centrifuge (Osterode am Harz Germany). An aliquot of 200 µL was transferred from the supernatant solution after centrifugation to a 2 mL Agilent HPLC screw-cap vial. The samples were then dried down using Martin Christ Alpha RVC (Osterode am Harz Germany) at successively reduced pressure of 180, 120, 80, 50, 20 and 5 mbar each at one-hour intervals, while 5 mbar was kept overnight. The dried samples were resuspended in 1.5 mL water containing 1.17 µmol mL^{-1} glucotropaeolin (a GSL not found in Brassicas) as internal standard. The tubes were mixed by inverting several times. Eight individual GSLs were quantified, including sinigrin (SIN), gluconapin (GNP), progroitrin (PGT), epi-pogroitrin (EPI), glucoiberin (GIB), glucoraphanin (GRF), glucobrassicin (GBS) and gluconarturtiin (GNT), using an Agilent 1260 Infinity II High Performance LC-MS instrument (Agilent Technologies, Palo Alto, CA, USA). HPLC-MS parameters used are detailed in Supplemental File S3. Total GSLs is the sum of the eight GSLs measured.

4.3. Bioinformatic Analyses and Data Processing

Illumina's FastQ sequence outputs were demultiplexed using Axe [73]. Both reads from the paired-end data were aligned against the *B. juncea* var. *tumida* T84-66 V1.5 genome reference (http://39.100.233.196:82/download_genome/Brassica_Genome_data/Braju_tum_V1.5, accessed on 15 January 2022) [53]. SNP calling was carried out using the Stacks pipeline [74], using default parameters and a low-level filter by looking for a minimum allele frequency of 5% for an SNP to be considered. Among the duplicated samples, the sample with the lower call rate was removed. Filtering of the resulting variant table for SNPs with a 60% call rate, non-minor allele frequency filtered and 10% maximum marker heterozygosity was done using TASSEL 5.2.73 [75]. Missing marker states for all remaining unique genotypes were imputed using Beagle 5.2 [76] with default parameters and the effective population size (Ne) set to 500,000.

4.4. Statistical Analysis

Residual distribution and quantile–quantile plots were visualised using Genstat 64-bit Release 18.1 (VSN International Ltd., Hemel Hempstead, England UK) to assess the normality and homoscedasticity of the phenotype data. Data were log10 (x + 0.01) transformed for subsequent estimation of the variance components and heritability values using REML Restricted Maximum Likelihood (REML) implemented in Genstat 64, as well as input for GWA. Best Linear Unbiased Predictions (BLUPs), calculated using genotype and replicate effects in REML, were used as phenotype input in GWAS. Correlations among GSL traits using raw mean values were computed using the 'ggpubr' package [77], implemented in the R environment.

4.5. Genome-Wide Association Analysis

Marker–trait association was performed using the Genome Association and Prediction Integrated Tool (GAPIT Version 3) [78,79]. To select the best models, an initial analysis using the four most recommended models as discussed in the GAPIT manual based on statistical power was conducted [79]. These were multiple locus mixed linear model (MLMM) [80], Settlement of MLM Under Progressively Exclusive Relationship (SUPER) [81], Fixed and random model circulating probability unification (FarmCPU) [57], Bayesian-information and Linkage-disequilibrium Iteratively Nested Keyway (BLINK) [58]. The best models were selected based on the resulting q-q plots, which reflected how well each model accounted for population structure and familial relatedness. Manhattan plots were visualised using R package 'CMplot' [82].

4.6. Cluster, Population Structure and Principal Components Analysis

A separate set of 1174 higher-confidence SNPs imputed and filtered for >80% call rate, 5% minor allele frequency (maf) and 10% maximum heterozygosity, covering pseudochromosomes, and linkage-disequilibrium (LD) pruned, was used for cluster, population and principal components analyses. LD pruning was done using Plink [83] (version 1.07) with the following parameters: window of 50 SNPs, step size of five markers and an r^2 threshold of 0.4 [84]. An UPGMA (unweighted pair group method with arithmetic mean) tree was built for cluster analysis of all 158 lines. The genetic distance input for tree building was simple matching coefficients calculated in TASSEL (version 5.2.72) [75] and UPGMA was visualised using ITOLv6 (https://itol.embl.de/, accessed on 3 September 2021). A maximum likelihood estimate for population structure was carried out in ADMIXTURE [85] and barplots for Q matrix (probability of group membership) were visualised using package 'pophelper' [86] implemented in the R environment. The analysis was done for K = 1 to K = 10, and a ten-fold cross-validation procedure was used to determine the "best" K. PCA was conducted in TASSEL (version 5.2.72) and plotted using the 'ggplot2' [87] R package.

4.7. Candidate Genes within Significant SNPs

Predicted candidate genes within 250 kb upstream and downstream of each significantly associated SNP were identified using the *B. juncea* BRAD v.1.5 annotation. The BRAD V1.5 genes were annotated for putative function by alignment to the *Arabidopsis* TAIR10 release using NCBI BLASTP [88,89], integrated into the in-house SCPS Galaxy (http://lr-scps5-rh7v.scps.scu.edu.au:8080, accessed on 9 September 2021), and associating the annotation of the *Arabidopsis* genes in the top-scoring hits. All these annotations and genome information were integrated into the SCPS Galaxy. Next, we matched the *Arabidopsis* locus identifiers from our BLAST+ list and that of "known" and "potential GSL genes" curated in SuCCombase (https://plant-scc.org, accessed on 7 September 2021) for listing our candidate genes. Top hits identified as either "known" or "potential GSL genes" based on SuCCombase were prioritised as candidate genes. In a few cases, we chose the "known" or "potential GSL gene" even if they ranked second to third in BLASTP, provided that the percent identity was more than 60% across more than 50% of the total length alignment.

Supplementary Materials: The following supporting information can be downloaded at: https://www.mdpi.com/article/10.3390/plants11030364/s1, Supplemental File S1. Glucosinolate measurements of the accessions in the diversity panel; Table S1. List of accessions and country of origin in the diversity panel; Figure S1. Residual distribution and normal plots for total GSLs, sinigrin and gluconapin; Table S2. Variance components analysis and descriptive statistics for total GSLs, sinigrin and gluconapin evaluated in 158 diverse *B. juncea* L. accessions; Figure S2. Correlations of major GSLs, sinigrin and gluconapin, and total GSLs in ADMIXTURE clusters; Figure S3. Quantile–quantile plots reflecting correspondence between observed and expected $-\log 10$ (p) values from association analyses using four models (SUPER, MLMM, FarmCPU, BLINK) for total GSLs, sinigrin and gluconapin; Supplemental File S2. List of SNPs passing the Bonferroni threshold from four models; Figure S4. Linkage disequilibrium (LD) depicted based on squared correlation coefficient of pairwise markers in a sliding window of 100 SNP markers; Supplemental File S3. HPLC-MS parameters used for glucosinolate analysis.

Author Contributions: E.T. and T.K. conceptualised and designed the experiment. E.T. performed the experiments. P.B. advised and helped in glucosinolate quantification. R.M. advised and assisted in bioinformatic analyses and built the Galaxy instance used in candidate gene analysis. E.T. prepared the manuscript. P.B., R.M. and T.K. edited the manuscript. All authors have read and agreed to the published version of the manuscript.

Funding: This study was funded by the Australian Research Council (ARC) linkage grant LP170101062.

Institutional Review Board Statement: Not applicable.

Informed Consent Statement: Not applicable.

Data Availability Statement: All relevant research data is included as supplementary material. The genotype data can be accessed through the following DOI: DOI 10.25918/data.186.

Acknowledgments: We acknowledge Carolyn Raymond for performing REML on the phenotype data. We would like to thank LP170101062 co-chief investigators Graham King, Bronwyn Barkla and Terry Rose for their overall support in this study.

Conflicts of Interest: The authors declare no conflict of interest.

References

1. Halkier, B.A.; Gershenzon, J. Biology and biochemistry of glucosinolates. *Annu. Rev. Plant Biol.* **2006**, *57*, 303–333. [CrossRef]
2. Burow, M.; Halkier, B.A.; Kliebenstein, D.J. Regulatory networks of glucosinolates shape *Arabidopsis thaliana* fitness. *Curr. Opin. Plant Biol.* **2010**, *13*, 347–352. [CrossRef] [PubMed]
3. Bakhtiari, M.; Rasmann, S. Genotypic variation in below-to aboveground systemic induction of glucosinolates mediates plant fitness consequences under herbivore attack. *J. Chem. Ecol.* **2019**, 317–329. [CrossRef]
4. Harun, S.; Abdullah-Zawawi, M.R.; Goh, H.H.; Mohamed-Hussein, Z.A. A comprehensive gene inventory for glucosinolate biosynthetic pathway in *Arabidopsis thaliana*. *J. Agric. Food Chem.* **2020**, *68*, 7281–7297. [CrossRef] [PubMed]
5. Sønderby, I.E.; Geu-Flores, F.; Halkier, B.A. Biosynthesis of glucosinolates—Gene discovery and beyond. *Trends Plant Sci.* **2010**, *15*, 283–290. [CrossRef] [PubMed]

6. Augustine, R.; Mukhopadhyay, A.; Bisht, N.C. Targeted silencing of BjMYB28 transcription factor gene directs development of low glucosinolate lines in oilseed Brassica juncea. *Plant Biotechnol. J.* **2013**, *11*, 855–866. [CrossRef]
7. Dixon, G.R. Vegetable brassicas and related crucifers: Origins and diversity of brassica and its relatives. *Veg. Brassicas Relat. Crucif.* **2006**, 1–33. [CrossRef]
8. Raymer, P.L. Canola: An emerging oilseed crop. *Trends New Crop. New Uses* **2002**, *1*, 122–126.
9. Love, H.K.; Rakow, G.; Raney, J.P.; Downey, R.K. Development of low glucosinolate mustard. *Can. J. Plant Sci.* **1990**, *70*, 419–424. [CrossRef]
10. Potts, D.A.; Rakow, G.W.; Males, D.R. Canola quality Brassica juncea, a new oilseed crop for the canadian prairies. In Proceedings of the 10th International Rapeseed Congress, Canberra, Australia, 26–29 September 1999.
11. Norton, R.; Potter, T.; Haskins, B.; Mccaffery, D.; Bambach, R. Juncea Canola in the Low Rainfall Zones of Victoria and South Australia. Juncea Canola Growers Guide–Victoria and South Australia. Available online: http://anz.ipni.net/ipniweb/region/anz.nsf/0/CE50267DC5CD6D5385257AA10052C4E0/$FILE/ViCSAGrowersGuide.pdf (accessed on 26 September 2021).
12. Woods, D.L.; Capcara, J.J.; Downey, R.K. The potential of mustard (*Brassica juncea* (L.) Coss) as an edible oil crop on the Canadian prairies. *Can. J. Plant Sci.* **1991**, *71*, 195–198. [CrossRef]
13. Burton, W.; Pymer, S.; Salisbury, P.; Kirk, J.; Oram, R. Performance of Australian canola quality Brassica juncea breeding lines. In Proceedings of the 10th International Rapeseed Congress, Canberra, Australia, 26–29 September 1999; pp. 2–7.
14. Merah, O. Genetic Variability in glucosinolates in seed of *Brassica juncea*: Interest in mustard condiment. *J. Chem.* **2015**, *2015*, 606142. [CrossRef]
15. Misiewicz, I.; Skupińska, K.; Kowalska, E.; Lubiński, J.; Kasprzycka-Guttman, T. Sulforaphane-mediated induction of a phase 2 detoxifying enzyme NAD(P)H:Quinone reductase and apoptosis in human lymphoblastoid cells. *Acta Biochim. Pol.* **2004**, *51*, 711–721. [CrossRef] [PubMed]
16. Ullah, M. Sulforaphane (SFN): An isothiocyanate in a cancer chemoprevention paradigm. *Medicines* **2015**, *2*, 141–156. [CrossRef] [PubMed]
17. Yang, J.; Wang, J.; Li, Z.; Li, X.; He, Z.; Zhang, L.; Sha, T.; Lyu, X.; Chen, S.; Gu, Y.; et al. Genomic signatures of vegetable and oilseed allopolyploid *Brassica juncea* and genetic loci controlling the accumulation of glucosinolates. *Plant Biotechnol. J.* **2021**, *19*, 2619–2628. [CrossRef]
18. Sharma, M.; Mukhopadhyay, A.; Gupta, V.; Pental, D.; Pradhan, A.K. BjuB.CYP79F1 regulates synthesis of propyl fraction of aliphatic glucosinolates in oilseed mustard *Brassica juncea*: Functional validation through genetic and transgenic approaches. *PLoS ONE* **2016**, *11*, e0150060. [CrossRef]
19. Moose, S.P.; Mumm, R.H. Molecular plant breeding as the foundation for 21st century crop improvement. *Plant Physiol.* **2008**, *147*, 969–977. [CrossRef]
20. Harper, A.L.; He, Z.; Langer, S.; Havlickova, L.; Wang, L.; Fellgett, A.; Gupta, V.; Kumar Pradhan, A.; Bancroft, I. Validation of an associative transcriptomics platform in the polyploid crop species Brassica juncea by dissection of the genetic architecture of agronomic and quality traits. *Plant J.* **2020**, *103*, 1885–1893. [CrossRef]
21. Rout, K.; Sharma, M.; Gupta, V.; Mukhopadhyay, A.; Sodhi, Y.S.; Pental, D.; Pradhan, A.K. Deciphering allelic variations for seed glucosinolate traits in oilseed mustard (*Brassica juncea*) using two bi-parental mapping populations. *Theor. Appl. Genet.* **2015**, *128*, 657–666. [CrossRef]
22. Rout, K.; Yadav, B.G.; Yadava, S.K.; Mukhopadhyay, A.; Gupta, V.; Pental, D.; Pradhan, A.K. QTL landscape for oil content in *Brassica juncea*: Analysis in multiple bi-parental populations in high and "0" erucic background. *Front. Plant Sci.* **2018**, *9*, 871. [CrossRef]
23. Ramchiary, N.; Bisht, N.C.; Gupta, V.; Mukhopadhyay, A.; Arumugam, N.; Sodhi, Y.S.; Pental, D.; Pradhan, A.K. qtl analysis reveals context-dependent loci for seed glucosinolate trait in the oilseed *Brassica juncea*: Importance of recurrent selection backcross scheme for the identification of "true" QTL. *Theor. Appl. Genet.* **2007**, *116*, 77–85. [CrossRef]
24. Raman, H.; Raman, R.; Kilian, A.; Detering, F.; Carling, J.; Coombes, N.; Diffey, S.; Kadkol, G.; Edwards, D.; McCully, M.; et al. Genome-wide delineation of natural variation for pod shatter resistance in *Brassica napus*. *PLoS ONE* **2014**, *9*, e101673. [CrossRef] [PubMed]
25. Liu, S.; Huang, H.; Yi, X.; Zhang, Y.; Yang, Q.; Zhang, C.; Fan, C.; Zhou, Y. Dissection of genetic architecture for glucosinolate accumulations in leaves and seeds of *Brassica napus* by genome-wide association study. *Plant Biotechnol. J.* **2020**, *18*, 1472–1484. [CrossRef] [PubMed]
26. Qu, C.; Jia, L.; Fu, F.; Zhao, H.; Lu, K.; Wei, L.; Xu, X.; Liang, Y.; Li, S.; Wang, R.; et al. Genome-wide association mapping and identification of candidate genes for fatty acid composition in *Brassica napus* L. using SNP Markers. *BMC Genom.* **2017**, *18*, 232. [CrossRef]
27. Wang, B.; Wu, Z.; Li, Z.; Zhang, Q.; Hu, J.; Xiao, Y.; Cai, D.; Wu, J.; King, G.J.; Li, H.; et al. Dissection of the genetic architecture of three seed-quality traits and consequences for breeding in *Brassica napus*. *Plant Biotechnol. J.* **2018**, *16*, 1336–1348. [CrossRef] [PubMed]
28. Kaur, S.; Akhatar, J.; Kaur, H.; Atri, C.; Mittal, M.; Goyal, A.; Pant, U.; Kaur, G.; Banga, S.S. Genome-wide association mapping for key seed metabolites using a large panel of natural and derived forms of *Brassica rapa* L. *Ind. Crops Prod.* **2021**, *159*, 113073. [CrossRef]

29. Sudan, J.; Singh, R.; Sharma, S.; Salgotra, R.K.; Sharma, V.; Singh, G.; Sharma, I.; Sharma, S.; Gupta, S.K.; Zargar, S.M. DdRAD sequencing-based identification of inter-genepool SNPS and association analysis in *Brassica juncea*. *BMC Plant Biol.* **2019**, *19*, 594. [CrossRef]
30. Paritosh, K.; Gupta, V.; Yadava, S.K.; Singh, P.; Pradhan, A.K.; Pental, D. RNA-Seq based snps for mapping in *Brassica juncea* (AABB): Synteny analysis between the two constituent genomes A (from *B. rapa*) and B (from *B. nigra*) shows highly divergent gene block arrangement and unique block fragmentation patterns. *BMC Genom.* **2014**, *15*, 396. [CrossRef]
31. Yang, J.; Zhang, C.; Zhao, N.; Zhang, L.; Hu, Z.; Chen, S.; Zhang, M. Chinese root-type mustard provides phylogenomic insights into the evolution of the multi-use diversified allopolyploid *Brassica juncea*. *Mol. Plant* **2018**, *11*, 512–514. [CrossRef]
32. Akhatar, J.; Singh, M.P.; Sharma, A.; Kaur, H.; Kaur, N.; Sharma, S.; Bharti, B.; Sardana, V.K.; Banga, S.S. Association mapping of seed quality traits under varying conditions of nitrogen application in *Brassica juncea* L. Czern & Coss. *Front. Genet.* **2020**, *11*, 744. [CrossRef]
33. Hirai, M.Y.; Klein, M.; Fujikawa, Y.; Yano, M.; Goodenowe, D.B.; Yamazaki, Y.; Kanaya, S.; Nakamura, Y.; Kitayama, M.; Suzuki, H.; et al. Elucidation of gene-to-gene and metabolite-to-gene networks in *Arabidopsis* by integration of metabolomics and transcriptomics. *J. Biol. Chem.* **2005**, *280*, 25590–25595. [CrossRef]
34. Hirai, M.Y.; Sugiyama, K.; Sawada, Y.; Tohge, T.; Obayashi, T.; Suzuki, A.; Araki, R.; Sakurai, N.; Suzuki, H.; Aoki, K.; et al. Omics-based identification of *Arabidopsis* Myb transcription factors regulating aliphatic glucosinolate biosynthesis. *Proc. Natl. Acad. Sci. USA* **2007**, *104*, 6478–6483. [CrossRef] [PubMed]
35. Wentzell, A.M.; Rowe, H.C.; Hansen, B.G.; Ticconi, C.; Halkier, B.A.; Kliebenstein, D.J. Linking metabolic QTLs with network and cis-eqtls controlling biosynthetic pathways. *PLoS Genet.* **2007**, *3*, 1687–1701. [CrossRef] [PubMed]
36. Lee, S.; Kaminaga, Y.; Cooper, B.; Pichersky, E.; Dudareva, N.; Chapple, C. Benzoylation and sinapoylation of glucosinolate R-groups in Arabidopsis. *Plant J.* **2012**, *72*, 411–422. [CrossRef] [PubMed]
37. Skirycz, A.; Reichelt, M.; Burow, M.; Birkemeyer, C.; Rolcik, J.; Kopka, J.; Zanor, M.I.; Gershenzon, J.; Strnad, M.; Szopa, J.; et al. DOF transcription factor AtDof1.1 (OBP2) is part of a regulatory network controlling glucosinolate biosynthesis in *Arabidopsis*. *Plant J.* **2006**, *47*, 10–24. [CrossRef]
38. Gigolashvili, T.; Yatusevich, R.; Berger, B.; Müller, C.; Flügge, U.I. The R2R3-MYB transcription factor HAG1/MYB28 is a regulator of methionine-derived glucosinolate biosynthesis in *Arabidopsis thaliana*. *Plant J.* **2007**, *51*, 247–261. [CrossRef]
39. Sønderby, I.E.; Hansen, B.G.; Bjarnholt, N.; Ticconi, C.; Halkier, B.A.; Kliebenstein, D.J. A systems biology approach identifies a R2R3 MYB gene subfamily with distinct and overlapping functions in regulation of aliphatic glucosinolates. *PLoS ONE* **2007**, *2*, e1322. [CrossRef]
40. Celenza, J.L.; Quiel, J.A.; Smolen, G.A.; Merrikh, H.; Silvestro, A.R.; Normanly, J.; Bender, J. The *Arabidopsis* ATR1 Myb transcription factor controls indolic glucosinolate homeostasis. *Plant Physiol.* **2005**, *137*, 253–262. [CrossRef]
41. Frerigmann, H.; Gigolashvili, T. Update on the role of R2R3-MYBs in the regulation of glucosinolates upon sulfur deficiency. *Front. Plant Sci.* **2014**, *5*, 626. [CrossRef]
42. Kroymann, J.; Textor, S.; Tokuhisa, J.G.; Falk, K.L.; Bartram, S.; Gershenzon, J.; Mitchell-Olds, T. A gene controlling variation in *Arabidopsis* glucosinolate composition is part of the methionine chain elongation pathway. *Plant Physiol.* **2001**, *127*, 1077–1088. [CrossRef]
43. Benderoth, M.; Pfalz, M.; Kroymann, J. Methylthioalkylmalate synthases: Genetics, ecology and evolution. *Phytochem. Rev.* **2009**, *8*, 255–268. [CrossRef]
44. Li, J.; Hansen, B.G.; Ober, J.A.; Kliebenstein, D.J.; Halkier, B.A. Subclade of flavin-monooxygenases involved in aliphatic glucosinolate biosynthesis. *Plant Physiol.* **2008**, *148*, 1721–1733. [CrossRef] [PubMed]
45. Li, J.; Kristiansen, K.A.; Hansen, B.G.; Halkier, B.A. Cellular and subcellular localization of flavin-monooxygenases involved in glucosinolate biosynthesis. *J. Exp. Bot.* **2011**, *62*, 1337–1346. [CrossRef] [PubMed]
46. Huseby, S.; Koprivova, A.; Lee, B.R.; Saha, S.; Mithen, R.; Wold, A.B.; Bengtsson, G.B.; Kopriva, S. Diurnal and light regulation of sulphur assimilation and glucosinolate biosynthesis in *Arabidopsis*. *J. Exp. Bot.* **2013**, *64*, 1039–1048. [CrossRef] [PubMed]
47. Sul, J.H.; Martin, L.S.; Eskin, E. Population structure in genetic studies: Confounding factors and mixed models. *PLoS Genet.* **2018**, *14*, e1007309. [CrossRef] [PubMed]
48. Akhatar, J.; Goyal, A.; Kaur, N.; Atri, C.; Mittal, M.; Singh, M.P.; Kaur, R.; Rialch, I.; Banga, S.S. Genome-wide association analyses to understand genetic basis of flowering and plant height under three levels of nitrogen application in *Brassica juncea* (L.) Czern & Coss. *Sci. Rep.* **2021**, *11*, 4278. [CrossRef]
49. Vaughan, J.G.; Gordon, E.I. A taxonomic study of *Brassica juncea* using the techniques of electrophoresis, gas-liquid chromatography and serology. *Ann. Bot.* **1973**, *37*, 167–184. [CrossRef]
50. Gland, A.; Röbbelen, G.; Thies, W. Variation of alkenyl glucosinolates in seeds of Brassica species. *Z. Pflanzenzüchtg* **1981**, *87*, 96–110.
51. Sodhi, Y.S.; Mukhopadhyay, A.; Arumugam, N.; Verma, J.K.; Gupta, V.; Pental, D.; Pradhan, A.K. Genetic analysis of total glucosinolate in crosses involving a high glucosinolate indian variety and a low glucosinolate line of *Brassica Juncea*. *Plant Breed.* **2002**, *121*, 508–511. [CrossRef]
52. Velasco, L.; Becker, H.C. Variability for seed glucosinolates in a germplasm collection of the genus brassica. *Genet. Resour. Crop Evol.* **2000**, *47*, 231–238. [CrossRef]

53. Yang, J.; Liu, D.; Wang, X.; Ji, C.; Cheng, F.; Liu, B.; Hu, Z.; Chen, S.; Pental, D.; Ju, Y.; et al. The genome sequence of allopolyploid *Brassica juncea* and analysis of differential homoeolog gene expression influencing selection. *Nat. Genet.* **2016**, *48*, 1225–1232. [CrossRef] [PubMed]
54. Kaler, A.S.; Gillman, J.D.; Beissinger, T.; Purcell, L.C. Comparing different statistical models and multiple testing corrections for association mapping in soybean and maize. *Front. Plant Sci.* **2020**, *10*, 1794. [CrossRef] [PubMed]
55. Muhammad, A.; Li, J.; Hu, W.; Yu, J.; Khan, S.U.; Khan, M.H.U.; Xie, G.; Wang, J.; Wang, L. Uncovering genomic regions controlling plant architectural traits in hexaploid wheat using different GWAS models. *Sci. Rep.* **2021**, *11*, 6767. [CrossRef] [PubMed]
56. Zhong, H.; Liu, S.; Meng, X.; Sun, T.; Deng, Y.; Kong, W.; Peng, Z.; Li, Y. Correction to: Uncovering the genetic mechanisms regulating panicle architecture in rice with GPWAS and GWAS. *BMC Genom.* **2021**, *22*, 86. [CrossRef] [PubMed]
57. Liu, X.; Huang, M.; Fan, B.; Buckler, E.S.; Zhang, Z. Iterative usage of fixed and random effect models for powerful and efficient genome-wide association studies. *PLoS Genet.* **2016**, *12*, e1005767. [CrossRef] [PubMed]
58. Huang, M.; Liu, X.; Zhou, Y.; Summers, R.M.; Zhang, Z. BLINK: A package for the next level of genome-wide association studies with both individuals and markers in the millions. *Gigascience* **2018**, *8*, giy154. [CrossRef]
59. Harun, S.; Abdullah-Zawawi, M.R.; A-Rahman, M.R.A.; Muhammad, N.A.N.; Mohamed-Hussein, Z.A. SuCComBase: A manually curated repository of plant sulfur-containing compounds. *Database* **2019**, *2019*, 1–9. [CrossRef] [PubMed]
60. Kang, H.G.; Singh, K.B. Characterization of salicylic acid-responsive, *Arabidopsis* Dof domain proteins: Overexpression of OBP3 leads to growth defects. *Plant J.* **2000**, *21*, 329–339. [CrossRef] [PubMed]
61. Kliebenstein, D.J.; Lambrix, V.M.; Reichelt, M.; Gershenzon, J.; Mitchell-Olds, T. Gene duplication in the diversification of secondary metabolism: Tandem 2-oxoglutarate-dependent dioxygenases control glucosinolate biosynthesis in *Arabidopsis. Plant Cell* **2001**, *13*, 681–693. [CrossRef]
62. Burow, M.; Atwell, S.; Francisco, M.; Kerwin, R.E.; Halkier, B.A.; Kliebenstein, D.J. The glucosinolate biosynthetic gene *AOP2* mediates feed-back regulation of jasmonic acid signaling in *Arabidopsis. Mol. Plant* **2015**, *8*, 1201–1212. [CrossRef]
63. Benstein, R.M.; Ludewig, K.; Wulfert, S.; Wittek, S.; Gigolashvili, T.; Frerigmann, H.; Gierth, M.; Flügge, U.I.; Krueger, S. Arabidopsis phosphoglycerate dehydrogenase1 of the phosphoserine pathway is essential for development and required for ammonium assimilation and tryptophan biosynthesis. *Plant Cell* **2013**, *25*, 5011–5029. [CrossRef]
64. Watanabe, M.; Tohge, T.; Fernie, A.R.; Hoefgen, R. The effect of single and multiple serat mutants on serine and sulfur metabolism. *Front. Plant Sci.* **2018**, *9*, 702. [CrossRef] [PubMed]
65. Niemiro, A.; Cysewski, D.; Brzywczy, J.; Wawrzyńska, A.; Sieńko, M.; Poznański, J.; Sirko, A. Similar but not identical—Binding properties of LSU (Response to Low Sulfur) proteins from *Arabidopsis Thaliana. Front. Plant Sci.* **2020**, *11*, 1246. [CrossRef] [PubMed]
66. Aarabi, F.; Naake, T.; Fernie, A.R.; Hoefgen, R. Coordinating sulfur pools under sulfate deprivation. *Trends Plant Sci.* **2020**, *25*, 1227–1239. [CrossRef] [PubMed]
67. Ohkubo, Y.; Tanaka, M.; Tabata, R.; Ogawa-Ohnishi, M.; Matsubayashi, Y. Shoot-to-Root Mobile Polypeptides Involved in Systemic Regulation of Nitrogen Acquisition. *Nat. Plants* **2017**, *3*, 17029. [CrossRef]
68. Grubb, C.D.; Zipp, B.J.; Ludwig-Müller, J.; Masuno, M.N.; Molinski, T.F.; Abel, S. *Arabidopsis* glucosyltransferase UGT74B1 functions in glucosinolate biosynthesis and auxin homeostasis. *Plant J.* **2004**, *40*, 893–908. [CrossRef]
69. Grubb, C.D.; Zipp, B.J.; Kopycki, J.; Schubert, M.; Quint, M.; Lim, E.K.; Bowles, D.J.; Pedras, M.S.C.; Abel, S. Comparative analysis of *Arabidopsis* UGT74 glucosyltransferases reveals a special role of UGT74C1 in glucosinolate biosynthesis. *Plant J.* **2014**, *79*, 92–105. [CrossRef]
70. Romero, L.C.; Aroca, M.Á.; Laureano-Marín, A.M.; Moreno, I.; García, I.; Gotor, C. Cysteine and cysteine-related signaling pathways in *Arabidopsis thaliana. Mol. Plant* **2014**, *7*, 264–276. [CrossRef]
71. Aarabi, F.; Kusajima, M.; Tohge, T.; Konishi, T.; Gigolashvili, T.; Takamune, M.; Sasazaki, Y.; Watanabe, M.; Nakashita, H.; Fernie, A.R.; et al. Sulfur deficiency-induced repressor proteins optimize glucosinolate biosynthesis in plants. *Sci. Adv.* **2016**, *2*, e1601087. [CrossRef]
72. Borpatragohain, P.; Rose, T.J.; Liu, L.; Barkla, B.J.; Raymond, C.A.; King, G.J. Remobilization and fate of sulphur in mustard. *Ann. Bot.* **2019**, *124*, 471–480. [CrossRef]
73. Murray, K.D.; Borevitz, J.O. Axe: Rapid, Competitive sequence read demultiplexing using a trie. *Bioinformatics* **2018**, *34*, 3924–3925. [CrossRef]
74. Catchen, J.; Hohenlohe, P.A.; Bassham, S.; Amores, A.; Cresko, W.A. Stacks: An analysis tool set for population genomics. *Mol. Ecol.* **2013**, *22*, 3124–3140. [CrossRef]
75. Bradbury, P.J.; Zhang, Z.; Kroon, D.E.; Casstevens, T.M.; Ramdoss, Y.; Buckler, E.S. TASSEL: Software for association mapping of complex traits in diverse samples. *Bioinformatics* **2007**, *23*, 2633–2635. [CrossRef] [PubMed]
76. Browning, B.L.; Zhou, Y.; Browning, S.R. A one-penny imputed genome from next-generation reference panels. *Am. J. Hum. Genet.* **2018**, *103*, 338–348. [CrossRef] [PubMed]
77. Kassambara, A. Ggpubr: 'Ggplot2' Based Publication Ready Plots 2020. 2021. Available online: https://github.com/kassambara/ggpubr (accessed on 1 September 2021).
78. Lipka, A.E.; Tian, F.; Wang, Q.; Peiffer, J.; Li, M.; Bradbury, P.J.; Gore, M.A.; Buckler, E.S.; Zhang, Z. GAPIT: Genome association and prediction integrated tool. *Bioinformatics* **2012**, *28*, 2397–2399. [CrossRef] [PubMed]

79. Buckler, E.; Zhang, Z. User Manual for Genomic Association and Prediction Integrated Tool (GAPIT). 2018. Version 3. Available online: https://zzlab.net/GAPIT/gapit_help_document.pdf (accessed on 27 August 2021).
80. Segura, V.; Vilhjálmsson, B.J.; Platt, A.; Korte, A.; Seren, Ü.; Long, Q.; Nordborg, M. An efficient multi-locus mixed-model approach for genome-wide association studies in structured populations. *Nat. Genet.* **2012**, *44*, 825–830. [CrossRef] [PubMed]
81. Wang, Q.; Tian, F.; Pan, Y.; Buckler, E.S.; Zhang, Z. A SUPER powerful method for genome wide association study. *PLoS ONE* **2014**, *9*, e0107684. [CrossRef]
82. Yin, L.; Zhang, H.; Tang, Z.; Xu, J.; Yin, D.; Zhang, Z.; Yuan, X.; Zhu, M.; Zhao, S.; Li, X.; et al. RMVP: A memory-efficient, visualization-enhanced, and parallel-accelerated tool for genome-wide association study. *Genom. Proteom. Bioinform.* **2021**. [CrossRef]
83. Purcell, S.; Neale, B.; Todd-Brown, K.; Thomas, L.; Ferreira, M.A.R.; Bender, D.; Maller, J.; Sklar, P.; De Bakker, P.I.W.; Daly, M.J.; et al. PLINK: A tool set for whole-genome association and population-based linkage analyses. *Am. J. Hum. Genet.* **2007**, *81*, 559–575. [CrossRef]
84. Niu, S.; Song, Q.; Koiwa, H.; Qiao, D.; Zhao, D.; Chen, Z.; Liu, X.; Wen, X. Genetic diversity, linkage disequilibrium, and population structure analysis of the tea plant (*Camellia sinensis*) from an origin center, guizhou plateau, using genome-wide SNPs developed by genotyping-by-sequencing. *BMC Plant Biol.* **2019**, *19*, 328. [CrossRef]
85. Alexander, D.H.; Novembre, J.; Lange, K. Fast model-based estimation of ancestry in unrelated individuals. *Genome Res.* **2009**, *19*, 1655–1664. [CrossRef]
86. Francis, R.M. Pophelper: An R package and web app to analyse and visualize population structure. *Mol. Ecol. Resour.* **2017**, *17*, 27–32. [CrossRef] [PubMed]
87. Wickham, H. *Ggplot2: Elegant Graphics for Data Analysis*; Springer: New York, NY, USA, 2016; pp. 1–69. ISBN 978-0-387-98141-3.
88. Camacho, C.; Coulouris, G.; Avagyan, V.; Ma, N.; Papadopoulos, J.; Bealer, K.; Madden, T.L. BLAST+: Architecture and Applications. *BMC Bioinform.* **2009**, *10*, 1–9. [CrossRef] [PubMed]
89. Cock, P.J.A.; Chilton, J.M.; Grüning, B.; Johnson, J.E.; Soranzo, N. NCBI BLAST+ Integrated into Galaxy. *Gigascience* **2015**, *4*, s13742-015-0080-7. [CrossRef] [PubMed]

plants

Article

Portuguese Common Bean Natural Variation Helps to Clarify the Genetic Architecture of the Legume's Nutritional Composition and Protein Quality

Francisco A. Mendes [1], Susana T. Leitão [1,*], Verónica Correia [1,2], Elsa Mecha [1,3], Diego Rubiales [4], Maria Rosário Bronze [1,2,3] and Maria Carlota Vaz Patto [1]

1 Instituto de Tecnologia Química e Biológica António Xavier, Universidade Nova de Lisboa, Av. da República, 2780-157 Oeiras, Portugal; famendes@itqb.unl.pt (F.A.M.); vero_correi@hotmail.com (V.C.); emecha@itqb.unl.pt (E.M.); mbronze@ibet.pt (M.R.B.); cpatto@itqb.unl.pt (M.C.V.P.)

2 Faculdade de Farmácia, Universidade de Lisboa, 1649-019 Lisboa, Portugal

3 iBET—Instituto de Biologia Experimental e Tecnológica, Av. da República, 2780-157 Oeiras, Portugal

4 Instituto de Agricultura Sostenible, CSIC, Av. Menéndez Pidal, 14004 Cordova, Spain; diego.rubiales@ias.csic.es

* Correspondence: sleitao@itqb.unl.pt

Abstract: Common bean is a nutritious food legume widely appreciated by consumers worldwide. It is a staple food in Latin America, and a component of the Mediterranean diet, being an affordable source of protein with high potential as a gourmet food. Breeding for nutritional quality, including both macro and micronutrients, and meeting organoleptic consumers' preferences is a difficult task which is facilitated by uncovering the genetic basis of related traits. This study explored the diversity of 106 Portuguese common bean accessions, under two contrasting environments, to gain insight into the genetic basis of nutritional composition (ash, carbohydrates, fat, fiber, moisture, protein, and resistant starch contents) and protein quality (amino acid contents and trypsin inhibitor activity) traits through a genome-wide association study. Single-nucleotide polymorphism-trait associations were tested using linear mixed models accounting for the accessions' genetic relatedness. Mapping resolution to the gene level was achieved in 56% of the cases, with 102 candidate genes proposed for 136 genomic regions associated with trait variation. Only one marker-trait association was stable across environments, highlighting the associations' environment-specific nature and the importance of genotype × environment interaction for crops' local adaptation and quality. This study provides novel information to better understand the molecular mechanisms regulating the nutritional quality in common bean and promising molecular tools to aid future breeding efforts to answer consumers' concerns.

Keywords: ash; amino acids; carbohydrates; fat; fiber; GWAS; nutritional quality; *Phaseolus vulgaris*; resistant starch; trypsin inhibitor

check for updates

Citation: Mendes, F.A.; Leitão, S.T.; Correia, V.; Mecha, E.; Rubiales, D.; Bronze, M.R.; Vaz Patto, M.C. Portuguese Common Bean Natural Variation Helps to Clarify the Genetic Architecture of the Legume's Nutritional Composition and Protein Quality. *Plants* **2022**, *11*, 26. https://doi.org/10.3390/plants11010026

Academic Editor: Abdelmajid Kassem

Received: 29 November 2021
Accepted: 17 December 2021
Published: 22 December 2021

1. Introduction

Consumers are increasingly more health-conscious and striving to have greater diversity and healthier foods in their diets [1]. Common bean, or bean (*Phaseolus vulgaris* L.), is an important and affordable source of protein, dietary fiber, essential vitamins, and minerals [2]. It is one of the most important food legumes cultivated and consumed worldwide [3], being a staple food in Eastern Africa and Latin America [4], and a component of the Mediterranean diet [5]. Consequently, bean is an important crop to fight malnourishment, particularly protein malnutrition [3], as well as an important food for the prevention of a variety of non-communicable diseases due to its diversity of health-promoting compounds, such as resistant starch and dietary fiber (reviewed in [6]). In addition to its nutritive composition, common bean can be consumed in a variety of culinary forms and specific high-quality landraces can attract high market prices (gourmet foods) [7].

Despite its compositional richness, common bean production and consumption have been declining in Europe. Several factors contributed to this, such as low productivity, changes in consumers' preferences, and little investment in breeding and food innovation [8]. Nevertheless, breeding programs could be instrumental to reverse this decline through the development of bean varieties, first with higher and stable yields, increasing the attractiveness of this crop to farmers, but also with higher nutritional quality, more adapted to present consumers' demands [8]. Although particularly important to increase bean consumption and market demand, the traits related to nutritional quality, like the contents of particular compounds, are normally complex, controlled by many genes with small effects, and highly influenced by environmental factors [9]. Consequently, nutritional quality is hard to handle by conventional plant breeding.

The complexity of food nutritional quality breeding comes also from potential hidden interactions among the quality-related compounds. In particular, legumes possess bioactive compounds which can act simultaneously as health-promoting compounds and anti-nutritional factors, impairing nutrients' bioavailability, and influencing both taste and consumers' food acceptability [10]. Protease inhibitors are among these bioactive compounds, as they are anti-inflammatory and anti-carcinogenic compounds, and simultaneously interfere with digestion through the irreversible inhibition of trypsin and chymotrypsin (reviewed in [6]). Trypsin inhibitors interfere for instance with protein digestibility, reducing protein quality, a nutritional quality-related trait that has gained importance in recent years [11]. Protein quality is also evaluated in terms of amino acid composition. Although considered valuable sources of protein, the nutritional quality of legumes protein is often lower than that of animal protein, due to their reduced content on sulfur amino acids (methionine and cysteine) [12]. Nevertheless, variation exists in nature or may be generated, and higher protein quality in terms of improved amino acid composition and digestibility is presently regarded as an important target for legume breeding [13].

A common aspect of the mentioned nutritional quality-related targets is that many of these nutritional compounds' contents are laborious and expensive to measure and so difficult to routinely implement in breeding programs. However, the use of genomics-assisted breeding allows a considerable time and cost reduction in the development of crop varieties with improved nutritional contents as compared to conventional breeding [9]. Nevertheless, genomics-assisted breeding requires the clarification of the genetic basis of the target traits to be applied. Until now, studies on the genetic control of common bean seed composition have mainly focused on specific minerals such as iron and zinc. These studies used linkage mapping approaches, resorting to segregating populations from controlled crosses (e.g., [14,15]), or genome-wide association study (GWAS) approaches (e.g., [16,17]) to understand the genetic basis of these traits. However, only a reduced number of linkage and association mapping studies focused on macronutrients and protein quality. Examples are the study of Casañas et al. [18] on the genetic basis of ash, dietary fiber, starch, and protein among other traits, using a recombinant inbred line (RIL) population to perform a linkage mapping approach, or the Katuuramu et al. [19] study, on the genetic basis of protein and mineral contents using an association mapping approach.

Genome-wide association studies (GWAS) using populations of unrelated individuals to examine associations between genotypic polymorphisms and phenotypes are presently regarded as a good alternative to linkage mapping approaches to identify quantitative trait loci (QTL) responsible for complex traits variation [20]. Coupled with other recent advancements, such as the sequencing of reference genomes (e.g., the common bean genome [21]) or high-throughput next-generation sequencing (NGS) approaches, GWAS uncovers functional loci/genes underlying the genetic variation of complex traits with higher mapping resolution and broader genetic basis [22].

More than five centuries of bean cultivation in Portugal have produced a very diverse germplasm. Indeed, Portugal is considered a secondary center of origin of common bean diversity [23]. This germplasm diversity is expressed at agronomic, nutritional, and molecular levels [24,25] and proofed ideal for association genetic studies such as the identification of fusarium wilt resistance-associated candidate genes, and the identification of SNP alleles and candidate genes affecting photosynthesis under contrasting water regimes [26,27].

The present study aims to identify the genomic regions and/or candidate genes associated with common bean nutritional composition (ash, carbohydrates, fat, fiber, moisture, protein, and resistant starch contents) and protein quality (amino acid contents and trypsin inhibitor activity) within a diverse Portuguese germplasm collection, using a GWAS approach. We will test SNP-trait associations using linear mixed models accounting for the genetic relatedness between accessions and compare the SNP-traits association profile under two contrasting environments. By incorporating a heat stress environment in the study, we will be able to assess QTL stability under an expected climate change scenario for future exploitation of potential genotype × environment (G × E) for local adaptation. This study will also be useful for the development of molecular tools to facilitate routine evaluations on nutritional composition and protein quality, increasing the efficiency of common bean breeding for improved nutritional quality.

2. Results

The present study was carried out to clarify the genetic architecture of nutritional composition and protein quality-related traits in common beans by making use of the Portuguese germplasm natural variation. For that, previously collected data [25] on the nutritional composition and protein quality of a diverse collection of 106 Portuguese common bean accessions, cropped in two contrasting environments (Cabrela, central Portugal, with a mild climate, and Córdoba, southern Spain, a heat stress prone region), was complemented with the resistant starch quantification of the same samples prior to a genetic analysis through a genome-wide association study.

The total phenotypic data analyzed included nutritional composition related traits such as macronutrient contents (protein, carbohydrates (CH), fat, fiber, ash, moisture, and resistant starch (RS)), and protein quality-related traits such as amino acids contents (Ala-alanine; Arg-arginine; Asp-aspartic acid; Glu-glutamic acid; Gly-glycine; His-histidine; Ile-isoleucine; Leu-leucine; Lys-lysine; Met-methionine; Phe-phenylalanine; Pro-proline; Ser-serine; Thr-threonine; Tyr-tyrosine; Val-valine), and trypsin inhibitor activity (TIA). Traits regarding nutritional composition were measured in the two contrasting environments, whereas protein quality-related traits were only measured in the samples harvested in the most stressed environment, Córdoba.

This phenotypic data was then integrated with previously obtained single-nucleotide polymorphism (SNP)-based genotypic data (16,689 SNPs before quality control), screened through Illumina Infinium BARCBean6k_3 BeadChip™ assay and DArTseq™ analysis [26]. Genomic regions associated with the traits were highlighted, through GWAS, taking into consideration the population structure.

2.1. Phenotypic Trait Variation

Most nutritional composition-related traits showed a similar range of phenotypic variation between environments (Figure S1). Nevertheless, for all these traits, apart from fat, significant differences were detected between the two environments, and most traits showed higher coefficients of variation, or variability, in Córdoba than in Cabrela (Table 1). Fat and resistant starch showed the highest variability among the nutritional composition-related traits, in both environments.

Table 1. Average ± standard deviation (and coefficient of variation (%)) of ash, fat, fiber, carbohydrates (CH), protein, moisture, and resistant starch (RS) contents (g/100 g) in a collection of 106 Portuguese common bean accessions grown in two contrasting environments (Cabrela with a mild climate, and Córdoba, a heat stress prone region). Data calculated from best linear unbiased estimators (BLUEs). In each column different letters indicate significant differences ($p < 0.05$).

	Ash	Fat	Fiber	CH	Moisture	Protein	RS
Cabrela	3.16 ± 0.08 [a] (2.6)	1.44 ± 0.24 [a] (16.4)	5.75 ± 0.45 [a] (7.8)	60.57 ± 1.53 [a] (2.5)	13.55 ± 0.47 [a] (3.5)	21.28 ± 1.44 [a] (6.8)	30.74 ± 3.49 [a] (11.4)
Córdoba	3.25 ± 0.13 [b] (4.1)	1.49 ± 0.32 [a] (21.7)	6.77 ± 0.72 [b] (10.7)	56.65 ± 1.75 [b] (3.1)	14.48 ± 0.51 [b] (3.5)	24.14 ± 1.66 [b] (6.9)	45.23 ± 11.36 [b] (25.1)

Of the traits measured only in Córdoba, trypsin inhibitor activity showed the highest variability (29.6%). Among the amino acid contents, Methionine stood out showing the highest variability (21.2%). The remaining amino acid contents had all very similar coefficients of variation. The amino acid present, on average, in lower amounts was Methionine (1.06 g/100 g) and the one present in higher amounts was Glutamic acid (20.44 g/100 g) (Table S1).

Variance components were estimated for the nutritional composition-related data obtained from the Córdoba and Cabrela trials taken together. A high influence of the environment (E) was observed for the majority of the traits (Figure 1). The effect of E ranged from 0 to 70.4% and the effect of G×E ranged from 12.8% to 60%. From the seven analyzed traits, ash and fat were the only ones that did not have the environment as the largest variance component. The biggest variance component for ash was G×E (60%), followed by E (20.3%), whereas the biggest variance component for fat was the genotype (G) (51%), followed by G×E (39.7%). Fat was thus an exception as the remaining six traits had variance components values inferior to 9% for G. Since the G×E effect was generally larger than G effects on the phenotypic variability, subsequent analyses were performed separately for the two contrasting environments with the exception of fat.

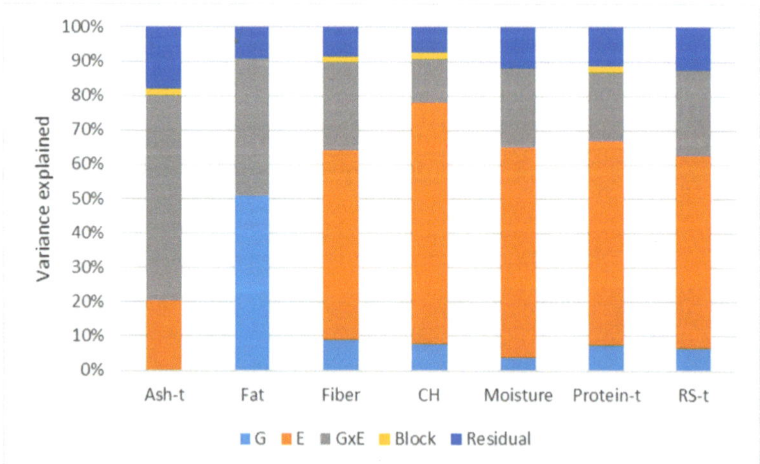

Figure 1. Variance components for the nutritional composition related traits measured in a collection of 106 Portuguese common bean accessions grown in two contrasting environments (Cabrela with a mild climate, and Córdoba, a heat stress prone region). The "-t" after the trait's name indicates that data was transformed following a Box-Cox transformation. Genotype (G), environment (E), genotype by environment interaction (G×E), block, and residual (error), carbohydrates (CH), resistant starch (RS).

Traits' broad-sense heritabilities (Tables S2–S4) were, in general high, with values between 53% and 98%. The nutritional composition-related traits and trypsin inhibitor activity showed, in general, higher heritabilities than the amino acid contents. Also, the nutritional composition-related traits showed higher heritabilities for Córdoba than for Cabrela environment. Wald tests (Tables S2 and S3) indicated that there were significant differences among genotypes and no block effects for most of the measured traits. Wald test for fat (Table S4), calculated with genotype and environment terms fixed, indicated that there were significant differences among genotypes but not between environments.

A strong negative correlation between carbohydrates and protein (Pearson correlation coefficient, $r = -0.97$, Cabrela; $r = -0.95$, Córdoba; Figures S2 and S3), and a moderately strong negative correlation between carbohydrates and ash ($r = -0.51$, Cabrela; $r = -0.63$, Córdoba), were observed. Considerable positive correlations were identified between protein and ash ($r = 0.62$, Cabrela; $r = 0.74$ Córdoba) and between moisture and ash ($r = 0.56$, Cabrela; $r = 0.40$ Córdoba). There were strong positive correlations among the different amino acid contents, but only small correlations between these and the trypsin inhibitor activity. The correlations between protein quality-related traits and nutritional composition-related traits were, overall, small (Figure S2).

With the best linear unbiased estimators (BLUEs or adjusted means), two principal component analyses (PCAs) were computed: one with the nutritional composition data from both environments, and another with the nutritional composition and protein quality data measured only at Córdoba environment.

The PCA with nutritional composition data collected in Cabrela and Córdoba environments (Figure 2A) depicted the impact of the environment on common beans' nutritional composition as well as highlighted interesting quality accessions considering the measured traits. The two first principal components explained 70.79% of the total variability. The biplot demonstrates a clear separation of environments, with accessions collected in Córdoba showing higher variability and, in general, higher contents of ash, protein, moisture, fiber, and resistant starch. Examples of bean accessions grown in Córdoba with high contents of ash and protein are accessions 1636 and 1644. Complementing these, accessions 587, 1952, 4049, 4073, 5367, and 5370 showed some of the highest contents of protein and fiber in Córdoba but also high contents of ash and resistant starch in the same environment. Accession 4100 showed one of the highest values of fat, the highest value of carbohydrates in Córdoba, and one of the lowest values for protein, ash, fiber, and moisture. Interestingly, despite having the highest content of fiber among samples grown in Córdoba and one of the highest contents of protein, accession 5370 showed the opposite phenotype in the samples grown in Cabrela where it had one of the highest contents of carbohydrates. On the other hand, accessions 4049 and 5367 showed some of the highest contents of ash, protein, and fiber in Cabrela, maintaining a similar nutritional composition in Córdoba. Accessions 600, 1631, 4081, 4100, and 5381, grown in Cabrela, showed the highest contents of carbohydrates overall.

The second PCA (Figure 2B) included protein quality-related traits in addition to the nutritional composition traits measured in common bean accessions grown in the heat stress environment (Córdoba). The first two principal components explained 64.94% of the total variability, with the first principal component explaining most of the variability (51.91%). This PCA showed that accession 5370 was among the accessions with the highest protein contents and with the highest content of the various amino acids. On the other hand, accession 5371 had even higher amino acid contents, but only average contents of protein and remaining nutritional composition-related traits.

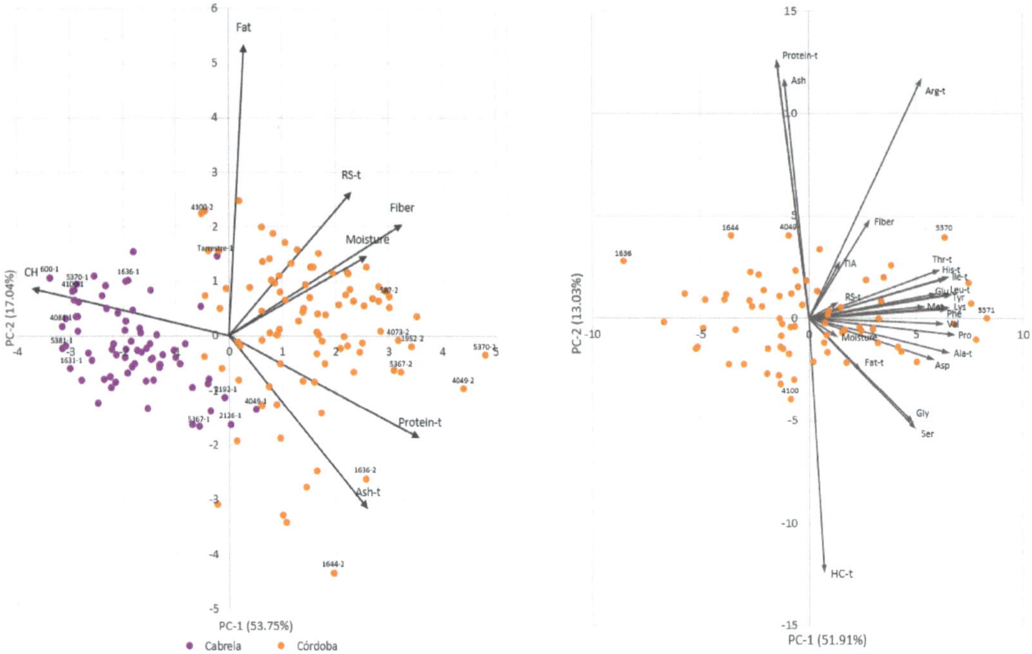

(A) (B)

Figure 2. Principal component analysis based on BLUEs of nutritional composition and protein quality-related traits measured in a collection of 106 Portuguese common bean accessions. (**A**) Biplot representing accessions grown in Cabrela (purple) and accessions grown in Córdoba (orange). Trait loading vectors of the seven nutritional composition-related traits are represented by arrows. Relevant accessions are identified by their accession numbers followed by 1 or 2 according to the corresponding environment: 1-Cabrela; 2-Córdoba. (**B**) Biplot representing accessions grown in Córdoba (orange). Trait loading vectors of the 24 nutritional composition and protein quality-related traits are represented by arrows. CH-carbohydrates; RS-resistant starch; Ala-alanine; Arg-arginine; Asp-aspartic acid; Glu-glutamic acid; Gly-glycine; His-histidine; Ile-isoleucine; Leu-leucine; Lys-lysine; Met-methionine; Phe-phenylalanine; Pro-proline; Ser-serine; Thr-threonine; Tyr-tyrosine; Val-valine; TIA-trypsin inhibitor activity. The "-t" after the trait's name indicates that data was transformed following a Box-Cox transformation.

2.2. *SNP-Trait Associations*

Accessions adjusted means for each trait were tested for association with SNP data taking population structure or familial relatedness into consideration. Manhattan plots depicting GWAS results for protein are shown in Figure 3, all the remaining traits' Manhattan plots are shown in Figures S4–S6. For most traits, the best model for association analysis, with an inflation factor closer to 1 (Table S5) and Q–Q plots showing fewer P-values deviating from the expected uniform distribution that holds under the null hypothesis (Figure S7), was the model using a different kinship matrix per chromosome. For His, Lys, Phe, Ser, and Thr a model using 15 principal components to control for population structure was used.

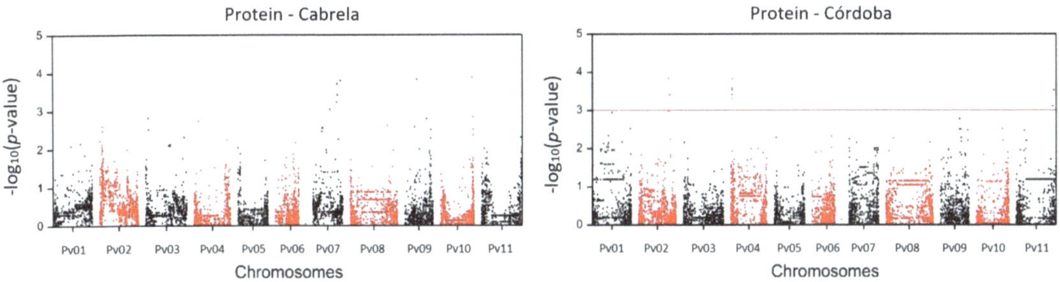

Figure 3. Manhattan plot depicting the genome-wide association results for protein content in common bean using 78 Portuguese accessions grown in the Cabrela environment (left) and 94 Portuguese accessions grown in the Córdoba environment (right). The y-axis represents the $-\log_{10}$ (*p*-value) of 9601 SNPs and the x-axis shows their chromosomal positions across the common bean genome. The horizontal red line indicates the significance threshold (*p*-value = 10^{-3}).

A total of 224 marker-trait associations were identified for the 24 nutritional composition- and protein quality-related traits studied, with the defined threshold $-\log_{10}$ (*p*-value) = 3. The conservative Benjamini-Yekutieli *p*-value adjustment supported the significance of 151 (67%) of these associations (Table S6). Marker-trait associations were found for all traits except for Glu, Gly, His, Ile, Leu, Met, Pro, and Thr. A total of 181 unique SNPs was responsible for the 224 marker-trait detected associations, indicating that some of the SNPs were associated with more than one trait. The 181 associated SNPs were organized in 136 unique genomic regions, each genomic region englobing the markers within a linkage disequilibrium (LD) block. One hundred and five SNP markers were significantly associated with the nutritional composition traits measured in Cabrela and 37 in Córdoba, being one SNP marker (DART03724) commonly associated with the same trait (moisture) in both environments. Moisture was the trait with the biggest number of associations detected in one environment (62 in Cabrela but only 6 in Córdoba). However, a significant portion of the SNPs associated with moisture in Cabrela was located within the same LD blocks (62 SNPs were located within 36 genomic regions). Among protein quality-related traits, Arg content had the biggest number of associations detected (32). However, similarly to moisture, a large portion of these SNPs was within the same LD blocks (32 SNPs were located within 13 genomic regions). The association of the same SNP with different traits occurred mainly among amino acid content traits, being eight markers associated with more than one trait. SNP04308 and SNP09201 were the markers associated with the highest number of traits, each associated with four different amino acid contents. Except for Arg, out of the eight amino acid traits with associated markers, all shared markers with another trait.

Marker-trait associations were identified across all the common bean chromosomes. Amino acid contents, in particular, were mostly associated with markers belonging either to chromosome Pv02 or Pv09.

Most marker-trait associations explained only a small percentage of the observed phenotypic variance, with an average of 16.2%. Moisture (measured in Cabrela) was an exception to this, with markers explaining up to 70.6% (DART04597) of the observed phenotypic variance. Other SNPs explaining larger proportions of variance, besides the previously referred, were SNP01084 (45.6%; resistant starch, Cabrela), DART02462 (31.3%; fat, both environments), SNP01413 (29.7%; ash, Cabrela) DART11240 (24.0%; carbohydrates, Cabrela), and SNP00787 (24.1%; Arg, Córdoba) (Table S6). The variant allele for 64% of the associated SNP markers had a positive effect on the trait in this association panel.

2.3. Candidate Gene Identification

A gene was considered a putative candidate for the phenotypic trait under analysis if it contained an associated SNP or was in linkage disequilibrium (LD) with an SNP associated with the trait, observing a strict LD-decay threshold ($r^2 > 0.2$). This was investigated using the JBrowse tool in *Phaseolus vulgaris* v2.1 genome in Phytozome v12 portal. One hundred and two different candidate genes were identified within 136 genomic regions responsible for trait variation.

The gene-trait network (Figure 4) established for the 102 identified candidate genes showed a clear separation between groups of traits and environments. Candidate genes for protein quality-related traits did not connect to candidate genes for nutritional composition-related traits. Similarly, with one exception, candidate genes for traits measured in Cabrela did not connect to candidate genes for traits measured in Córdoba. The exception to this was gene Phvul.004G045900, encoding for a galacturonosyltransferase 9, which linked moisture measured in Córdoba to moisture, protein, and carbohydrates measured in Cabrela. Most connections of the same gene to various traits occurred among amino acids, with four different genes connecting three different amino acids. There were also various genes connected to both carbohydrates and protein, in both environments.

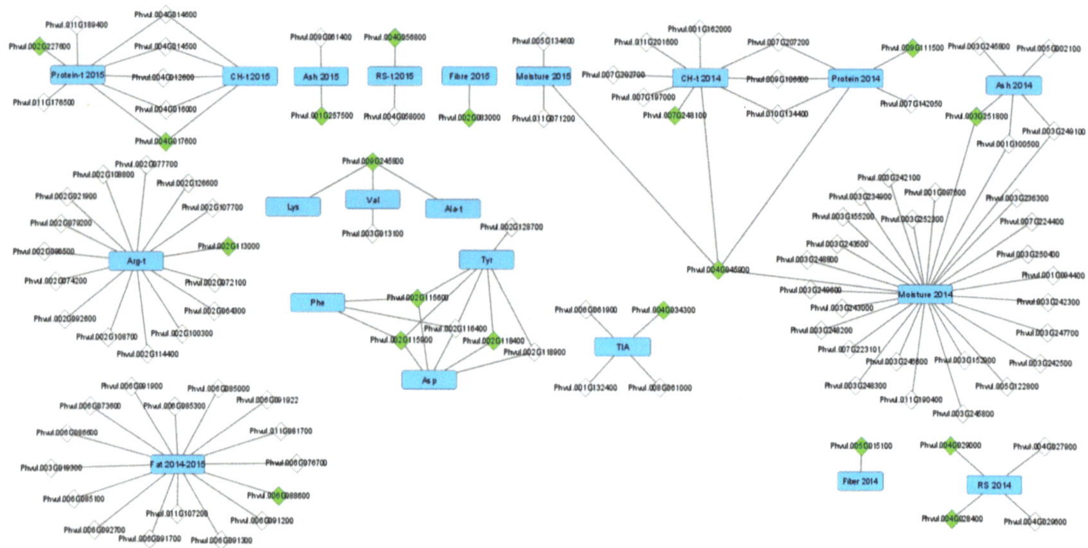

Figure 4. Network analysis of the candidate genes proposed for the nutritional quality traits using 106 Portuguese common bean accessions grown in two contrasting environments (Cabrela and Córdoba), using Cytoscape software. Traits represented as rectangles; genes represented as diamonds. Genes identified by green diamonds correspond to candidate genes for the SNP markers associated with the highest P-value for each trait. Traits are identified as: 2014-Cabrela; 2015-Córdoba; 2014–2015 — both environments; "-t" — trait data transformed following a Box-Cox transformation. CH-carbohydrates; RS-resistant starch; Ala-alanine; Arg-arginine; Asp-aspartic acid; Lys-lysine; Phe-phenylalanine; Tyr-tyrosine; Val—valine; TIA-trypsin inhibitor activity.

Functional categorization of the candidate genes was obtained using MapMan (Mercator) web tools to better understand the involvement of the candidate genes in different metabolic pathways. From the candidate genes identified, 41.7% had a functional category assigned (Figure 5, Table S7). The assigned categories showed some diversity within each trait and a high diversity overall, with 15 different categories assigned to the 102 candidate genes. The most common functional categories attributed by MapMan across all traits were "enzyme classification" (12%), "RNA biosynthesis" (4.6%), "RNA processing" (3.7%), and "vesicle trafficking" (2.8%).

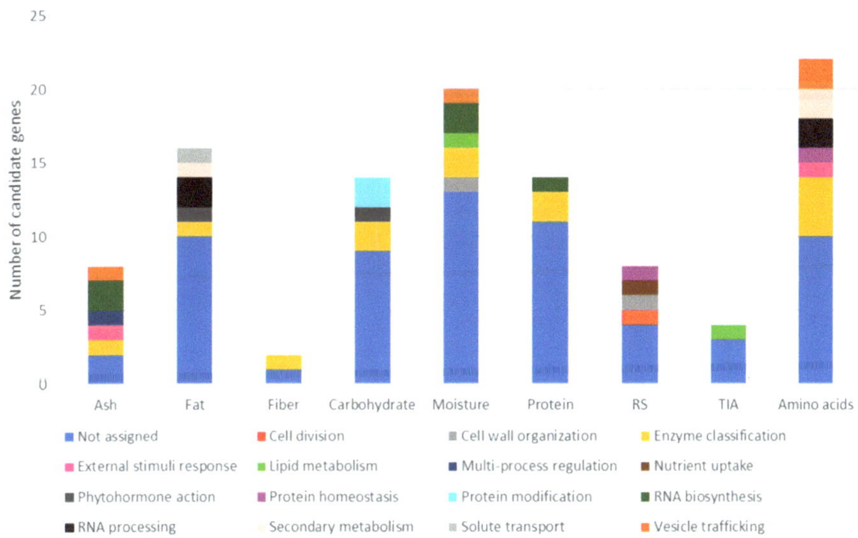

Figure 5. MapMan functional categories of the candidate genes associated with the nutritional composition and protein quality-related traits in 106 Portuguese common bean accessions grown in two contrasting environments. Candidate genes for amino acid contents were pooled together. The nine bar charts represent the number of candidate genes of a given functional category associated with ash, fat, fiber, carbohydrate, moisture, protein, resistant starch (RS), trypsin inhibitor activity (TIA), and amino acid contents.

In the frame of this work, it was not possible to highlight all candidate genes located within the associated genomic regions in detail. We, therefore, restrict ourselves to highlight those that were (1) located within regions associated with multiple quality-related traits, and with (2) a biological annotation related to the studied trait. The candidate gene associated with the biggest number of quality related-traits, which also had a biological annotation related to the traits, was Phvul004G045900, encoding a galacturonosyltransferase 9, which was associated with carbohydrates, protein, and moisture (in both environments). Another candidate gene highlighted due to its association with more than one quality-related trait was Phvul010G13440, encoding a hydroxyproline-rich glycoprotein family protein, which was associated with protein and carbohydrate contents. Finally, the genes highlighted due to their biological annotation were Phvul004G056800, encoding an ankyrin repeat family protein, Phvul009G061400, encoding a cytochrome P450, family 82, subfamily C, polypeptide 4, and Phvul002G113000, encoding a transmembrane amino acid transporter family protein, respectively associated with resistant starch, ash, and Arg.

3. Discussion

This study explored the natural variation of 106 accessions from the highly diverse and underused Portuguese common bean germplasm grown under contrasting environments (traditional and heat stress), using a GWAS approach to unveil the genetic architecture of 24 nutritional compositions and protein quality-related traits. The generated knowledge will allow a better understanding of the molecular mechanisms and pathways regulating the nutritional composition and protein quality in common beans. Further, it will assist the development of promising molecular tools to help breeders answer consumers' diet concerns and to support farmers to better exploit G×E quality interactions under future climate constraints. A total of 136 common bean genomic regions controlling the natural variation of the analyzed traits were identified in this association panel. Additionally, 102 putative candidate genes for the trait-associated regions were proposed.

3.1. Genomic Regions and Candidate Genes Associated with Common Bean Nutritional Composition and Protein Quality-Related Traits

SNP marker-trait associations were identified for all the nutritional composition and protein quality-related traits analyzed except for Glu, Gly, His, Ile, Leu, Met, Pro, and Thr. A mean of 9.5 marker-trait associations was identified per trait, each SNP explaining, on average, a small percentage of the phenotypic variation (around 16%). Mapping resolution to the gene level was achieved in 55.9% of the cases (LD blocks where a single gene was identified), demonstrating the complex genetic nature of common bean nutritional quality-related traits and validating the use of GWAS to harness the diversity of this Portuguese common bean germplasm.

The protein content is one of the most relevant food grain legume traits for breeding, as the interest in plant-based protein increases in developed countries to provide healthier diets, and the need for cheap protein sources to fight malnourishment in developing countries remains [28,29]. Previous studies identified Quantitative Trait Loci (QTL) for seed protein content on common bean chromosomes Pv05 and Pv07 using a Xana × Cornell 49242 RIL population, with parental lines belonging to the Andean and Mesoamerican gene pools, respectively [18], and on Pv03, Pv06, and Pv07 using a subset of the Andean Diversity Panel [19]. Our study identified significant marker-trait associations for protein content on chromosomes Pv02, Pv04, Pv07, Pv09, Pv10, and Pv11. The LD blocks around the five associated markers identified in Pv07 (DART06714, DART06845, DART06856, DART03216, and DART03273) were such that each marker was located in an independent genomic region. Of these five marker-trait associations one was at 23Mb, three between 32 and 33 Mb, and the fifth at 37 Mb. The QTL identified on Pv07 by Casañas et al. [18] was located 5Mb away from the QTL identified by Katuuramu et al. [19] which was located at 7.6 Mb. Therefore, both previously identified QTLs were located more than 10 Mb away from the presently identified QTLs, and therefore out of our LD windows. This suggests that in common bean several regions control seed protein concentration on chromosome Pv07. The different genotypic resources used between these three genetic studies may probably explain these findings. The use of association mapping populations, such as the one used in the present study, characterized by unrelated accessions of Andean, Mesoamerican, and of admixed genetic origin, or of Andean origin as in Katuuramu et al. [19], allows the exploration of a larger allelic diversity (broader genetic basis), with higher mapping resolution. This contrasts with the use of bi-parental linkage mapping populations which have a narrower genetic basis, resulting in a smaller potential identification of genomic regions associated with the trait of interest [22]. This might explain the smaller amount of genomic regions associated with protein content identified by Casañas et al. [18] using a RILs population developed from the cross of only two parental lines although from the different Andean and Mesoamerican gene pools.

Several of the markers associated with protein content were simultaneously associated with carbohydrate content (in both environments) but with contrasting effects on trait variation, reflecting the expected strong negative correlation between these related traits. One of these markers was SNP04726, which was the second most significantly associated with protein content (explaining 9.4% of the variability) and was also associated with carbohydrates content, explaining 10.8% of its variability (data from Cabrela field trial). A candidate gene including this marker sequence was Phvul.010G134400, which encodes a hydroxyproline-rich glycoprotein family protein (HRGP). HRGPs are known to accumulate in cell walls as a general response of dicotyledons to infection by biotrophic and necrotrophic fungi, bacteria, and viruses [30]. In particular, this reaction was described in common beans as a response to infection by the causal agent of anthracnose [31]. Selecting for this marker/gene candidate could allow an increase in protein content to the detriment of the carbohydrate content in new bean varieties. Nevertheless, several of the other markers associated with these traits in the present study were only associated with either protein or carbohydrate contents. Therefore, selecting for those markers could allow an increase or decrease of the contents of either trait (protein and carbohydrates) independently of the other trait.

Apart from the genes associated with both protein and carbohydrates contents, very few other SNPs/candidate genes related to nutritional composition were connected to more than one trait, reflecting the small correlations among most nutritional composition-related traits. Indeed, most SNPs and subsequent candidate genes for all the studied nutritional quality-related traits were only associated with one trait, as can be seen in the network analysis of candidate genes (Figure 4).

Of note, among the genes connected to more than one trait, is Phvul.004G045900, which encodes a galacturonosyltransferase 9. Two markers led to the identification of this gene, DART03724 which was associated with carbohydrate contents measured in seeds from Cabrela and moisture contents measured in both environments, and DART06845 which was associated with protein content measured in seeds from Cabrela. This gene was the only identified candidate associated with more than three traits, the only gene connected to the same trait in both environments, and with a corresponding variant allele responsible for the largest effect on trait variability for carbohydrates, protein, and moisture (data from Córdoba). Galacturonosyltransferases are required for the synthesis of pectin, a family of complex polysaccharides present in the cell walls of all land plants [32]. Moreover, pectin is involved in the regulation of ion transport and porosity of cell walls and is consequently involved in the control of cell wall permeability and determination of water holding capacity [33]. Therefore, galacturonosyltransferase 9 seems to be an interesting candidate for the variation of both carbohydrate and moisture contents in common bean seeds.

Moisture measured in seeds from Cabrela showed the biggest number of marker-trait associations detected (62) and the largest percentage of variation explained by a marker (70.6%). However, a significant portion of the associated SNPs was located within the same LD blocks, as 62 SNPs were located within 36 genomic regions. Two chromosomes, Pv07 and Pv03, are of particular interest and could be responsible for most of the observed variation. Pv07 contains one genomic region which includes the marker responsible for 70.6% of the moisture phenotypic variation. Pv03 contains most of the markers associated with moisture (43) as well as the marker associated with this trait at the highest statistical significance. Most of the associated markers within this chromosome (37) were enclosed within a relatively small section of the chromosome (3Mb) and were thus relatively close physically. Despite the physical closeness, 25 genomic regions (LD blocks) were still detected within this section of the chromosome suggesting that various QTL may be responsible for the phenotypic variation of moisture within this chromosomal region. Moisture from Cabrela contrasted with most other traits in the amount of explained variability per QTL, including moisture measured in seeds from Córdoba. In the Córdoba environment, moisture was associated with six markers, each explaining less than 9% of the phenotypic variance. The difference in the proportion of variance explained between environments was mainly due to higher effects of each marker on the phenotypic variation in Cabrela (up to 0.7) than in Córdoba (up to 0.3). This difference might indicate some inflation of the effect attributed to each marker. This inflation could be explained by the relatively small mapping population used since smaller populations lead to inflated QTL effects [34].

QTLs for ash, fiber, and starch in common bean were also identified through linkage mapping in the previously mentioned study of Casañas et al. [18]. In that study, QTLs for ash were identified on chromosomes Pv01 and Pv07, for fiber on Pv06 and Pv07, and for starch on Pv01, Pv02, Pv04, and Pv07. Similarly, our study identified a marker-trait association for ash on chromosome Pv01, and for resistant starch on Pv01 and Pv04. A candidate gene for DART03741, associated with resistant starch on Pv04, is Phvul.004G056800, encoding an Ankyrin repeat family protein. This family of proteins includes members implicated in carbohydrate allocation and has been associated with reduced starch in tobacco transformants carrying a specific ankyrin repeat family gene [35].

Also associated with ash content was the SNP03984 in Pv09. A candidate gene for this associated SNP is Phvul.009G061400 which encodes a cytochrome 450, family 82, subfamily C, polypeptide 4 (CYP82C4). CYP82C4 is a heme-containing enzyme that is strongly correlated with genes involved in metal uptake/transport and is possibly involved in the early iron deficiency responses in *Arabidopsis thaliana*, as was proposed by Murgia et al. [36]. The involvement of this gene in the uptake/transport of iron becomes particularly interesting when considering that iron deficiency is the most common and widespread nutritional disorder in the world [36].

The strong correlations among amino acid contents were reflected by the association of several SNPs to more than one of these traits. As an example, four molecular markers and the respective candidate genes were associated with three different amino acid contents: DART01478, SNP00739, SNP00741 were associated with Asp, Phe and Tyr; and SNP04308 was associated with Ala, Lys, and Val contents. Interestingly, Arg was the only amino acid content that did not share SNPs and subsequently candidate genes with other amino acid contents. Of the several SNP markers associated with Arg, the one within the candidate gene Phvul.002G113000 was the most strongly associated and the one that explained the highest proportion of variance (20.2%). This gene encodes a protein of the transmembrane amino acid transporter superfamily. These are integral membrane proteins involved in the absorption of amino acids from the soil, load and transport of amino acids in the phloem, absorption of amino acids in seeds, and long-distance transport and distribution of amino acids in seeds [37]. Functional analysis of this gene would be relevant to understand the usefulness of this gene and whether the content of Arg is the only affected by this variant allele, as the marker has not been considered as associated with the other amino acid contents due to the chosen $-\log_{10}$ (P) threshold.

3.2. Portuguese Common Bean Germplasm Nutritional Quality Richness and the Influence of Environment

As has been previously shown, the environment, in particular heat and drought stress, influence the maturing process of seeds, affecting various nutritional quality traits in the process (reviewed in [38]). Accordingly, environment and G×E had large effects on the nutritional quality traits variation observed in the Portuguese common bean germplasm collection and scored across field trials. Common bean accessions grown in Córdoba (heat stress environment) showed higher variability, as well as higher average contents of ash, fiber, moisture, protein, and resistant starch than accessions grown in Cabrela (milder environment). On the other hand, carbohydrates and moisture contents were on average lower in Córdoba than in Cabrela. As discussed by Mecha et al. [25], despite the general connection between reduced yield and heat stress [39], no differences in yield were observed in the two environments used in this study, likely due to the presence of artificial irrigation. Therefore, the differences in nutritional composition observed between environments were likely due to heat stress. In particular, heat stress has been associated with alterations in carbohydrate metabolism, affecting synthesis but also the accumulation of carbohydrates during seed filling [38,40], which could explain the reduction of carbohydrate contents in the seeds grown in Córdoba. Unlike the remaining nutritional composition traits, fat was the only trait with a reduced effect of the environment on trait variability.

In addition, G×E was also significant on the traits analysed in both environments, particularly for ash and fat contents. G×E occurs when genotypes differ in their relative performance across environments and corresponds to the presence of genetic factors with environment-specific effects [41]. The magnitude of G×E varies among traits and accessions and in some cases can lead to crossover interactions as can be seen in the present study with accession 5370, for example, which showed one of the highest contents of protein when grown in Córdoba and one of the lowest contents of protein when grown in Cabrela. Another example is accession 1636, which showed one of the highest contents of fat when grown in Cabrela and lowest contents of fat in Córdoba, and the opposite concerning ash (one of the highest contents of ash in Córdoba and one of the lowest in Cabrela). In other cases, crossover interactions do not occur, and the relative performance of the genotypes is maintained across environments, as can be seen with accession 4049, for example, which showed one of the highest contents of protein, ash, and fiber in both environments.

As previously stated, DART03724 with Phvul.004G045900 as a putative candidate gene, was the only molecular marker/gene associated with the same trait (moisture) in both environments. The lack of more markers stably associated with traits across environments is a consequence of the environment-specific nature of the genomic regions associated with the studied traits. Constitutive QTLs, which show consistent effects across environments, are the main targets for breeding programs as they can improve crop performance across various regions where the crop can be grown. Nevertheless, it is possible to take advantage of significant G×E by breeding for local adaptation [41] and also increase the efficiency of this selection using the specifically associated molecular markers instead of performing time-consuming and expensive phenotyping. Examples of promising associated markers explaining considerable amounts of trait variability are DART11240 for carbohydrates content (in Cabrela) and SNP00732 for Arg (in Córdoba) that explained 24.0% and 20.2% of these traits' phenotypic variation. The fact that the great majority of the presently identified QTLs are environment-specific indicates that breeding efforts for nutritional quality traits in the Portuguese common bean collection should focus on developing varieties adapted to location-specific growing conditions, as proposed by Vaz Patto and Araújo [8]. The higher variability within traits observed in the stressed environment of Córdoba highlights the genetic richness of the Portuguese germplasm collection and its potential for location-specific breeding, namely for production under heat stress conditions.

Under these more stressful conditions, particular Portuguese accessions stood out as promising sources of protein quality-related traits. Traits such as bioactive compounds that influence protein digestibility and limiting amino acids are factors used to determine protein quality [42]. Limiting amino acids are the essential amino acids, or the amino acids that must necessarily be provided by the diet, that are in shortest supply [6]. As expected in a legume species, Met was the limiting amino acid (1.06 g/100 g) in the Portuguese common bean accessions analyzed. Pearson correlations showed that amino acid contents were strongly correlated among themselves and weakly correlated with TIA and the nutritional composition traits. As a consequence of the strong correlation among amino acids contents, targeted conventional breeding for increased Met should not be feasible. This can be explained by the low Met content of phaseolin, the main storage protein of common bean (40–50% of the total protein content), which despite being deficient in Met is still the major source of this amino acid [43]. Montoya et al. [43] proposed instead the exploration of the natural variability of phaseolin in terms of their protein digestibility, to increase the availability of Met and improve the protein quality of common bean. Although highly reduced by the processing, namely boiling (reviewed in [42,44]), the presence of trypsin inhibitors compromises protein digestibility and amino acid absorption, thus affecting protein quality (reviewed in [6]). In general, the accessions that showed in the present study the highest contents of protein and amino acids were also the accessions with the highest TIA. For example, accession 5371 displayed one of the highest contents of amino acids but also high contents of TIA. However, the reduced correlation of TIA with the remaining protein quality-related traits at the genetic level (different genomic regions involved in their control) suggests the possibility of altering the contents of amino acids and TIA independently through conventional breeding, which is especially facilitated by the help of the detected associated molecular markers.

Due to the detailed and expensive nature of the analysis developed to measure some of the nutritional quality traits, the size of the population and the number of tested environments had to be restricted in the present study. Smaller populations lead to lower QTL detection power as well as inflated estimates of QTL effects [34]. Nevertheless, the relatively high heritability observed in the analyzed traits compensates for the small population size to some extent, as the power for detecting QTLs is a function of population size x heritability of the trait [34]. Nevertheless, we were able to detect the most relevant

QTL for breeding in this association panel, as QTL with minor effects are the ones harder to detect [34]. Additionally, due to the previously referred constraints, protein quality-related traits were only measured in one of the environments. This decision was based on the higher variability observed in that environment but allowed us also to collect phenotypic data in the heat stress environment mimicking future climate changes expected in the Mediterranean region.

In conclusion, this study further characterized the Portuguese common bean germplasm through the clarification of its genetic architecture of nutritional quality traits. The functional categorization of the proposed 102 candidate genes for 24 nutritional composition and protein quality-related traits demonstrated the involvement of a variety of metabolic pathways in the determination of common bean nutritional quality, corroborating the genetic complexity of these traits. Additionally, this study provided a unique resource of molecular markers associated with common bean nutritional and protein quality traits, which will help to answer consumers' nutritional demands as well as broader vs. local adaptation on future breeding efforts. In particular, the inclusion of data from Córdoba, a heat stress environment, allowed the identification of markers relevant for quality breeding in a context of climate change with increasing temperature scenarios, such as the one being experienced in the Mediterranean area.

4. Materials and Methods

4.1. Plant Material and Growing Conditions

A collection of 106 common bean accessions, from the Portuguese plant germplasm bank (BPGV, INIAV, Braga, Portugal) was used for the present genetic study. This collection was the same as described by Leitão et al. [26,27] to identify genomic regions controlling fusarium wilt resistance and photosynthetic efficiency-related traits in common bean. Based on genotypic data, it is known that this collection is mainly composed of accessions belonging to the Andean gene pool and a smaller proportion to the Mesoamerican gene pool. Additionally, one-third of the accessions have an admixed origin and might represent putative hybrids between the Andean and Mesoamerican gene pools [24].

Seeds from the bean accessions were sown in two different environments following a randomized complete block design, with two replicates at each environment, as described by Mecha et al. [25]. The field trials were developed in two different years and locations. The first field trial took place from May to September 2014 in Cabrela, Portugal, and the second from March to July 2015 in Córdoba, Spain. Cabrela represents a standard common bean production area in Portugal, characterized by temperature average ranges of 18–21 °C and 66–80% of relative humidity during the growing season, whereas Córdoba represents a heat stress prone area, characterized by temperature average ranges of 15–32 °C and 31–63% of relative humidity during the growing season. The two field trials were established under artificial irrigation. Mature dried seeds were collected from a total of 106 accessions, 66 in both environments, 12 exclusively in Cabrela (environment with a total of 78 accessions) and 28 exclusively in Córdoba (environment with a total of 94 accessions). The mature dried seeds were milled to a particle size of 0.8 mm and stored at −20 °C until chemical analysis.

4.2. Phenotypic Data Acquisition

Twenty-three traits related to nutritional composition and protein quality were measured in the common bean harvested from the two field trials as described by Mecha et al. [25] and retrieved for the present genetic association study. Total protein, total carbohydrates (CH), fat, fiber, moisture and ash contents, were measured in the samples from Córdoba and Cabrela, and the contents of 16 different amino acids (Ala-alanine; Arg-arginine; Asp-spartic acid; Glu-glutamic acid; Gly-glycine; His-histidine; Ile-isoleucine; Leu-leucine; Lys-lysine; Met-methionine; Phe-phenylalanine; Pro-proline; Ser-serine; Thr-threonine; Tyr-tyrosine; Val-valine) and trypsin inhibitor activity (TIA), were measured only in the samples from the most stressful environment (Córdoba). In addition, in the present study, resistant starch (RS) content was measured in all the harvested samples, completing a total of 24 traits analyzed.

Briefly, Mecha et al. [25] determined total protein, fat, fiber, moisture and ash (%) content using a near-infrared (NIR) analyzer (MPA; Bruker, Billerica, MA, USA). Total carbohydrates were calculated following Equation (1):

$$\text{total carbohydrates} = 100 - (\text{total protein} + \text{total fat} + \text{moisture} + \text{ash}). \quad (1)$$

A LC-MS/MS system Waters Alliance 2695 HPLC system coupled to a triple quadrupole mass spectrometer, Micromass®Quattro micro API (Micromass, Waters, Milford, MA, USA), equipped with an electrospray ionization source (ESI) was used to determine the content of the 16 amino acids [25]. Protein was hydrolyzed using a solution of HCL 6 M with 0.1% of phenol and the final amino acid extract resulted from the resuspension of the hydrolysates in HCL 0.1M after the evaporation to dryness of the initial solution.

The chromatographic separation was performed in a Mediterranean Sea 18, 5 μm 20 × 0.21 cm, 1.8 μm, (Teknokroma®, Barcelona, Spain) column. The amino acids were analyzed by multiple reaction monitoring (MRM) mode, using an ESI source operating in ion positive mode. Amino acids were identified by comparison with the amino acids' standard retention time and corresponding m/z values.

Trypsin inhibitors were extracted from 0.5 g of common bean flour to which 25 mL of NaOH 0.01M were added, and the pH was adjusted to 9.5 ± 0.1. Inhibition percentage and trypsin inhibitor activity were calculated according to Mecha et al. [25]. Resistant starch was quantified following the methods AACC 32-40.01 [45] and AOAC 2002.02 [46]. The method was performed using a Resistant Starch Assay Kit (K-RSTAR, Megazyme, Bray, Ireland). A buffer solution of sodium maleate (pH 6.0), containing pancreatic α-amylase and amyloglucosidase, was added to the thawed samples of common bean flour. The samples were incubated in a water bath with horizontal agitation (100 rpm) at 37 °C for 36 h. The reaction was interrupted with the addition of ethanol (99%). The solution was centrifuged at 3000 rpm for 10 min, and the pellet, containing resistant starch, was washed two additional times with ethanol (50%) followed by centrifugation. The resulting pellet was resuspended in potassium hydroxide (2M) under continuous agitation in an ice and water bath for 20 min. The solution was then neutralized with a buffer solution of sodium acetate (pH 3.8). The existing starch was hydrolyzed to glucose through the action of the amyloglucosidase in a water bath at 50 °C for 30 min. The samples were centrifugated, and two 0.1 mL aliquots were collected from the liquid phase. Simultaneously, a blank sample was prepared with 0.1 mL of sodium acetate 0.1 M (pH 4.5) as well as four standard glucose solutions with 0.1 mL of glucose solution (1 mg/mL). For glucose quantification, through spectrophotometry, 3 mL of glucose oxidase/peroxidase reagent were added to each tube, followed by a 50 °C incubation for 20 min. The absorbance of the samples was evaluated at 510 nm against the blank sample. The average glucose content of each sample was compared to the absorbance values of the standard glucose solutions to obtain the concentration of resistant starch.

4.3. Phenotypic Data Analysis

Quality control of phenotypic data was performed separately for each environment. Technical repetitions were averaged for each accession to minimize technical error. A descriptive statistical analysis was performed using the summary statistics option of Genstat®software, 21st edition [47]. Histograms and boxplots were generated per trait to analyze data distribution and to identify outliers. The normality of residuals was assessed for each trait using the Shapiro-Wilk test. A Box-Cox transformation was applied when needed to meet normality assumptions.

A linear mixed model was fitted per trait as trait = genotype + block + error for the analysis of traits measured in a single environment and as trait = genotype + environment + genotype x environment + block + error for the analysis of traits measured in both environments using the restricted maximum likelihood (REML) variance component analysis framework of Genstat software, where environment identifies the two field trials and block identifies the two plot replicates within each trial.

Models were initially fitted with all terms as random to obtain the best linear unbiased predictors (BLUPs), estimate variance components, broad-sense heritability and Pearson correlation coefficients between traits. In a second step, genotype and block were fitted as fixed terms to obtain the best linear unbiased estimates (BLUEs), for each trait and accession. BLUEs were determined for fat with genotype and environment fitted as fixed terms. Wald tests for the significance of fixed effects were performed. BLUEs were used for principal component analysis (PCA) and as input phenotypic data for GWAS.

To estimate how much of the variation of accessions' nutritional composition could be explained by the environment or the interaction with the environment, the previously defined model was fitted considering environment and genotype × environment as fixed terms, and a Wald test was performed to test for the significance of the fixed effects.

4.4. Genotypic Data

4.4.1. Association-Mapping Analysis

Genome-wide association studies were conducted for all the 24 traits using the QTL library procedures from Genstat software. Adjusted means (BLUEs) for each trait were tested for association with a previously collected genotypic dataset [26] retrieved for the present study. This genotypic dataset was constituted of 9601 single nucleotide polymorphisms (SNPs) after quality control (removal of SNP markers and accessions with >25% missing data, as well as SNPs with a minor allele frequency <0.01 from the 16,689 SNPs originally screened) and was obtained through two different approaches, the Illumina Infinium BARCBean6k_3 BeadChipTM assay and DArTseqTM analysis [26]. SNPs called heterozygous were set as missing data. Furthermore, some accessions were not phenotyped for some of the traits. Thus, association mapping was performed using 72 accessions for the amino acid contents and trypsin inhibitor activity and 94 accessions for the remaining traits measured in seeds from Córdoba; and for 78 accessions for all the traits measured in Cabrela.

GWAS was performed separately for Cabrela and Córdoba environments for the traits that showed a higher variance component of genotype by environment (GxE) interaction than of genotype (G). Otherwise, the association study was performed with data from both environments together. GWAS was performed in the mixed model framework of Genstat software, fitting SNP as fixed and genotype as random terms using REML [48]. Four models were tested to detect significant marker-trait association: a null model [Phenotype = SNP + Error], which does not account for any population structure or familial relatedness; a model accounting for population structure (Q) [Phenotype = Q + SNP + Error], using 15 principal components from the principal component analysis (PCA); and two models accounting for familial relatedness (K) [Phenotype = SNP + genotype + Error], one with genotype random effects structured following a kinship matrix K [48,49]; and another using a different kinship matrix calculated for each chromosome using only the SNPs located on the remaining 10 chromosomes, as proposed by Cheng et al. [50]. The kinship matrices to account for familial relatedness per chromosome among genotypes were previously calculated by Leitão et al. [26] and retrieved to perform the present association studies. Inflation factor values near 1 and quantile-quantile (Q–Q) plots of the respective p-values with lower deviations from the expected uniform distribution under the null hypothesis were the considered parameters to select the best model accounting for genetic structure/relatedness among genotypes.

The observed $-\log_{10}$ (p-value) of each SNP was plotted against its chromosomal position to produce a Manhattan plot. Significant SNP-trait associations were detected at a threshold of $-\log_{10}$ (p-value) = 3. This threshold was set taking into consideration two aspects: the size of the association panel used and the background noise of the obtained Manhattan plots. Similar criteria were already described in other works with comparable panel sizes and a similar number of markers, focusing on resistance/tolerance traits [26] to avoid losing potentially interesting regions while applying a conservative type of adjustment such as Bonferroni correction. However, as a "conservative" guidance, adjusted *p*-values following the Benjamini and Yekutieli (BY) false discovery rate (FDR) method [51] were calculated, with $\alpha = 0.2$ and $k = 520$ (the effective number of independent tests was set as the number of LD blocks per chromosome [52]), to control type I errors due to multiple testing.

For every SNP significantly associated with a trait, the effect of the minor-frequency SNP variant was calculated. The proportion of variance explained by each SNP-trait association was estimated using the formula V_{QTL}/V_{pheno}, where $V_{QTL} = 2freq(1\text{-}freq)effect^2$ and V_{pheno} is the phenotypic variance of the adjusted means of each trait [53].

4.4.2. Candidate Gene Identification

A gene was considered a putative candidate for the phenotypic trait under analysis if it contained an associated SNP or was in linkage disequilibrium (LD) with an associated SNP observing a strict LD-decay threshold ($r^2 > 0.2$). LD was previously calculated for each common bean chromosome using the squared coefficient of the correlation between marker pairs r^2 [26], and retrieved for the present study. Neighbouring SNPs showing $r^2 > 0.2$ in relation to the associated SNPs were considered to be within the same LD block or genomic region. Putative candidate genes were searched for using the JBrowse tool in the *Phaseolus vulgaris* v2.1 genome (DOE-JGI and USDA-NIFA, http://phytozome.jgi.doe.gov/, accessed on 16 December 2021), available at the Phytozome v12 portal [54]. The annotation of the candidate genes was obtained from the file "Pvulgaris_442_v2.1.annotation_info.txt", available in the previously referred portal.

Candidate genes were also assigned to MapMan bins, which described biological contexts/concepts, using Mercator4 V2.0 [55]. Cytoscape software [56], version 3.8.2, was used to visualize the candidate genes associated with each trait as a network.

Supplementary Materials: The following are available online at https://www.mdpi.com/article/10.3390/plants11010026/s1, Figure S1: Histograms of the 24 nutritional composition and protein quality traits measured in a collection of 106 Portuguese common bean accessions. Figure S2: Pearson's correlations between the 24 nutritional composition and protein quality traits measured in the seeds of a collection of 72 Portuguese common bean accessions grown in the Córdoba environment. Figure S3: Pearson's correlations between the seven nutritional composition traits measured in the seeds of a collection of 78 Portuguese common bean accessions grown in the Cabrela environment. Figure S4: Manhattan plot depicting the genome-wide association results for ash, fiber, carbohydrates, moisture, and resistant starch content in common bean using 78 Portuguese accessions grown in the Cabrela environment. Figure S5: Manhattan plot depicting the genome-wide association results for ash, fiber, carbohydrates, moisture, protein, resistant starch, Alanine, Arginine, Aspartic acid, Glutamic acid, Glycine, Histidine, Isoleucine, Leucine, Lysine, Methionine, Phenylalanine, Proline, Serine, Threonine, Tyrosine, Valine and Trypsin inhibitor activity content in common bean using 94 Portuguese accessions grown in the Córdoba environment. Figure S6: Manhattan plot depicting the genome-wide association results for fat content in common bean using 106 Portuguese accessions grown in the Cabrela and Córdoba environments. Figure S7: Quantile-quantile (Q-Q) plots for the SNP-trait associations of the 24 nutritional composition and protein quality-related traits measured in Cabrela, Córdoba, or both environments. Table S1: Average, standard deviation, and coefficient of variation (%) of 16 amino acids and trypsin inhibitor activity (g/100 g) measured in the seeds of 72 Portuguese common bean accessions grown in Córdoba. Table S2: Wald test statistics and broad-sense heritability for several nutritional composition traits measured in the seeds of 78 Portuguese common bean accessions grown in the Cabrela environment. Table S3: Wald test statistics and broad sense heritability for several nutritional composition and protein quality traits measured in the seed of 94 Portuguese common bean accessions grown in the Córdoba environment. Table S4: Wald test statistics and broad sense heritability for fat measured in the seed of 106 Portuguese common bean accessions grown in two contrasting environments. Table S5: Inflation factors for the linear mixed models tested for genome-wide association of 24 nutritional composition and protein quality-related traits measured in common bean seeds collected in Cabrela and Córdoba. Table S6: SNP associations ($-\log_{10}$ (*p*-value) ≥ 3) with 24 nutritional composition and protein quality traits under two contrasting environments (Cabrela and Córdoba), marker position within chromosomes, allelic reference and allelic variant for the associated SNP, minor allele frequency, the effect of the allelic variant (rare allele), and the proportion of phenotypic variance explained by each associated SNP detected using a panel of 106 Portuguese common bean accessions. Table S7: Putative candidate genes for the 16 nutritional composition and protein quality traits for the significant ($-\log_{10}$ (*p*-value) ≥ 3) SNP-trait associations under two contrasting environments (Cabrela and Córdoba).

Author Contributions: Conceptualization, M.C.V.P., M.R.B. and D.R.; resources, M.C.V.P., M.R.B. and D.R.; methodology, F.A.M., S.T.L., V.C. and E.M.; formal analysis, F.A.M. and S.T.L., investigation, F.A.M., S.T.L., V.C., E.M., D.R., M.R.B. and M.C.V.P.; writing-original draft preparation, F.A.M.; writing—review and editing, F.A.M., S.T.L., E.M., D.R. and M.C.V.P.; supervision, S.T.L., M.R.B. and M.C.V.P.; funding acquisition M.C.V.P. and M.R.B.; project administration, M.C.V.P. and M.R.B. All authors have read and agreed to the published version of the manuscript.

Funding: Financial support by FCT—Fundação para a Ciência e a Tecnologia, I.P., Portugal, is acknowledged through research project BeGeQA (PTDC/AGR-TEC/3555/2012) and R&D Unit "GREEN-IT—Bioresources for Sustainability" (UIDB/04551/2020 and UIDP/04551/2020). The European Union Rural Development Program PDR2020 project (PDR2020-784-042734) is also acknowledged.

Institutional Review Board Statement: Not applicable.

Informed Consent Statement: Not applicable.

Data Availability Statement: Data is contained within the article or Supplementary Material.

Conflicts of Interest: The authors declare no conflict of interest.

References

1. Boye, J.; Zare, F.; Pletch, A. Pulse Proteins: Processing, Characterization, Functional Properties and Applications in Food and Feed. *Food Res. Int.* **2010**, *43*, 414–431. [CrossRef]
2. Câmara, C.R.S.; Urrea, C.A.; Schlegel, V. Pinto Beans (*Phaseolus vulgaris* L.) as a Functional Food: Implications on Human Health. *Agriculture* **2013**, *3*, 90–111. [CrossRef]
3. Rawal, V.; Navarro, D.K. *The Global Economy of Pulses*, 2019 ed.; FAO: Rome, Italy, 2019.
4. Petry, N.; Boy, E.; Wirth, J.P.; Hurrell, R.F. Review: The Potential of the Common Bean (*Phaseolus vulgaris*) as a Vehicle for Iron Biofortification. *Nutrients* **2015**, *7*, 1144–1173. [CrossRef]
5. Kalogeropoulos, N.; Chiou, A.; Ioannou, M.; Karathanos, V.T.; Hassapidou, M.; Andrikopoulos, N.K. Nutritional Evaluation and Bioactive Microconstituents (Phytosterols, Tocopherols, Polyphenols, Triterpenic Acids) in Cooked Dry Legumes Usually Consumed in the Mediterranean Countries. *Food Chem.* **2010**, *121*, 682–690. [CrossRef]
6. Mecha, E.; Figueira, M.E.; Vaz Patto, M.C.; do Rosario Bronze, M. Two Sides of the Same Coin: The Impact of Grain Legumes on Human Health: Common Bean (*Phaseolus vulgaris* L.) as a Case Study. In *Legume Seed Nutraceutical Research;* Jimenez-Lopez, J.C., Clemente, A., Eds.; IntechOpen: London, UK, 2018; ISBN 978-1-78985-397-1.
7. Escribano, M.R.; Santalla, M.; de Ron, A.M. Genetic Diversity in Pod and Seed Quality Traits of Common Bean Populations from Northwestern Spain. *Euphytica* **1997**, *93*, 71–81. [CrossRef]
8. Vaz Patto, M.C.; Araújo, S.S. Positioning Portugal into the Context of World Production and Research in Grain Legumes. *Rev. Ciênc. Agrár.* **2016**, *39*, 471–489. [CrossRef]
9. Gaikwad, K.B.; Rani, S.; Kumar, M.; Gupta, V.; Babu, P.H.; Bainsla, N.K.; Yadav, R. Enhancing the Nutritional Quality of Major Food Crops Through Conventional and Genomics-Assisted Breeding. *Front. Nutr.* **2020**, *7*, 198. [CrossRef] [PubMed]
10. Vaz Patto, M.C.; Amarowicz, R.; Aryee, A.N.A.; Boye, J.I.; Chung, H.-J.; Martín-Cabrejas, M.A.; Domoney, C. Achievements and Challenges in Improving the Nutritional Quality of Food Legumes. *Crit. Rev. Plant Sci.* **2015**, *34*, 105–143. [CrossRef]
11. Vaz Patto, M.C. Grain Legume Protein Quality: A Hot Subject. *Arbor* **2016**, *192*, a314. [CrossRef]
12. Boye, J.; Wijesinha-Bettoni, R.; Burlingame, B. Protein Quality Evaluation Twenty Years after the Introduction of the Protein Digestibility Corrected Amino Acid Score Method. *Br. J. Nutr.* **2012**, *108*, S183–S211. [CrossRef] [PubMed]
13. Wang, T.L.; Domoney, C.; Hedley, C.L.; Casey, R.; Grusak, M.A. Can We Improve the Nutritional Quality of Legume Seeds? *Plant Physiol.* **2003**, *131*, 886–891. [CrossRef]
14. Blair, M.W.; Astudillo, C.; Grusak, M.A.; Graham, R.; Beebe, S.E. Inheritance of Seed Iron and Zinc Concentrations in Common Bean (*Phaseolus vulgaris* L.). *Mol. Breed.* **2009**, *23*, 197–207. [CrossRef]
15. Guzmán-Maldonado, S.H.; Martínez, O.; Acosta-Gallegos, J.A.; Guevara-Lara, F.; Paredes-López, O. Putative Quantitative Trait Loci for Physical and Chemical Components of Common Bean. *Crop Sci.* **2003**, *43*, 1029–1035. [CrossRef]
16. Delfini, J.; Moda-Cirino, V.; dos Santos Neto, J.; Zeffa, D.M.; Nogueira, A.F.; Ribeiro, L.A.B.; Ruas, P.M.; Gepts, P.; Gonçalves, L.S.A. Genome-Wide Association Study for Grain Mineral Content in a Brazilian Common Bean Diversity Panel. *Theor. Appl. Genet.* **2021**, *134*, 2795–2811. [CrossRef] [PubMed]
17. Gunjača, J.; Carović-Stanko, K.; Lazarević, B.; Vidak, M.; Petek, M.; Liber, Z.; Šatović, Z. Genome-Wide Association Studies of Mineral Content in Common Bean. *Front. Plant Sci.* **2021**, *12*, 305. [CrossRef]

18. Casañas, F.; Pérez-Vega, E.; Almirall, A.; Plans, M.; Sabaté, J.; Ferreira, J.J. Mapping of QTL Associated with Seed Chemical Content in a RIL Population of Common Bean (*Phaseolus vulgaris* L.). *Euphytica* **2013**, *192*, 279–288. [CrossRef]
19. Katuuramu, D.N.; Hart, J.P.; Porch, T.G.; Grusak, M.A.; Glahn, R.P.; Cichy, K.A. Genome-Wide Association Analysis of Nutritional Composition-Related Traits and Iron Bioavailability in Cooked Dry Beans (*Phaseolus vulgaris* L.). *Mol. Breed.* **2018**, *38*, 44. [CrossRef]
20. Mitchell-Olds, T. Complex-Trait Analysis in Plants. *Genome Biol.* **2010**, *11*, 113. [CrossRef]
21. Schmutz, J.; McClean, P.E.; Mamidi, S.; Wu, G.A.; Cannon, S.B.; Grimwood, J.; Jenkins, J.; Shu, S.; Song, Q.; Chavarro, C.; et al. A Reference Genome for Common Bean and Genome-Wide Analysis of Dual Domestications. *Nat. Genet.* **2014**, *46*, 707–713. [CrossRef] [PubMed]
22. Alqudah, A.M.; Sallam, A.; Stephen Baenziger, P.; Börner, A. GWAS: Fast-Forwarding Gene Identification and Characterization in Temperate Cereals: Lessons from Barley—A Review. *J. Adv. Res.* **2020**, *22*, 119–135. [CrossRef]
23. Santalla, M.; Rodiño, A.; De Ron, A. Allozyme Evidence Supporting Southwestern Europe as a Secondary Center of Genetic Diversity for the Common Bean. *Theor. Appl. Genet.* **2002**, *104*, 934–944. [CrossRef]
24. Leitão, S.T.; Dinis, M.; Veloso, M.M.; Šatović, Z.; Vaz Patto, M.C. Establishing the Bases for Introducing the Unexplored Portuguese Common Bean Germplasm into the Breeding World. *Front. Plant Sci.* **2017**, *8*, 1296. [CrossRef]
25. Mecha, E.; Natalello, S.; Carbas, B.; da Silva, A.B.; Leitão, S.T.; Brites, C.; Veloso, M.M.; Rubiales, D.; Costa, J.; do Fatima Cabral, M.; et al. Disclosing the Nutritional Quality Diversity of Portuguese Common Beans—The Missing Link for Their Effective Use in Protein Quality Breeding Programs. *Agronomy* **2021**, *11*, 221. [CrossRef]
26. Leitão, S.T.; Malosetti, M.; Song, Q.; van Eeuwijk, F.; Rubiales, D.; Vaz Patto, M.C. Natural Variation in Portuguese Common Bean Germplasm Reveals New Sources of Resistance Against *Fusarium oxysporum* f. sp. *Phaseoli and Resistance-Associated Candidate Genes. Phytopathology* **2020**, *110*, 633–647. [CrossRef] [PubMed]
27. Leitão, S.T.; Bicho, M.C.; Pereira, P.; Paulo, M.J.; Malosetti, M.; de Sousa Araújo, S.; van Eeuwijk, F.; Vaz Patto, M.C. Common Bean SNP Alleles and Candidate Genes Affecting Photosynthesis under Contrasting Water Regimes. *Hortic. Res.* **2021**, *8*, 4. [CrossRef] [PubMed]
28. Considine, M.J.; Siddique, K.H.M.; Foyer, C.H. Nature's Pulse Power: Legumes, Food Security and Climate Change. *J. Exp. Bot.* **2017**, *68*, 1815–1818. [CrossRef]
29. Magrini, M.-B.; Anton, M.; Cholez, C.; Corre-Hellou, G.; Duc, G.; Jeuffroy, M.-H.; Meynard, J.-M.; Pelzer, E.; Voisin, A.-S.; Walrand, S. Why Are Grain-Legumes Rarely Present in Cropping Systems despite Their Environmental and Nutritional Benefits? Analyzing Lock-in in the French Agrifood System. *Ecol. Econ.* **2016**, *126*, 152–162. [CrossRef]
30. Mazau, D.; Esquerré-Tugayé, M.T. Hydroxyproline-Rich Glycoprotein Accumulation in the Cell Walls of Plants Infected by Various Pathogens. *Physiol. Mol. Plant Pathol.* **1986**, *29*, 147–157. [CrossRef]
31. Showalter, A.M.; Bell, J.N.; Cramer, C.L.; Bailey, J.A.; Varner, J.E.; Lamb, C.J. Accumulation of Hydroxyproline-Rich Glycoprotein MRNAs in Response to Fungal Elicitor and Infection. *Proc. Natl. Acad. Sci. USA* **1985**, *82*, 6551–6555. [CrossRef] [PubMed]
32. Sterling, J.D.; Atmodjo, M.A.; Inwood, S.E.; Kumar Kolli, V.S.; Quigley, H.F.; Hahn, M.G.; Mohnen, D. Functional Identification of an Arabidopsis Pectin Biosynthetic Homogalacturonan Galacturonosyltransferase. *Proc. Natl. Acad. Sci. USA* **2006**, *103*, 5236–5241. [CrossRef] [PubMed]
33. Voragen, A.G.J.; Coenen, G.-J.; Verhoef, R.P.; Schols, H.A. Pectin, a Versatile Polysaccharide Present in Plant Cell Walls. *Struct. Chem.* **2009**, *20*, 263. [CrossRef]
34. Bernardo, R. *Breeding for Quantitative Traits in Plants*, 3rd ed.; Stemma Press: Woodbury, MN, USA, 2020; ISBN 978-0-9720724-3-4.
35. Wirdnam, C.; Motoyama, A.; Arn-Bouldoires, E.; van Eeden, S.; Iglesias, A.; Meins, F. Altered Expression of an Ankyrin-Repeat Protein Results in Leaf Abnormalities, Necrotic Lesions, and the Elaboration of a Systemic Signal. *Plant Mol. Biol.* **2004**, *56*, 717–730. [CrossRef] [PubMed]
36. Murgia, I.; Tarantino, D.; Soave, C.; Morandini, P. Arabidopsis CYP82C4 Expression Is Dependent on Fe Availability and Circadian Rhythm, and Correlates with Genes Involved in the Early Fe Deficiency Response. *J. Plant Physiol.* **2011**, *168*, 894–902. [CrossRef]
37. Ma, H.; Cao, X.; Shi, S.; Li, S.; Gao, J.; Ma, Y.; Zhao, Q.; Chen, Q. Genome-Wide Survey and Expression Analysis of the Amino Acid Transporter Superfamily in Potato (*Solanum tuberosum* L.). *Plant Physiol. Biochem.* **2016**, *107*, 164–177. [CrossRef]
38. Sehgal, A.; Sita, K.; Siddique, K.H.M.; Kumar, R.; Bhogireddy, S.; Varshney, R.K.; HanumanthaRao, B.; Nair, R.M.; Prasad, P.V.V.; Nayyar, H. Drought or/and Heat-Stress Effects on Seed Filling in Food Crops: Impacts on Functional Biochemistry, Seed Yields, and Nutritional Quality. *Front. Plant Sci.* **2018**, *9*, 1705. [CrossRef]
39. Deryng, D.; Conway, D.; Ramankutty, N.; Price, J.; Warren, R. Global Crop Yield Response to Extreme Heat Stress under Multiple Climate Change Futures. *Environ. Res. Lett.* **2014**, *9*, 034011. [CrossRef]
40. Sita, K.; Sehgal, A.; HanumanthaRao, B.; Nair, R.M.; Vara Prasad, P.V.; Kumar, S.; Gaur, P.M.; Farooq, M.; Siddique, K.H.M.; Varshney, R.K.; et al. Food Legumes and Rising Temperatures: Effects, Adaptive Functional Mechanisms Specific to Reproductive Growth Stage and Strategies to Improve Heat Tolerance. *Front. Plant Sci.* **2017**, *8*, 1658. [CrossRef] [PubMed]
41. El-Soda, M.; Malosetti, M.; Zwaan, B.J.; Koornneef, M.; Aarts, M.G.M. Genotype × Environment Interaction QTL Mapping in Plants: Lessons from Arabidopsis. *Trends Plant Sci.* **2014**, *19*, 390–398. [CrossRef]
42. Nosworthy, M.G.; House, J.D. Factors Influencing the Quality of Dietary Proteins: Implications for Pulses. *Cereal Chem.* **2017**, *94*, 49–57. [CrossRef]

43. Montoya, C.A.; Lallès, J.-P.; Beebe, S.; Leterme, P. Phaseolin Diversity as a Possible Strategy to Improve the Nutritional Value of Common Beans (*Phaseolus vulgaris*). *Food Res. Int.* **2010**, *43*, 443–449. [CrossRef]
44. Wang, N.; Hatcher, D.W.; Tyler, R.T.; Toews, R.; Gawalko, E.J. Effect of Cooking on the Composition of Beans (*Phaseolus vulgaris* L.) and Chickpeas (*Cicer Arietinum* L.). *Food Res. Int.* **2010**, *43*, 589–594. [CrossRef]
45. American Association of Cereal Chemists (AACC). *Method 32-40.01 Resistant Starch Assay Procedure*; AACC: St. Paul, MN, USA, 2009.
46. Association of Analytical Chemists (AOAC). *Method 2002.02 Resistant Starch Assay Procedure*; AACC: St. Paul, MN, USA, 2002.
47. VSN International. *Genstat for Windows*, 21st ed.; VSN International: Hemel Hempstead, UK, 2020.
48. Malosetti, M.; van der Linden, C.G.; Vosman, B.; van Eeuwijk, F.A. A Mixed-Model Approach to Association Mapping Using Pedigree Information with an Illustration of Resistance to Phytophthora Infestans in Potato. *Genetics* **2007**, *175*, 879–889. [CrossRef] [PubMed]
49. Alves, M.L.; Carbas, B.; Gaspar, D.; Paulo, M.; Brites, C.; Mendes-Moreira, P.; Brites, C.M.; Malosetti, M.; van Eeuwijk, F.; Vaz Patto, M.C. Genome-Wide Association Study for Kernel Composition and Flour Pasting Behavior in Wholemeal Maize Flour. *BMC Plant Biol.* **2019**, *19*, 123. [CrossRef] [PubMed]
50. Cheng, R.; Parker, C.C.; Abney, M.; Palmer, A.A. Practical Considerations Regarding the Use of Genotype and Pedigree Data to Model Relatedness in the Context of Genome-Wide Association Studies. *G3 Genes Genomes Genet.* **2013**, *3*, 1861–1867. [CrossRef]
51. Benjamini, Y.; Yekutieli, D. The Control of the False Discovery Rate in Multiple Testing under Dependency. *Ann. Stat.* **2001**, *29*, 1165–1188. [CrossRef]
52. Duggal, P.; Gillanders, E.M.; Holmes, T.N.; Bailey-Wilson, J.E. Establishing an Adjusted P-Value Threshold to Control the Family-Wide Type 1 Error in Genome Wide Association Studies. *BMC Genom.* **2008**, *9*, 516. [CrossRef]
53. Resende, R.T.; Resende, M.D.V.; Silva, F.F.; Azevedo, C.F.; Takahashi, E.K.; Silva-Junior, O.B.; Grattapaglia, D. Regional Heritability Mapping and Genome-Wide Association Identify Loci for Complex Growth, Wood and Disease Resistance Traits in Eucalyptus. *New Phytol.* **2017**, *213*, 1287–1300. [CrossRef]
54. Goodstein, D.M.; Shu, S.; Howson, R.; Neupane, R.; Hayes, R.D.; Fazo, J.; Mitros, T.; Dirks, W.; Hellsten, U.; Putnam, N.; et al. Phytozome: A Comparative Platform for Green Plant Genomics. *Nucleic Acids Res.* **2012**, *40*, D1178–D1186. [CrossRef] [PubMed]
55. Schwacke, R.; Ponce-Soto, G.Y.; Krause, K.; Bolger, A.M.; Arsova, B.; Hallab, A.; Gruden, K.; Stitt, M.; Bolger, M.E.; Usadel, B. MapMan4: A Refined Protein Classification and Annotation Framework Applicable to Multi-Omics Data Analysis. *Mol. Plant* **2019**, *12*, 879–892. [CrossRef]
56. Shannon, P.; Markiel, A.; Ozier, O.; Baliga, N.S.; Wang, J.T.; Ramage, D.; Amin, N.; Schwikowski, B.; Ideker, T. Cytoscape: A Software Environment for Integrated Models of Biomolecular Interaction Networks. *Genome Res.* **2003**, *13*, 2498–2504. [CrossRef]

plants

MDPI

Article

Using Crop Databases to Explore Phenotypes: From QTL to Candidate Genes

Anne V. Brown [1], David Grant [1,2] and Rex T. Nelson [1,*]

[1] United States Department of Agriculture-Agricultural Research Service, Corn Insects and Crop Genetics Research Unit, Ames, IA 50011, USA; anne.brown@usda.gov (A.V.B.); dgrant@iastate.edu (D.G.)
[2] Department of Agronomy, Iowa State University, Ames, IA 50011, USA
* Correspondence: rex.nelson@usda.gov; Tel.: +1-515-294-1297

Abstract: Seeds, especially those of certain grasses and legumes, provide the majority of the protein and carbohydrates for much of the world's population. Therefore, improvements in seed quality and yield are important drivers for the development of new crop varieties to feed a growing population. Quantitative Trait Loci (QTL) have been identified for many biologically interesting and agronomically important traits, including many seed quality traits. QTL can help explain the genetic architecture of the traits and can also be used to incorporate traits into new crop cultivars during breeding. Despite the important contributions that QTL have made to basic studies and plant breeding, knowing the exact gene(s) conditioning each QTL would greatly improve our ability to study the underlying genetics, biochemistry and regulatory networks. The data sets needed for identifying these genes are increasingly available and often housed in species- or clade-specific genetics and genomics databases. In this demonstration, we present a generalized walkthrough of how such databases can be used in these studies using SoyBase, the USDA soybean Genetics and Genomics Database, as an example.

Keywords: QTL; GWAS; candidate gene; genomics; genetics; database; SoyBase

Citation: Brown, A.V.; Grant, D.; Nelson, R.T. Using Crop Databases to Explore Phenotypes: From QTL to Candidate Genes. *Plants* **2021**, *10*, 2494. https://doi.org/10.3390/plants10112494

Academic Editor: Abdelmajid Kassem

Received: 15 October 2021
Accepted: 13 November 2021
Published: 18 November 2021

1. Introduction

Since the introduction of bi-parental QTL analysis in plants [1] in the early 1980s, QTL regions have been described in both plant and animal species [2]. In early QTL analyses, the number of markers used and the limited number of progeny examined meant that the genetic regions encompassed by a QTL were usually large. These regions could include dozens, if not hundreds of genes, making candidate gene identification for the trait measured tedious, if not impossible (reviewed in [3]). Fine mapping with more markers is necessary to further limit the genetic region containing the gene conditioning the trait. This process would be aided if a naturally occurring or synthetic mutant in the gene conditioning the trait existed [4].

In previous years, fine-mapping was both a time consuming and expensive process that was not routinely performed to identify candidate genes. More recently, with the drop in sequencing costs, identification of vast numbers of single nucleotide polymorphisms (SNPs) and relatively inexpensive analysis technologies, it has become feasible to both identify smaller QTL regions and generate sequence information for those regions [5]. Additionally, Genome-Wide Association Studies (GWAS) utilizing SNP allele information have been employed to identify sequence regions associated with phenotypic traits and tools have been developed to integrate GWAS studies with QTL data such as QTLtools [6].

As more genomic data become easily accessible by quick and easy data sharing [7], some clade and species genome databases are now actively curating both bi-parental QTL and GWAS QTL information. This information can be used to identify candidate regions, although these regions typically contain many candidate genes. The list of candidate genes can often be reduced by considering molecular function annotations and tissue expression patterns. To illustrate this process, we will use, as an example, the information curated in the species database SoyBase [8].

SoyBase is the United States Department of Agriculture, Agricultural Research Service (USDA-ARS) soybean genetics and genomics database [8] and has been actively curated since its inception in the early 1990s. In 2010, the first assembly of a soybean genome (CV. Williams 82) was released [9]. Since then, SoyBase has been curating genomic information and presenting these data in the context of the original genetic data. We will demonstrate how genetic and genomic data can be used in silico to help identify candidate gene(s) that might condition a phenotype of interest. This process has often been referred to as phenotype to genotype (P2G) or field to genes (F2G).

2. Example Walkthrough

This demonstration on using a genomic/genetic database in P2G research was developed using SoyBase. Although the specific examples presented are for soybean, most species- or clade-specific databases will have somewhat equivalent data; however, the tools to display that data vary. In this demonstration, we present a series of steps that demonstrate how the various data types in SoyBase can be used together to identify a candidate gene controlling a trait. We do not intend to imply that the path through the database we present is the only one that would accomplish this, only that this is one way of solving the problem that highlights some of the important data sets available.

Seed oil is a major product extracted from soybeans, and seed oil composition is a significant factor in determining the price of oil paid by processors. Oil that contains reduced linolenic content is more stable during storage [10] and as a frying oil [11]. Thus, determining the genes and regulatory networks of linolenic synthesis is an important step in developing improved varieties, and this will be the trait used in this demonstration. The first step in identification of the gene(s) controlling seed linolenic acid content is to identify QTL for this trait, i.e., region(s) of the genetic map that have been associated with the phenotype.

In this example, we will use the SoyBase Search function to obtain a list of QTL for the search term "linolenic". SoyBase contains information for 68 bi-parental QTL related to seed linolenic acid content that have been reported in 14 papers. Further examination of these results shows that there is a region on molecular linkage group B2 (chromosome 14) that has a large number of bi-parental QTL for seed oil traits, including several for seed linolenic acid content (Figure 1).

The SoyBase genetic map viewer is composed of two panes (Figure 1). The left shows a representation of the soybean physical or sequence map based on the Williams 82 genome sequence. This view of the chromosome shows the positions of molecular markers, the gene models (Glyma.14gxxxxxx) and the GWAS QTL identified in soybean. On the right is the soybean Composite Genetic Map, which shows the genetically mapped molecular markers along with the QTL identified in soybean.

The hand-curated Composite Genetic Map is based on the reported QTL mapping studies in soybean and allows QTL from different publications to be displayed using a common coordinate system. Markers present on both the genetic and sequence maps are connected by a blue line. These two views of a chromosome allow the easy identification of regions with relatively high or low recombination as well as where the genetic and sequence maps are not congruent. In addition, comparing the locations of the bi-parental and GWAS QTL can provide information that is not available if used individually. Note that these two views of a chromosome have an important difference: coordinates on the sequence map are in base pairs (bp, left) while those on the genetic map are in centi-Morgans (cM, right).

We will use Seed linolenic 11-2 as the QTL of interest in this example (Figure 2). Along with information about the cross used to identify this QTL and other related information, the QTL page for Seed linolenic 11-2 provides links to the QTL on the SoyBase Genetic Map and to the approximate region containing this QTL in the SoyBase Genome Sequence Browser. Seed linolenic 11-2 was originally identified as a bi-parental QTL where the inheritance of the trait was genetically associated with the molecular marker Satt063 (Figure 3).

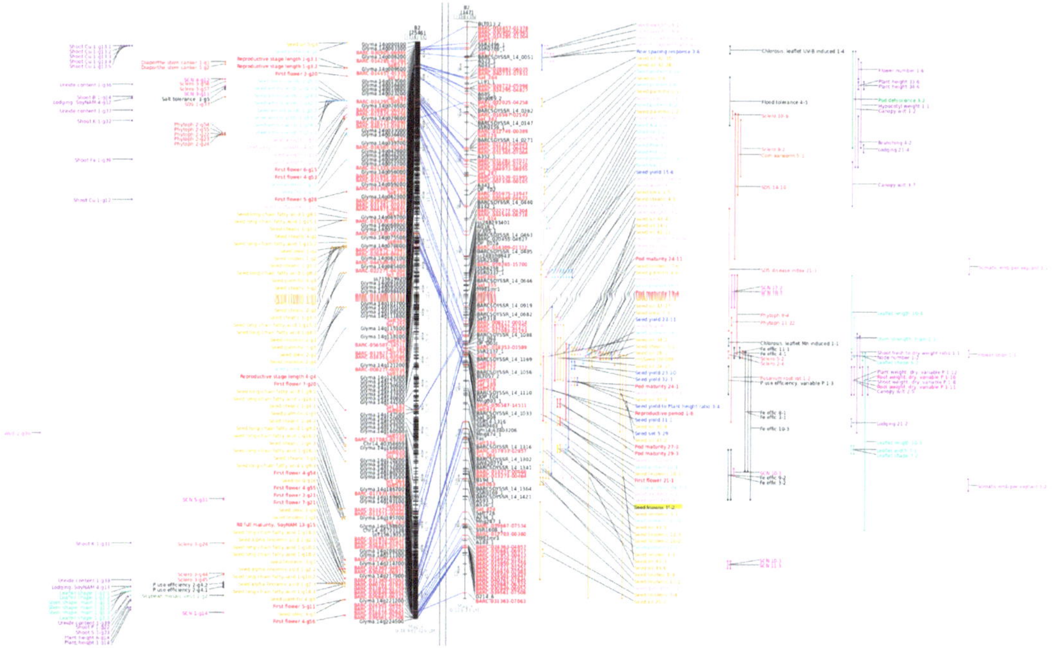

Figure 1. The composite genetic map and physical map of linkage group B2/chromosome 14. The left pane shows the physical or sequence map based on the soybean reference cultivar Williams 82. Genetically mapped molecular markers for which the sequence is available are shown on the physical map along with the gene models. The right pane shows the GmComposite2003 genetic map. This map was created in 2003 as the composite genetic map for soybean and is continually updated with new QTL and genetic markers. Markers in common between the two maps are connected by blue lines and shown in red text. Both bi-parental and GWAS QTL are grouped in columns by function or developmental category. Related QTL within categories are shown using the same color. Both QTL types use the same groupings and color to make correlations across the chromosome representations easier. The Seed linolenic 11-2 QTL is highlighted in yellow. Larger version.

For clarity, in this example, only seed related QTL are shown. Comparison of the physical and genetic maps indicates that not only have there been many seed oil and linolenic content bi-parental QTL identified in the region but also that a number of GWAS QTL for seed oil content, linolenic acid and long-chain fatty acids are present in the corresponding region of the physical map. As this region contains many genes, a useful first step to identifying potential candidate genes is to view this region of the chromosome in the SoyBase Sequence Browser where a short annotation is provided for each gene.

This region can be viewed by selecting the closest flanking markers around the QTL (BARC-013273-00464 and Sat_424, shown in red text) and showing this region in the Sequence Browser (Figure 4A, flanking markers highlighted in orange). This figure also includes tracks for the related GWAS QTL and genes. Zooming into this view shows the short annotations for each gene (Figure 4B). In this view, a track showing gene expression as revealed by RNA-seq has been added.

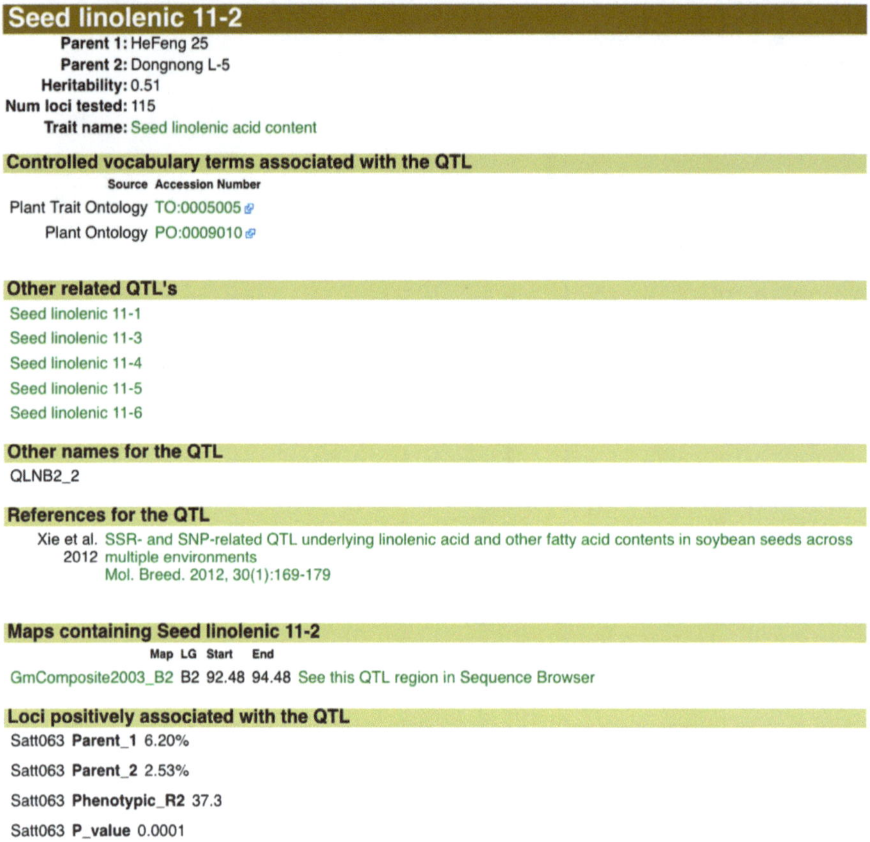

Seed linolenic 11-2

Parent 1: HeFeng 25
Parent 2: Dongnong L-5
Heritability: 0.51
Num loci tested: 115
Trait name: Seed linolenic acid content

Controlled vocabulary terms associated with the QTL

Source	Accession Number
Plant Trait Ontology	TO:0005005
Plant Ontology	PO:0009010

Other related QTL's

Seed linolenic 11-1
Seed linolenic 11-3
Seed linolenic 11-4
Seed linolenic 11-5
Seed linolenic 11-6

Other names for the QTL

QLNB2_2

References for the QTL

Xie et al. 2012 SSR- and SNP-related QTL underlying linolenic acid and other fatty acid contents in soybean seeds across multiple environments
Mol. Breed. 2012, 30(1):169-179

Maps containing Seed linolenic 11-2

Map	LG	Start	End	
GmComposite2003_B2	B2	92.48	94.48	See this QTL region in Sequence Browser

Loci positively associated with the QTL

Satt063 **Parent_1** 6.20%
Satt063 **Parent_2** 2.53%
Satt063 **Phenotypic_R2** 37.3
Satt063 **P_value** 0.0001

Figure 2. QTL report page for Seed linolenic 11-2. The QTL report for Seed linolenic 11-2 provides details on the QTL such as its heritability, parents and parental phenotype. It also lists any other phenotypes measured in the study (none in this example) and other QTLs for the trait identified in the study (Other Related QTLs). The map and location of the QTL is presented in the section "Maps containing Seed linolenic 11-2". Clicking on the link "See this QTL region in Sequence Browser" will take the user to the sequence browser view of the approximate QTL on the sequence map to allow browsing of the gene model annotations. Genetic loci that are associated with the QTL are listed in the "Loci positively associated with the QTL" section along with association values for the loci.

Figure 4B shows several lines of evidence that point to Glyma.14g194300 (highlighted in yellow) as a candidate for the gene conditioning seed linolenic content:

- Located physically close to Satt063 (highlighted in red), the molecular marker most associated with Seed linolenic 11-2.
- Located within the region of GWAS QTL Seed alpha-linolenic acid 1-g2 (highlighted in orange).
- Annotated as a Fatty Acid Desaturase.
- Preferentially expressed in developing seeds.

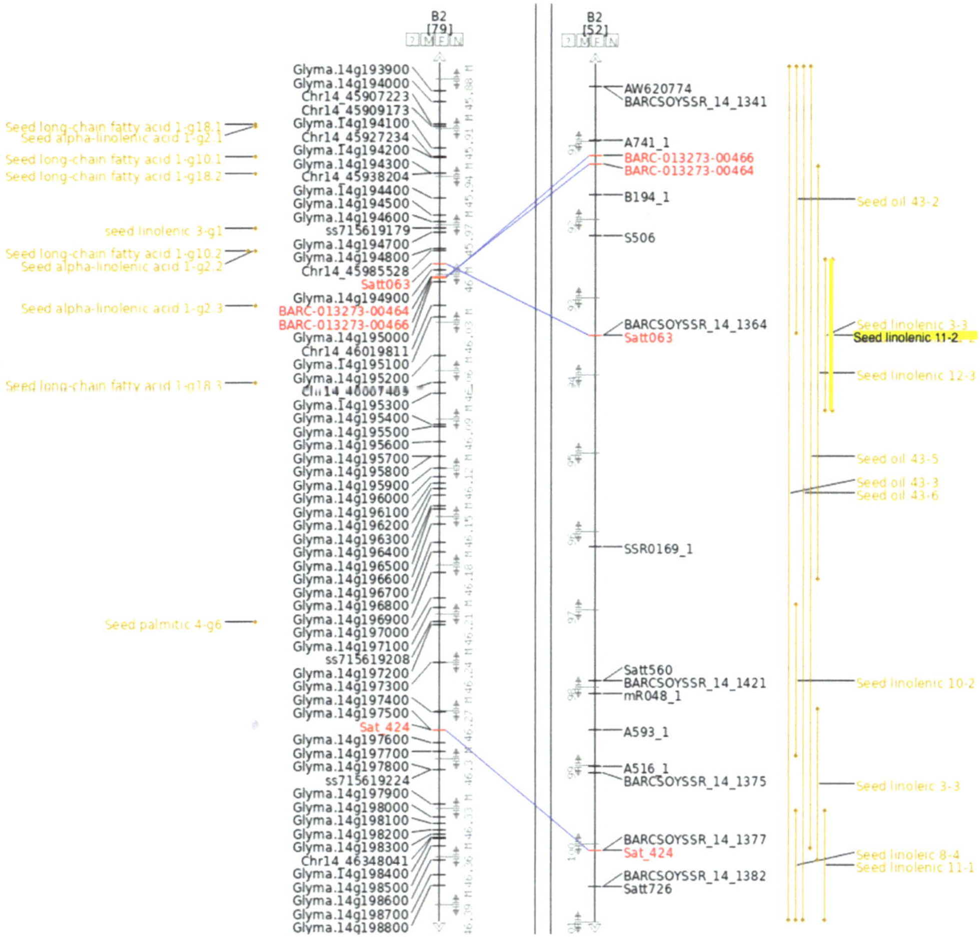

Figure 3. Genetic and physical map region containing Seed linolenic 11-2. Region of the physical and genetic map of MLG B2/Gm14 containing Seed linolenic 11-2. Only seed oil QTL are pictured for clarity. The physical map (**left**) includes the locations of gene models (Glyma.14g194300). The composite map (**right**) contains QTL regions for seed oil related QTL. Sequence based markers that have been genetically mapped are connected by blue lines. Larger version.

The information page for Glyma.14g194300 provides more information for this gene, parts of which are shown in Figure 5. Panel 5A gives the annotations from a number of sources for Glyma.14g194300. Panel 5B shows that the gene model is associated with the gene FAD3A, which is known to carry out a major step in linolenate biosynthesis and seed linolenic acid content [12]. Panel 5C presents a pictorial representation of the gene's expression in different tissues and steps in development [13]. Glyma.14g194300 has relatively high expression during seed development, which supports the conclusion above that it is a candidate gene for the Seed linolenic 11-2 and Seed alpha-linolenic acid 1-g2 QTL.

(**A**)

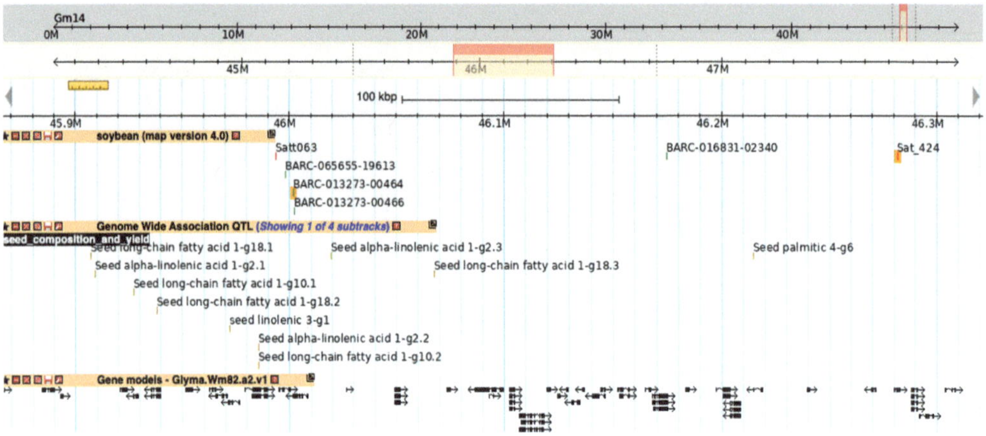

(**B**)

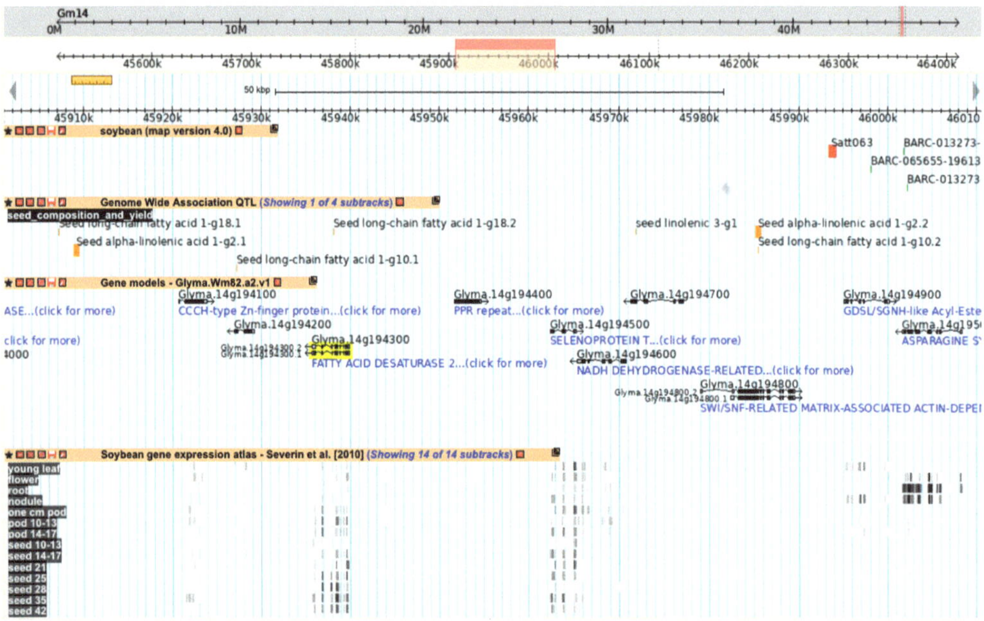

Figure 4. Identification of a candidate gene using the SoyBase Genome Browser. The region of the soybean physical map around Seed linolenic 11-2. (**A**) Magnification of the genomic region around Satt063. Molecular markers that flank Seed linolenic 11-2 are highlighted in orange. Tracks are also shown for GWAS QTL and genes. Larger version (**B**) Magnification of the chromosomal region in Panel A showing the short functional annotation for genes. The candidate gene Glyma.14g134300 is highlighted in yellow. The flanking GWAS QTL (orange) are indicated in the Genome Wide Association QTL track. Gene expression patterns indicating that the highlighted gene is preferentially expressed in seed tissue derived from RNA-seq are shown in the bottom track. Larger version.

(A)

Annotations for Glyma.14g194300

Database ID	Annotation Type	Annotation Description	Annotation Source	Match Score	Evidence Code
AT5G05580.1	AT	fatty acid desaturase 8	JGI	N/A	IEA
GO:0006629	GO-bp	lipid metabolic process	EnsemblGenomes	N/A	IEA
GO:0006629	GO-bp	lipid metabolic process	JGI	N/A	IEA
GO:0055114	GO-bp	oxidation-reduction process	EnsemblGenomes	N/A	IEA
GO:0055114	GO-bp	oxidation-reduction process	JGI	N/A	IEA
GO:0016020	GO-cc	membrane	EnsemblGenomes	N/A	IEA
GO:0016021	GO-cc	integral component of membrane	EnsemblGenomes	N/A	IEA
GO:0016717	GO-mf	oxidoreductase activity, acting on paired donors, with oxidation of a pair of donors resulting in the reduction of molecular oxygen to two molecules of water	EnsemblGenomes	N/A	IEA
GO:0016717	GO-mf	oxidoreductase activity, acting on paired donors, with oxidation of a pair of donors resulting in the reduction of molecular oxygen to two molecules of water	JGI	N/A	IEA
PTHR19353	Panther	FATTY ACID DESATURASE 2	JGI	N/A	IEA
PTHR19353:SF7	Panther		JGI	N/A	IEA
PF00487	PFAM	Fatty acid desaturase	JGI	N/A	IEA
PF11960	PFAM	Domain of unknown function (DUF3474)	JGI	N/A	IEA
PWY-5997	SoyCyc9	α-linolenate biosynthesis I (plants and red algae)	Plant Metabolic Network		ISS
PWY-762	SoyCyc9	phospholipid desaturation	Plant Metabolic Network		ISS
PWY-782	SoyCyc9	glycolipid desaturation	Plant Metabolic Network		ISS
GN7V-49191	SoyCyc9-rxn	1-18:2-2-trans-16:1-phosphatidylglycerol desaturase	Plant Metabolic Network		ISS

(B)

Proteins Associated with Glyma.14g194300

Locus	Gene Symbol	Protein Name
	FAD3a	omega-3-fatty acid desaturase 3 gene 1
	FAD3A	microsomal omega-3-fatty acid desaturase

(C)

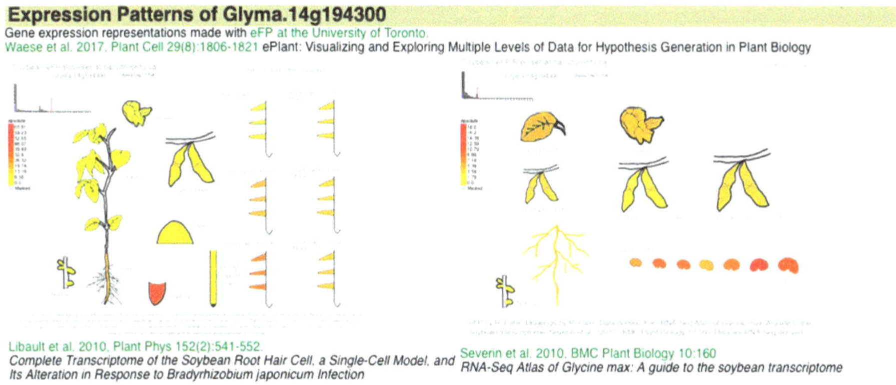

Figure 5. Detailed gene report for Glyma.14g194300. Details of the SoyBase gene report for the candidate gene Glyma.14g194300. (**A**) Functional and biochemical pathway annotation of the candidate indicates that it is a fatty acid desaturase and functions in the α-linolenate biosynthesis I pathway of plants and algae. Evidence codes are described at the GO evidence code page. (**B**) The protein product of this gene has been identified as FAD3A, a microsomal ω-3-fatty acid desaturase gene known to be involved in seed linolenic acid biosynthesis in soybean. (**C**) Expression of this gene measured by RNAseq is elevated in seed and shoot apical meristem tissue.

In this example, there is a gene previously shown to be involved in the seed linolenic content phenotype. In cases where there is no obvious candidate gene in the region, other sources of information will be necessary to identify a strong candidate gene. Such supplementary information includes gene function (geneontology.org, accessed on 12 November 2021), protein structure (pfam.xfam.org, accessed on 12 November 2021), orthology (pantherdb.org, plants.ensembl.org, accessed on 12 November 2021), participation in biological pathways (plantreactome.gramene.org, plantcyc.org, accessed on 12 November 2021) and protein–protein interactions (string-db.org, accessed on 12 November 2021), which can be found in the respective databases.

Additionally, information regarding gene function can often be inferred from or to other species based on orthology or sequence similarity. Orthologs of Glyma.14g194300 in other species can identify genes that may also condition the seed linolenic content in those species. Orthologous genes in other species can be viewed by clicking the "View Gene Family" button on the Glyma.14g194300 report page. This will present a sequence similarity or ontology tree from the Legume Information System (LIS, legumeinfo.org, accessed on 12 November 2021) (Figure 6). It is often the case that other well-characterized species may appear in the tree. These can then be used as an additional source of information when inferring a candidate gene's function.

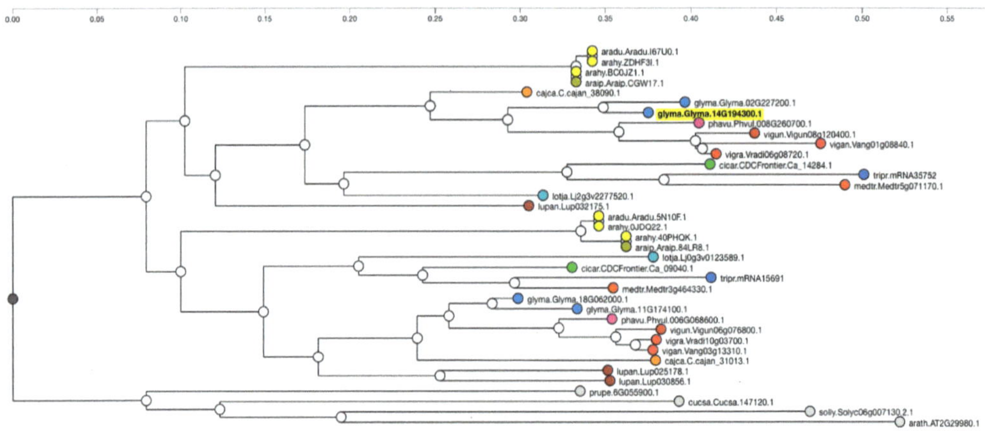

Figure 6. Orthologs of Glyma.14g194300. The phylogram derived from the Legume Information Service's Phylotree viewer. Sequences with high sequence similarity to Glyma.14g194300 (highlighted in yellow) are from Common Bean (phavu), Cowpean (vigun), Adzuki Bean (vigan) and Mung Bean (vigra). Larger version.

As an extra set of conformation of QTL, a new tool called the Genotype Comparison Visualization Tool (GCViT) [14], available on Github (https://github.com/LegumeFederation/gcvit, accessed on 12 November 2021) and SoyBase, can be of use. GCViT is a tool that can be used with any species and will plot SNPs from multiple accessions and display where the differences in alleles are. Therefore, we can confirm/and or identify new regions for linolenic QTL by comparing lines with high linolenics to lines with low linolenics. Another tool that can be used to confirm QTL locations are ZBrowse [15] and ZZBrowse (https://zzbrowse.legumeinfo.org/, accessed on 12 November 2021) [16]. ZBrowse is an interactive tool for the visualization of GWAS data across experiments within a single species, while ZZBrowse is an interactive web tool for the comparative analysis of GWAS and QTL between species [16].

3. Conclusions

In this exercise, we demonstrated how a genetics/genomics database can be used as a tool to help identify the gene(s) conditioning a QTL. Although we used SoyBase in this exercise, other species- or clade-specific databases may contain equivalent data and tools that can be used in concert to accomplish a similar investigation. While other databases may collect similar data, they are not focused on the same user experience that SoyBase tools are. Thus, the path a user takes to identify candidate genes is unique to each database.

A common theme of these databases is that they strive to collect what is known about a species' genetics, genomics, phenotypes, biochemistry and other data into a single repository that allows users to quickly identify the information relevant to the question of interest. The reader will still have to consult some of the external databases referred to above and to other primary literature to manually identify candidate genes as no single species or clade database can assemble all relevant data for a single gene.

Author Contributions: A.V.B., D.G. and R.T.N. contributed equally to the conceptualization, writing—original draft preparation and writing—review and editing of this manuscript. All authors have read and agreed to the published version of the manuscript.

Funding: This research was supported by the US. Department of Agriculture, Agricultural Research Service, project 5030-21000-069-00D. Mention of trade names or commercial products in this publication is solely for the purpose of providing specific information and does not imply recommendation or endorsement by the U.S. Department of Agriculture. USDA is an equal opportunity provider and employer.

Institutional Review Board Statement: Not Applicable.

Informed Consent Statement: Not applicable.

Data Availability Statement: All data on SoyBase.org (accessed on 9 November 2021) is publically available.

Acknowledgments: We would like to thank many previous biological curators that have extracted data from the primary literature for SoyBase and scientific programmers that have worked on components of the website.

Conflicts of Interest: The authors declare no conflict of interest.

References

1. Paterson, A.H.; Lander, E.S.; Hewitt, J.D.; Peterson, S.; Lincoln, S.E.; Tanksley, S.D. Resolution of quantitative traits into Mendelian factors by using a complete linkage map of restriction fragment length polymorphisms. *Nature* **1988**, *335*, 721–726. [CrossRef]
2. Tanksley, S.D. MAPPING POLYGENES. *Annu. Rev. Genet.* **1993**, *27*, 205–233. [CrossRef]
3. Kearsey, M.J.; Farquhar, A.G.L. QTL analysis in plants; where are we now? *Heredity* **1998**, *80*, 137–142. [CrossRef]
4. Koornneef, M.; Hanhart, C.J.; Van Der Veen, J.H. A genetic and physiological analysis of late flowering mutants in Arabidopsis thaliana. *Mol. Genet. Genom.* **1991**, *229*, 57–66. [CrossRef]
5. Cortes, L.T.; Zhang, Z.; Yu, J. Status and prospects of genome-wide association studies in plants. *Plant Genome* **2021**, *14*, e20077. [CrossRef]
6. Delaneau, O.; Ongen, H.; Brown, A.A.; Fort, A.; Panousis, N.; Dermitzakis, E.T. A complete tool set for molecular QTL discovery and analysis. *Nat. Commun.* **2017**, *8*, 15452. [CrossRef] [PubMed]
7. Brown, A.V.; Campbell, J.D.; Assefa, T.; Grant, D.; Nelson, R.T.; Weeks, N.T.; Cannon, S.B. Ten quick tips for sharing open genomic data. *PLoS Comput. Biol.* **2018**, *14*, e1006472. [CrossRef] [PubMed]
8. Brown, A.V.; Conners, S.I.; Huang, W.; Wilkey, A.P.; Grant, D.; Weeks, N.T.; Cannon, S.B.; Graham, M.A.; Nelson, R.T. A new decade and new data at SoyBase, the USDA-ARS soybean genetics and genomics database. *Nucleic Acids Res.* **2020**, *49*, D1496–D1501. [CrossRef] [PubMed]
9. Schmutz, J.; Cannon, S.B.; Schlueter, J.; Ma, J.; Mitros, T.; Nelson, W.; Hyten, D.L.; Song, Q.; Thelen, J.J.; Cheng, J.; et al. Genome sequence of the palaeopolyploid soybean. *Nature* **2010**, *463*, 178–183. [CrossRef] [PubMed]
10. Dutton, H.J.; Lancaster, C.R.; Evans, C.D.; Cowan, J.C. The flavor problem of soybean oil. VIII. Linolenic acid. *J. Am. Oil Chem. Soc.* **1951**, *28*, 115–118. [CrossRef]
11. Tompkins, C.; Perkins, E.G. Frying performance of low-linolenic acid soybean oil. *J. Am. Oil Chem. Soc.* **2000**, *77*, 223–229. [CrossRef]
12. Bilyeu, K.; Palavalli, L.; Sleper, D.; Beuselinck, P. Mutations in Soybean Microsomal Omega-3 Fatty Acid Desaturase Genes Reduce Linolenic Acid Concentration in Soybean Seeds. *Crop. Sci.* **2005**, *45*, 1830–1836. [CrossRef]
13. Severin, A.J.; Woody, J.L.; Bolon, Y.-T.; Joseph, B.; Diers, B.W.; Farmer, A.D.; Muehlbauer, G.J.; Nelson, R.T.; Grant, D.; Specht, J.E.; et al. RNA-Seq Atlas of Glycine max: A guide to the soybean transcriptome. *BMC Plant Biol.* **2010**, *10*, 160. [CrossRef] [PubMed]

14. Wilkey, A.P.; Brown, A.V.; Cannon, S.B.; Cannon, E.K.S. GCViT: A method for interactive, genome-wide visualization of resequencing and SNP array data. *BMC Genom.* **2020**, *21*, 822. [CrossRef]
15. Ziegler, G.R.; Hartsock, R.H.; Baxter, I. Zbrowse: An interactive GWAS results browser. *PeerJ Comput. Sci.* **2015**, *1*, e3. [CrossRef]
16. Berendzen, J.; Brown, A.V.; Cameron, C.T.; Campbell, J.D.; Cleary, A.M.; Dash, S.; Hokin, S.; Huang, W.; Kalberer, S.R.; Nelson, R.T.; et al. The legume information system and associated online genomic resources. *Legum. Sci.* **2021**, *3*, e74. [CrossRef]

plants

MDPI

Article

The Soybean High Density 'Forrest' by 'Williams 82' SNP-Based Genetic Linkage Map Identifies QTL and Candidate Genes for Seed Isoflavone Content

Dounya Knizia [1,2], Jiazheng Yuan [3], Nacer Bellaloui [4], Tri Vuong [5], Mariola Usovsky [5], Qijian Song [6], Frances Betts [3], Teresa Register [3], Earl Williams [3], Naoufal Lakhssassi [1], Hamid Mazouz [2], Henry T. Nguyen [5], Khalid Meksem [1], Alemu Mengistu [7] and My Abdelmajid Kassem [3,*]

1 Department of Plant, Soil, and Agricultural Systems, Southern Illinois University, Carbondale, IL 62901, USA; dounya.knizia@siu.edu (D.K.); naoufal.lakhssassi@siu.edu (N.L.); meksem@siu.edu (K.M.)
2 Laboratoire de Biotechnologies & Valorisation des Bio-Ressources (BioVar), Department de Biology, Faculté des Sciences, Université Moulay Ismail, Meknès 50000, Morocco; H.MAZOUZ@fs-umi.ac.ma
3 Plant Genomics and Biotechnology Laboratory, Department of Biological and Forensic Sciences, Fayetteville State University, Fayetteville, NC 28301, USA; jyuan@uncfsu.edu (J.Y.); fbetts@broncos.uncfsu.edu (F.B.); tregist2@broncos.uncfsu.edu (T.R.); ewilli17@broncos.uncfsu.edu (E.W.)
4 Crop Genetics Research Unit, USDA, Agriculture Research Service, 141 Experiment Station Road, Stoneville, MS 38776, USA; nacer.bellaloui@usda.gov
5 Division of Plant Science and Technology, University of Missouri, Columbia, MO 65211, USA; vuongt@missouri.edu (T.V.); klepadlom@missouri.edu (M.U.); nguyenhenry@missouri.edu (H.T.N.)
6 Soybean Genomics and Improvement Laboratory, USDA-ARS, Beltsville, MD 20705, USA; qijian.song@usda.gov
7 Crop Genetics Research Unit, USDA, Agricultural Research Service, Jackson, TN 38301, USA; alemu.mengistu@usda.gov
* Correspondence: mkassem@uncfsu.edu

Citation: Knizia, D.; Yuan, J.; Bellaloui, N.; Vuong, T.; Usovsky, M.; Song, Q.; Betts, F.; Register, T.; Williams, E.; Lakhssassi, N.; et al. The Soybean High Density 'Forrest' by 'Williams 82' SNP-Based Genetic Linkage Map Identifies QTL and Candidate Genes for Seed Isoflavone Content. *Plants* **2021**, *10*, 2029. https://doi.org/10.3390/plants10102029

Academic Editor: Toyoaki Anai

Received: 23 August 2021
Accepted: 21 September 2021
Published: 27 September 2021

Abstract: Isoflavones are secondary metabolites that are abundant in soybean and other legume seeds providing health and nutrition benefits for both humans and animals. The objectives of this study were to construct a single nucleotide polymorphism (SNP)-based genetic linkage map using the 'Forrest' by 'Williams 82' (F×W82) recombinant inbred line (RIL) population (n = 306); map quantitative trait loci (QTL) for seed daidzein, genistein, glycitein, and total isoflavone contents in two environments over two years (NC-2018 and IL-2020); identify candidate genes for seed isoflavone. The FXW82 SNP-based map was composed of 2075 SNPs and covered 4029.9 cM. A total of 27 QTL that control various seed isoflavone traits have been identified and mapped on chromosomes (Chrs.) 2, 4, 5, 6, 10, 12, 15, 19, and 20 in both NC-2018 (13 QTL) and IL-2020 (14 QTL). The six QTL regions on Chrs. 2, 4, 5, 12, 15, and 19 are novel regions while the other 21 QTL have been identified by other studies using different biparental mapping populations or genome-wide association studies (GWAS). A total of 130 candidate genes involved in isoflavone biosynthetic pathways have been identified on all 20 Chrs. And among them 16 have been identified and located within or close to the QTL identified in this study. Moreover, transcripts from four genes (*Glyma.10G058200*, *Glyma.06G143000*, *Glyma.06G137100*, and *Glyma.06G137300*) were highly abundant in Forrest and Williams 82 seeds. The identified QTL and four candidate genes will be useful in breeding programs to develop soybean cultivars with high beneficial isoflavone contents.

Keywords: soybean; RIL; Forrest; Williams 82; linkage map; isoflavone; daidzein; genistein; glycitein; SNP

1. Introduction

Soybean seeds are rich in secondary metabolites beneficial for human and animal consumption including tocopherols, phenolic compounds, saponins, and isoflavones such as genistein, daidzein, and glycitein that showed beneficial health and nutrition effects in animals and humans [1–3]. It is well established that isoflavones reduce menopausal symptoms, low density lipoprotein (LDL) cholesterol levels, breast and prostate cancers risks, improve the immune system [4–11], and play an important role in nitrogen fixation and defense against pathogens [12].

Due to these benefits and others, isoflavones, especially genistein, daidzein, and glycitein, have been widely studied during the past decades [13,14] and many studies tried to genetically map quantitative trait loci (QTL) that control seed genistein, daidzein, glycitein, and total isoflavone content as well as their precursors such as daidzin, glycitin, genistin, malonyldaidzin, malonylglycitin, malonylgenistin, etc., using different molecular markers such as AFLPs, RFLPs, SSRs, SNPs [15–27]. For example, using the 'Essex' by 'Forrest' recombinant inbred line (RIL) population (n = 100) and 250+ simple sequence repeat (SSR) markers, 11 QTL that control genistein, daidzein, glycitein, and total isoflavone contents have been identified on Chrs. 1, 3, 7, 8, 11, and 18 [15,16]. Likewise, Liang et al. (2010) used the 'Jindou 23' by 'Huibuzhi' RIL population (n = 474) and identified six QTL that control isoflavone contents and mapped them on soybean Chrs. 3, 16, 17, and 18 [18]. In another study, Smallwood et al. (2014) identified 3, 5, 7, and 6 QTL that control seed glycitein, daidzein, genistein, and total isoflavone contents, respectively [20]. Using the 'Zhongdou 27' by 'Jiunong 20' RIL population (n = 130) and 194 SSR markers, Han et al. (2015), identified 6, 5, 3, and 7 QTL that control seed glycitein, daidzein, genistein, and total isoflavone contents, respectively [24]. Akond et al. (2015) used the 'Hamilton' by 'Spencer' RIL population (n = 93), genotyped it with 1502 SNPs, and identified a major QTL that controls both seed daidzein and total isoflavone contents on Chr. 6 and a minor QTL that controls seed glycitein content on Chr. 18 [22]. Recently, the authors of [23] used 'Aokimame' by 'Fukuyutaka' and 'Kumaji-1' by 'Fukuyutaka' RIL populations and identified one QTL that controls malonylgenistin on Chr. 12 and two QTL that control malonylglycitin on Chrs. 11 and 15 [23]. Besides using biparental mapping populations, other researchers used natural populations and genome wide association studies (GWAS) to map QTL that control seed isoflavone contents and identified candidate genes within these QTL regions [28–32].

The objectives of this study were to construct a SNP-based genetic linkage map using the F×W82 RIL population (n = 306); map quantitative trait loci (QTL) for seed daidzein, genistein, glycitein, and total isoflavone contents in two environments over two years; identify candidate genes involved in soybean seed isoflavone biosynthesis.

2. Results and Discussion

2.1. The SNP-Based Genetic Map

A total of 5405 SNP markers were generated from the Infinium SNP6K BeadChips-based genotyping among 306 RILs, from which 2075 polymorphic SNPs were mapped on the 20 soybean chromosomes (Table 1, Figure 1). The F×W82 genetic map covered 4029.9 cM with an average marker density of 1.94 cM (Table 1). The genetic length ranged from 153.7 cM for Chr. 18 to 308.3 cM for Chr. 2 (Table 1). The polymorphism of SNPs in this RIL population (38.4%), number of linked SNPs, and map coverage were comparable to other reported SNP-based genetic linkage maps of soybean [33,34]. For example, in Akond et al. (2013) [33], only 27.33% of SNPs (1465/5361 × 100) have been used to construct the genetic map based on excluding missing data (~20%) and heterozygosity (3.99%). Polymorphic markers between parents (MD96-5722 and Spencer) among the 1465 SNPs used was 44.8% (657/1465 × 100) [33]. Likewise, in Kassem et al. (2012) [34], polymorphic markers between parents (PI 438489B and Hamilton) among the 1465 SNPs used was 44.2% (657/1465 × 100) [34].

Table 1. Distribution of SNP markers and their properties on the Chrs. Of Forrest by Williams 82 recombinant inbred line (RIL) population (n = 306).

Chr.	No. of SNP Markers	Length (cM)	Average Marker Density (cM)	Maximum Gap (cM)
1	110	190.1	1.73	48.7
2	161	308.3	1.91	59.0
3	92	173.9	1.89	22.9
4	71	214.9	3.03	57.4
5	138	167.2	1.21	38.1
6	114	253.7	2.23	43.4
7	117	224.0	1.91	18.4
8	71	211.1	2.97	45.3
9	109	179.2	1.64	62.6
10	100	216.5	2.17	48.5
11	95	168.9	1.78	41.3
12	73	192.6	2.64	31.5
13	156	265.7	1.70	57.7
14	50	158.9	3.18	25.8
15	94	219.1	2.33	68.4
16	95	169.0	1.78	46.7
17	79	185.4	2.35	46.5
18	144	153.7	1.07	23.1
19	125	190.4	1.52	53.3
20	81	187.3	2.31	25.7
Totals	2075	4029.9	Av. = 1.94	Av. = 43.2

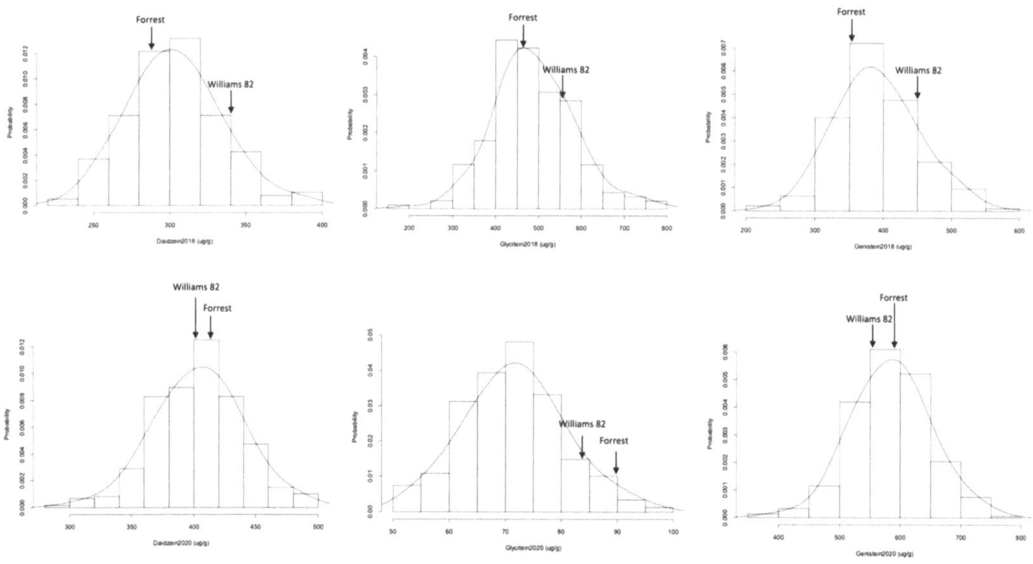

Figure 1. Frequency distribution of seed isoflavone contents (µg/g of seed weight) in the FxW82 RIL population. The seed daidzein, genistein, and glycitein contents were assessed in the RILs harvested in Spring Lake, NC (2018) and Carbondale, IL (2020).

2.2. Isoflavone Contents Frequency Distribution, Heritability, and Correlation

The seed isoflavone contents were normally distributed in the FxW82 RIL population based on Shapiro–Wilk's method for normality test, even though the positive or negative skewness and kurtosis value (>3) were observed in the RIL population (Table 2; Figure 1). The individual component of isoflavone also displayed small ranges of phenotypic variations in the seeds obtained from two geographically diverse field trials (Table 2, Figure 2). Daidzein 2018 in Spring Lake, NC had the highest coefficients of variation (CV) value (19.37%); however, the CV of this trait in Carbondale, IL (2020) was only 12.59% suggesting that phenotypic variability among isoflavone contents was impacted by different environmental conditions.

Table 2. Seed isoflavone means, ranges, CVs, skewness, and kurtosis in the FxW82 RIL population evaluated in Spring Lake, NC (2018) and Carbondale, IL (2020). Mean and range values are expressed in µg/g of seed weight.

Trait	Mean	Range	CV	SE	Skewness	Kurtosis	*p* Value ($p > 0.05$)
Daidzein 2018	303.22	171	10.11	2.23	0.26	3.11	0.99
Glycitein2018	490.47	610	19.37	7.01	0.29	3.65	0.98
Genistein2018	391	348	15.47	4.4	0.3	3.04	0.99
Daidzein 2020	14.48	8.08	13.72	0.42	−0.08	3.17	0.99
Glycitein2020	71.79	46	12.59	0.53	0.178	2.94	0.99
Genistein2020	584.88	383	10.94	3.73	−0.02	3.31	0.99

A. Unassorted correlogram.

B. Assorted correlogram.

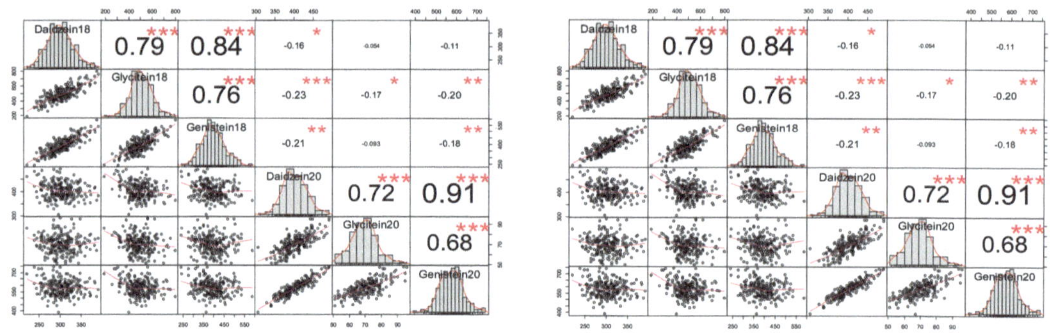

Figure 2. Correlations between daidzein, glycitein, and genistein in the two locations and years: Spring Lake, NC (2018) and Carbondale, IL (2020). (**A**). Unassorted correlogram, (**B**). Assorted correlogram. Significance level: * $p < 0.05$, ** $p < 0.01$, *** $p < 0.001$.

Azam et al. (2020) [35] reported that the total isoflavones ranged from 745 to 5253.98 µg/g, with highest mean of 2689.27 µg/g observed in some regions and up to 2518.91 and 1942.78 µg/g in others due to climatic conditions. Similar results have been reported by other studies [25–27,36–38]. Our results showed over 1000 µg/g and in some cases over 1100 µg/g. Therefore, the total concentrations of isoflavones are in the expected range of soybean seed. In addition, there is no premium to be given to growers for soybean seed isoflavone content and no docking is done at the grain elevator for seed isoflavone. Isoflavone concentrations vary depending on the year and environmental growing conditions. Although isoflavones are genetically controlled, environmental conditions including temperature, drought, absence or presence of diseases each year, many other biotic and abiotic factors can significantly affect the contents (by increasing or decreasing) and profile of isoflavone.

The broad sense heritability of percentage dry weight for daidzein, glycitein, and genistein across two different environments over two years appeared to be quite different. Glycitein had the highest heritability (72.4%) and the values for both daidzein and genistein were 42.8% and 42.5%, respectively, which displayed a similar fashion. The lower heritability of daidzein and genistein contents suggested that some portion of phenotypic variation was still not detected by the mapped QTL due to the complexity of these traits. The genotype–environment interactions still played a significant role in the molecular formation of daidzein and genistein molecules in soybean seeds based on our two-way ANOVA analysis because the σ_{GE2} is relatively high compared to that of glycitein (data not shown). It will certainly impact future breeding strategies for trait improvement based on the data we presented on these traits.

We used type I sum of squares (ANOVA (model)) function in R program to obtain the Sum Sq and Mean Sq and calculated $\sigma_G{}^2$ and $\sigma_{GE}{}^2$ for each trait (Table 3). However, $\sigma_e{}^2$ was 0 due to limited replicates. In this study, we only had three technical replicates due to cost effect of this student-centered project, but these replicates could only be considered as one biological replicate and hence, F value and probability could not be generated (Table 3). The FxW82 RIL population derived from two parental cultivars with different maturity groups (MGs). Forrest belongs to MG5-6 and Williams 82 to MG2-3 suggesting that the locations may play an important role on major agronomic traits including seed isoflavone. Based on our data (Figure 2), glycitein showed less correlation with daidzein and genistein which may indicate that its production may be less impacted by environment. Furthermore, Fayetteville, NC is a subtropic favorable weather for MG6-7 soybeans while Carbondale, IL is the favorable weather for MG 4-5. Therefore, further studies of the seed isoflavone in the FxW82 RIL population in different environments would be beneficial.

Table 3. Two-way ANOVA results for daidzein, genistein, and glycitein.

Response: Daidzein				
	Df	Sum Sq	Mean Seq	H^2
Line	301	541,800	1800	0.428
Year	1	974,711	974,711	
Line: Year	181	186,226	1029	
Residuals	0	0	NA	
Response: Glycitein				
	Df	Sum Sq	Mean Seq	H^2
Line	301	5,086,274	16,898	0.724
Year	1	16,033,506	16,033,506	
Line: Year	181	843,473	4660	
Residuals	0	0	NA	
Response: Genistein				
	Df	Sum Sq	Mean Seq	H^2
Line	301	1,922,339	6387	0.425
Year	1	3,630,207	3,630,207	
Line: Year	181	668,735	3695	
Residuals	0	0	NA	

The correlogram demonstrates a novel correlation among these assessed traits (Figure 2). Based on the unassorted data (all lines were included), each of the isoflavone components was positively correlated with the other sister isoflavones ($p < 0.001$) from the same geographical location but negatively correlated with the isoflavones from the other location inferring that the production of these isoflavones has been strongly impacted both by genotype and environmental conditions. The assorted data (lines tested in both locations) showed similar positivity, but the level of negative correlation was low (Figure 2). To the best of our knowledge, this observation has not been described in other studies.

2.3. Seed Isoflavone Contents QTL

Both interval mapping (IM) and composite interval mapping (CIM) methods of Win-QTL Cartographer 2.5 [39] were used to identify QTL for seed daidzein, genistein, glycitein, and total isoflavone contents in the present RIL population. A total of 27 QTL that control seed isoflavone contents have been identified in this population in both NC-2018 (13 QTL) and IL-2020 (14 QTL) (Table 4, Figure 3 and Figure S1).

Table 4. QTL that control seed isoflavone (daidzein, genistein, and glycitein) contents in two environments over two years (2A. 2018 and 2B. 2020). The two environments are Spring Lake, NC (2018) (2A) and Carbondale, IL (2020) (2B). Only QTL with LOD scores > 2.0 and identified by composite interval mapping (CIM) method of QTL Cartographer (Wang et al., 2012) are reported.

2A. QTL Identified in Spring Lake, NC (2018)								
Trait	**QTL**	**Chr.**	**Marker**	**Interval (cM)**	**LOD**	**R^2 (%)**	**Additive Effect**	**Environment**
Daidzein	*qDAID01*	5	Gm05_1705841	160.41	2.01	6.07	−7.61	Spring Lake, NC
Daidzein		5	Gm05_9012813	166.51	2.12	4.18	−6.35	Spring Lake, NC
Daidzein		5	Gm05_9097414	166.71	2.19	4.32	−6.46	Spring Lake, NC
Daidzein		5	Gm05_8916450	166.81	2.11	4.16	−6.34	Spring Lake, NC
Genistein	*qGEN03*	5	Gm05_1705841	152.41	2.06	9.37	−18.72	Spring Lake, NC
Genistein		5	Gm05_9012813	166.51	2.27	4.22	−12.53	Spring Lake, NC
Genistein		5	Gm05_9097414	166.71	2.36	4.39	−12.79	Spring Lake, NC
Genistein		5	Gm05_8916450	166.81	2.28	4.25	−12.57	Spring Lake, NC
Glycitein	*qGLY02*	5	Gm05_1705841	146.41	2.01	9.07	−29.89	Spring Lake, NC
Genistein	*qGEN01*	6	Gm06_5014399	64.41	2.58	8.52	21.15	Spring Lake, NC
Daidzein	*qDAID02*	6	Gm06_5014399	62.41	2.06	7.55	10.35	Spring Lake, NC
Daidzein		6	Gm06_3941524	78.21	2.02	7.24	8.67	Spring Lake, NC
Genistein	*qGEN04*	6	Gm06_5014399	60.41	2.26	9.06	22.94	Spring Lake, NC
Genistein		6	Gm06_3941524	70.21	2.11	3.98	13.86	Spring Lake, NC
Genistein	*qGEN02*	12	Gm12_915327	179.21	2.56	4.8	−16.87	Spring Lake, NC
Genistein		12	Gm12_1064727	179.41	2.58	4.85	−16.93	Spring Lake, NC
Genistein		12	Gm12_1229101	179.71	2.95	5.51	−17.79	Spring Lake, NC
Genistein		12	Gm12_1374970	179.91	2.95	5.5	−17.78	Spring Lake, NC
Genistein		12	Gm12_1433336	180.61	2.85	5.33	−17.4	Spring Lake, NC
Glycitein	*qGLY01*	12	Gm12_553862	177.31	2.76	5.82	−26.95	Spring Lake, NC
Glycitein		12	Gm12_915327	179.21	2.58	4.79	−27.05	Spring Lake, NC
Glycitein		12	Gm12_1064727	179.41	2.59	4.82	−27.12	Spring Lake, NC
Glycitein		12	Gm12_1229101	179.71	2.83	5.23	−27.76	Spring Lake, NC
Glycitein		12	Gm12_1374970	179.91	2.83	5.23	−27.76	Spring Lake, NC
Glycitein		12	Gm12_1433336	180.61	2.7	5.02	−27.03	Spring Lake, NC
Genistein	*qGEN05*	12	Gm12_553862	171.31	2.03	5.74	−15.18	Spring Lake, NC
Genistein		12	Gm12_975837	178.71	2.47	4.64	−16.47	Spring Lake, NC
Genistein		12	Gm12_1632399	181.31	2.32	4.38	−15.76	Spring Lake, NC
Glycitein	*qGLY03*	12	Gm12_553862	169.31	2.22	5.63	−23.71	Spring Lake, NC
Glycitein		12	Gm12_975837	178.71	2.46	4.58	−26.29	Spring Lake, NC
Glycitein		12	Gm12_1632399	181.31	2.2	4.1	−24.52	Spring Lake, NC
Daidzein	*qDAID03*	19	Gm19_4552537	109.51	2.14	4.17	6.92	Spring Lake, NC
Glycitein	*qGLY04*	19	Gm19_3010363	35.31	2.04	3.83	19.46	Spring Lake, NC
Genistein	*qGEN06*	20	Gm20_4657454	0.01	2.03	3.75	11.84	Spring Lake, NC

Table 4. *Cont.*

2B. QTL Identified in Carbondale, IL (2020)								
Trait	**QTL**	**Chr.**	**Marker**	**Interval (cM)**	**LOD**	**R2 (%)**	**Additive Effect**	**Environment**
Daidzein	*qDAID03*	2	Gm02_2282900	24.01	2.01	10.61	−11.55	Carbondale, IL
Genistein	*qGEN04*	4	Gm04_4461164	190.31	2.26	3.96	−12.97	Carbondale, IL
Genistein	*qGEN01*	10	Gm10_4670275	130.81	2.6	3.52	−12.38	Carbondale, IL
Genistein		10	Gm10_4035277	130.91	2.61	3.53	−12.39	Carbondale, IL
Daidzein	*qDAID04*	10	Gm10_4670275	130.81	2.18	2.97	−6.11	Carbondale, IL
Daidzein		10	Gm10_4035277	130.91	2.19	2.97	−6.12	Carbondale, IL
Genistein	*qGEN05*	10	Gm10_4670275	128.81	2.15	3.2	−11.75	Carbondale, IL
Genistein		10	Gm10_4035277	132.91	2.37	3.39	−12.11	Carbondale, IL
Daidzein	*qDAID01*	12	Gm12_9193994	53.21	2.56	4.07	7.18	Carbondale, IL
Daidzein		12	Gm12_1430950	61.71	3.99	5.61	8.49	Carbondale, IL
Daidzein		12	Gm12_1423120	62.31	4.12	5.78	8.62	Carbondale, IL
Daidzein		12	Gm12_1539402	63.01	4.53	6.34	9.01	Carbondale, IL
Daidzein		12	Gm12_1678702	63.11	4.59	6.45	9.1	Carbondale, IL
Daidzein		12	Gm12_3052701	64.11	4.89	6.82	9.42	Carbondale, IL
Daidzein		12	Gm12_2097199	64.41	4.4	6.18	8.97	Carbondale, IL
Daidzein		12	Gm12_2432082	65.31	4.26	5.97	8.77	Carbondale, IL
Daidzein		12	Gm12_1547239	65.51	3.77	5.31	8.27	Carbondale, IL
Daidzein		12	Gm12_1428801	65.91	3.36	4.75	7.87	Carbondale, IL
Genistein	*qGEN02*	12	Gm12_9193994	55.21	2.5	4.24	13.58	Carbondale, IL
Genistein		12	Gm12_1430950	61.71	3.78	5.28	15.31	Carbondale, IL
Genistein		12	Gm12_1423120	62.31	3.76	5.25	15.2	Carbondale, IL
Genistein		12	Gm12_1539402	63.01	4.39	6.1	16.4	Carbondale, IL
Genistein		12	Gm12_1678702	63.11	4.38	6.1	16.4	Carbondale, IL
Genistein		12	Gm12_3052701	64.11	4.66	6.46	16.99	Carbondale, IL
Genistein		12	Gm12_2097199	64.41	4.05	5.65	15.64	Carbondale, IL
Genistein		12	Gm12_2432082	65.31	3.83	5.35	15.25	Carbondale, IL
Genistein		12	Gm12_1547239	65.51	3.38	4.74	14.39	Carbondale, IL
Genistein		12	Gm12_1428801	65.91	3.16	4.44	13.96	Carbondale, IL
Daidzein	*qDAID05*	12	Gm12_9193994	51.21	2.03	2.91	6.05	Carbondale, IL
Daidzein		12	Gm12_1428801	73.91	2.31	4.8	7.83	Carbondale, IL
Genistein	*qGEN06*	12	Gm12_1428801	71.91	2.37	4.4	13.77	Carbondale, IL
Glycitein	*qGLY01*	15	Gm15_756303	212.31	2.07	2.86	−1.53	Carbondale, IL
Glycitein		15	Gm15_2072075	218.41	2.24	3.1	−1.6	Carbondale, IL
Glycitein		15	Gm15_2021199	218.81	2.06	2.84	−1.53	Carbondale, IL
Daidzein	*qDAID02*	20	Gm20_3804081	70.31	2.55	6.13	−8.81	Carbondale, IL
Genistein	*qGEN03*	20	Gm20_3804081	68.31	2.65	6.72	−16.81	Carbondale, IL
Daidzein	*qDAID06*	20	Gm20_3804081	66.31	2.22	5.58	−8.41	Carbondale, IL
Genistein	*qGEN07*	20	Gm20_3804081	64.31	2	5.08	−14.62	Carbondale, IL
Genistein		20	Gm20_3424023	80.01	2.44	3.3	−11.8	Carbondale, IL
Genistein		20	Gm20_3418121	80.51	2.1	2.85	−10.96	Carbondale, IL

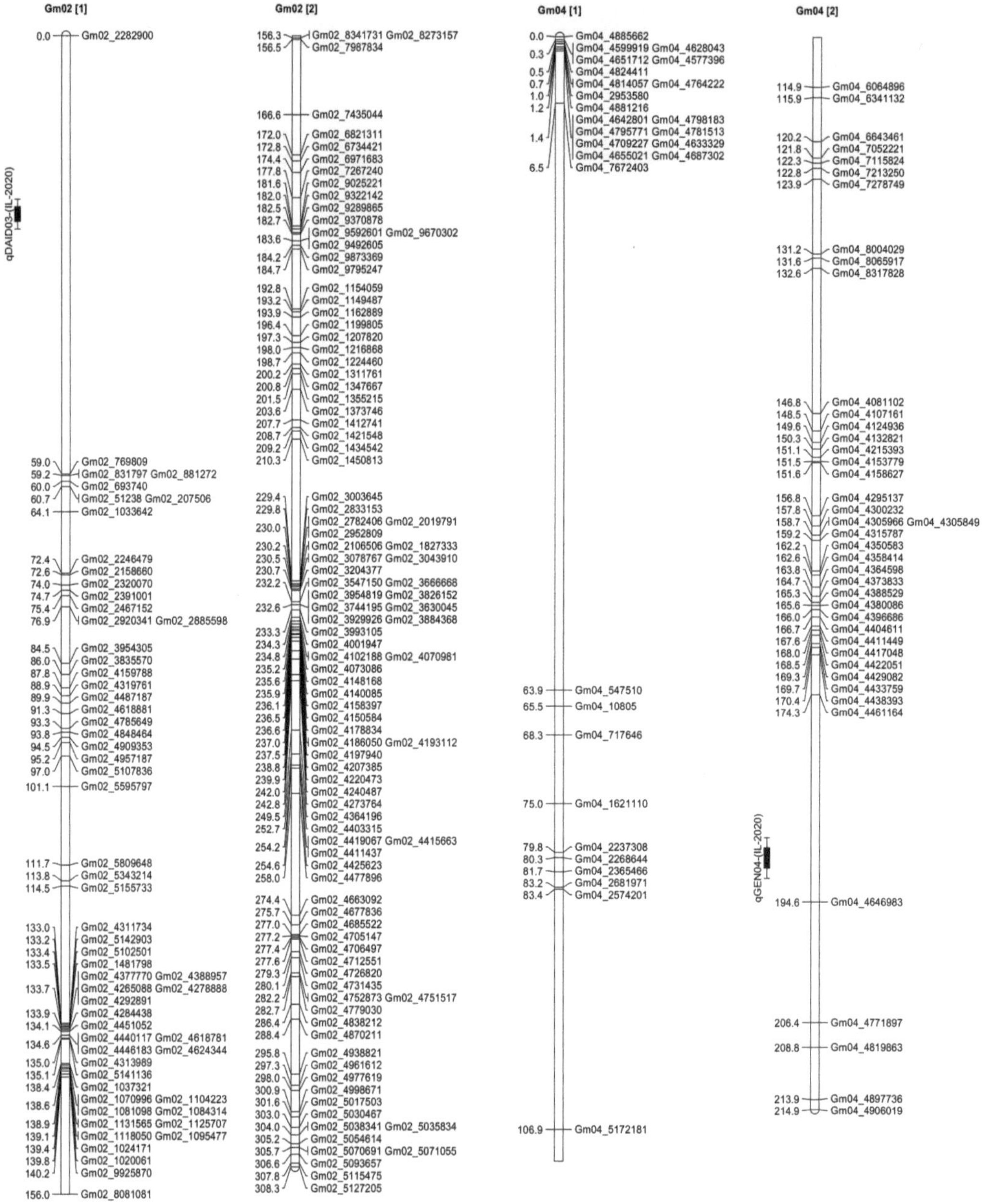

Figure 3. *Cont.*

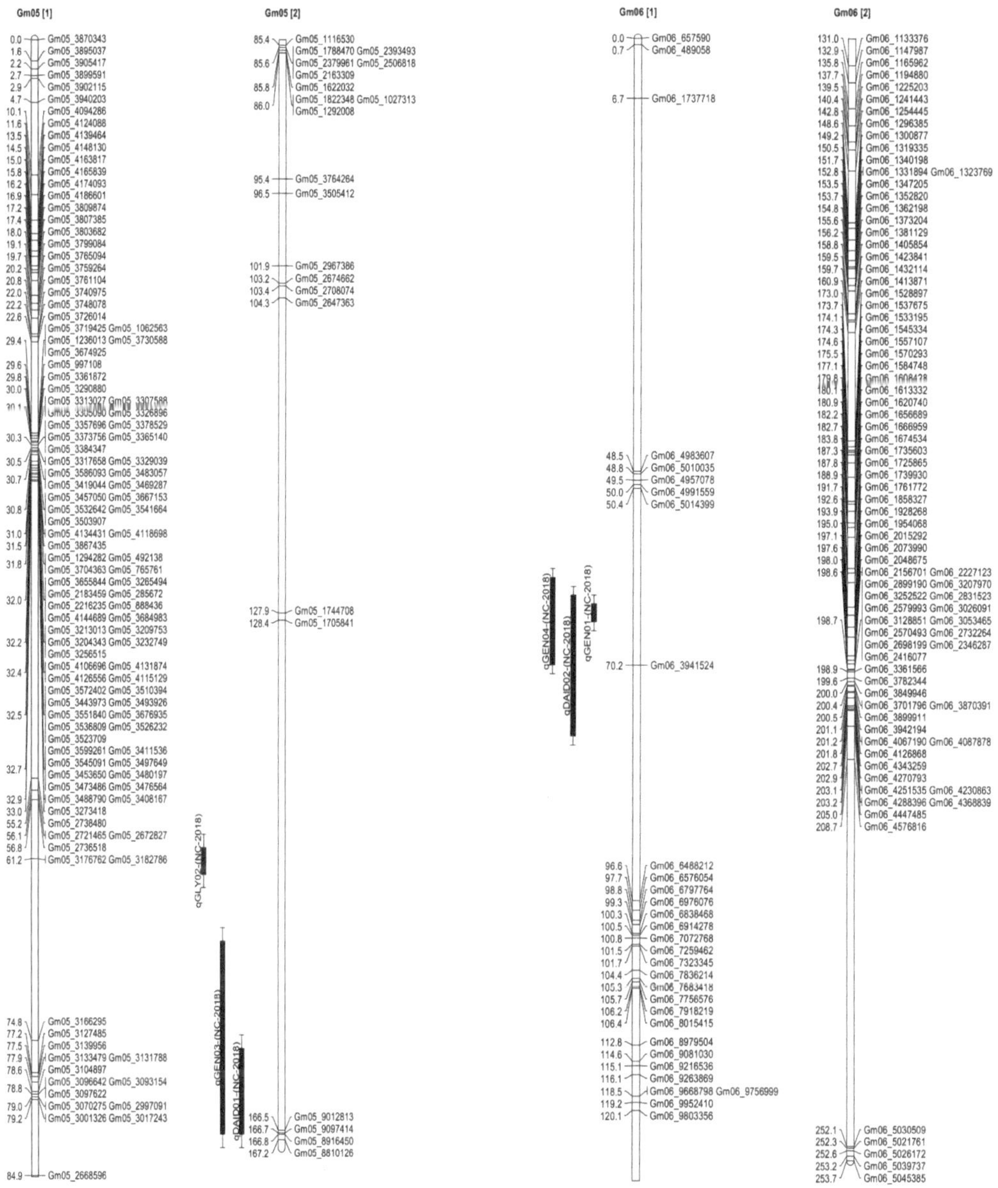

Figure 3. *Cont.*

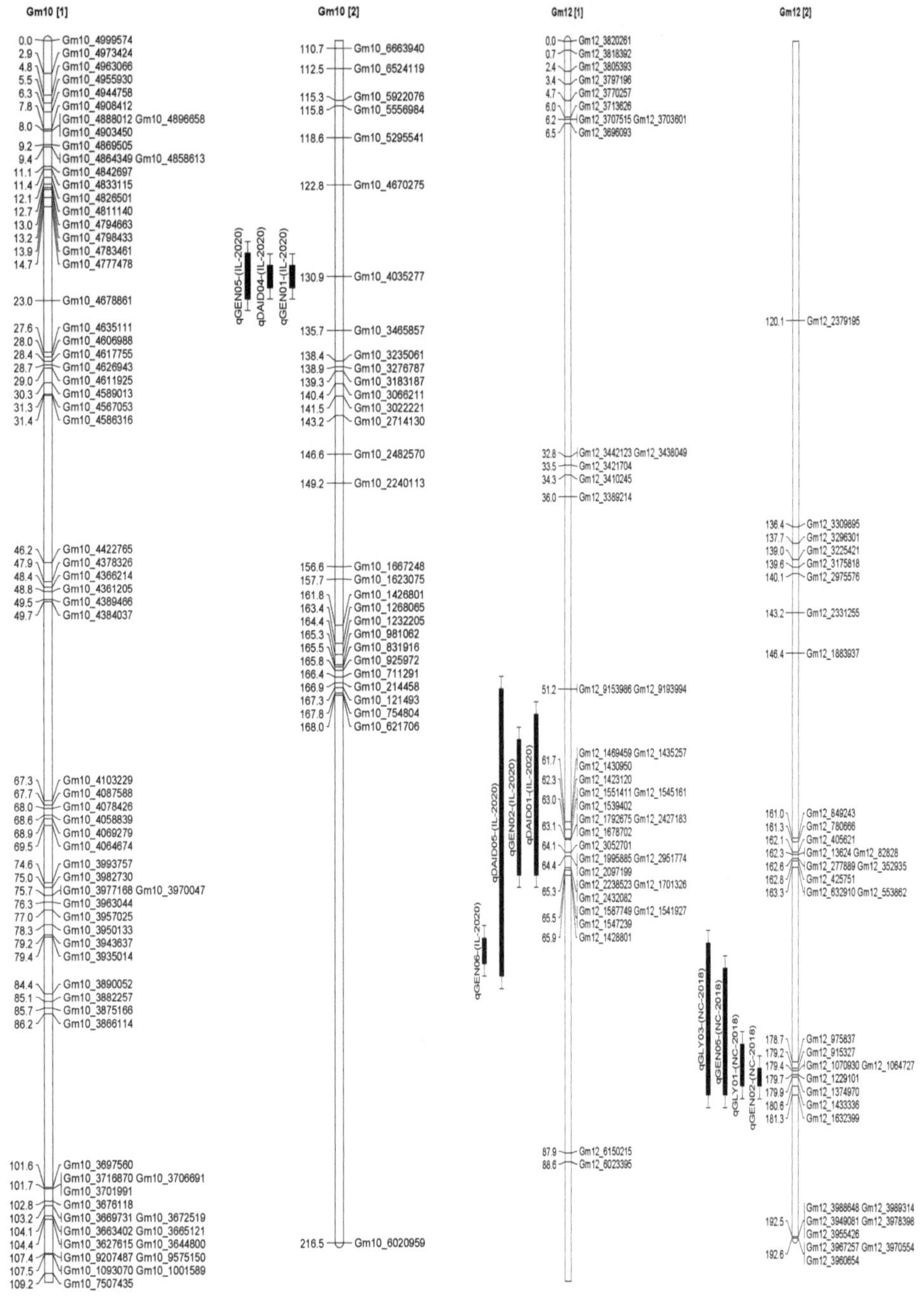

Figure 3. *Cont.*

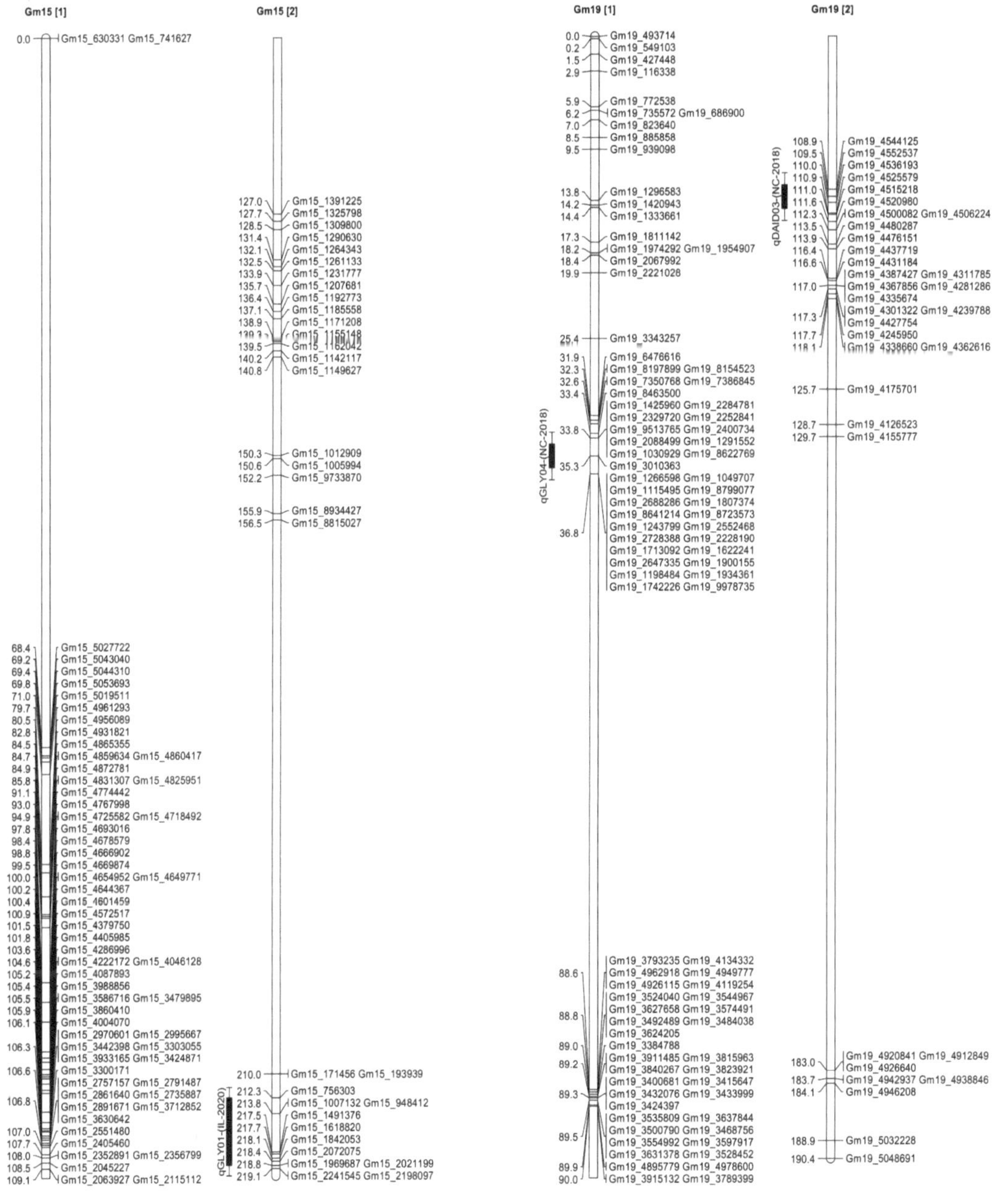

Figure 3. *Cont.*

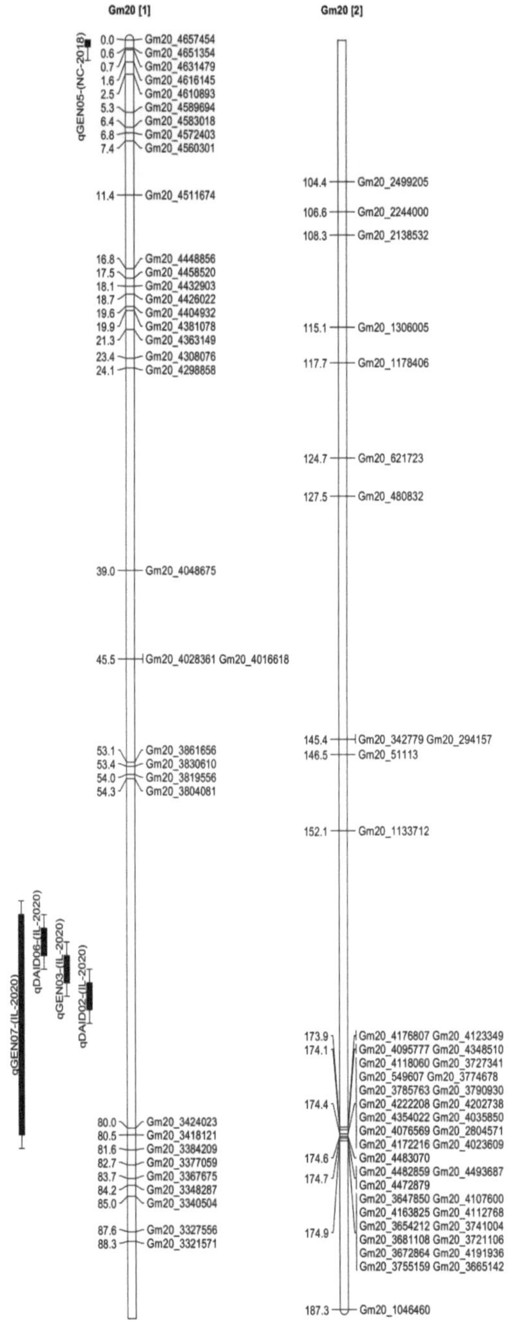

Figure 3. Positions of QTL that control seed genistein (qGEN), daidzein (qDAID), and glycitein (qGLY) contents on Chrs. 2, 4, 5, 6, 10, 12, 15, 19, and 20. QTL names are followed by a number, location, and year in which they are identified. For example, qGEN01-(NC-2018). The full SNP-based genetic linkage map of Forrest by Williams 82 recombinant inbred line (RIL) population (n = 306) of soybean is shown in Figure S2.

In Carbondale, IL (IL-2020), one QTL that controls seed daidzein content (*qDAID03*) has been identified and mapped on Chr. 2 and one QTL that controls seed genistein content (*qGEN04*) has been identified and mapped on Chr. 4 (Table 4, Figure 3 and Figure S1). One QTL that controls seed daidzein content (*qDAID04*) and two QTL that control seed genistein content (*qGEN01* and *qGEN05*) have been identified and mapped on Chr. 10. Two QTL that control each of seed daidzein (*qDAID01* and *qDAID05*) and seed genistein contents (*qGEN02* and *qGEN06*) have been identified and mapped on Chr. 12 (Table 4, Figure 3 and Figure S1). One QTL that controls seed glycitein (qGLY01) has been identified and mapped on Chr. 15 (Table 4, Figure 3 and Figure S1). Two QTL that control each of seed daidzein (*qDAID02* and *qDAID06*) and genistein contents (*qGEN03* and *qGEN07*) have been identified and mapped on Chr. 20 (Table 4, Figure 3 and Figure S1).

In Spring Lake, NC (NC-2020), one QTL that controls each of seed daidzein (*qDAID01*), genistein (*qGEN03*), and glycitein contents (*qGLY02*) have been identified and mapped on Chr. 5 (Table 4, Figure 3 and Figure S1). Two QTL that control seed genistein content (*qGEN01* and *qGEN04*) and one QTL that controls seed daidzein content (*qDAID02*) have been identified and mapped on Chr. 6 (Table 4, Figure 3 and Figure S1). Two QTL that control each of seed genistein (*qGEN02* and *qGEN05*) and glycitein contents (*qGLY01* and *qGLY03*) have been identified and mapped on Chr. 12 (Table 4, Figure 3 and Figure S1). One QTL that controls each of seed daidzein (*qDAID03*) and glycitein contents (*qGLY04*) have been identified and mapped on Chr. 19 (Table 4, Figure 3 and Figure S1). One QTL that controls seed genistein content (*qGEN06*) has been identified and mapped on Chr. 20 (Table 4, Figure 3 and Figure S1). No QTL that controls total seed isoflavone contents has been identified in both years and locations.

No previous studies identified QTL that control seed isoflavone contents in the QTL region identified on Chr. 2 (*qDAID03*-(IL-2020), 23–25 cM), indicating that this is a novel QTL region; however, other studies identified QTL that control seed calcium content, plant height, and few other traits [40,41]. Likewise, no other studies identified QTL that control seed isoflavone contents in the QTL region identified on Chr. 4 (*qGEN04*-(IL-2020), 189.3–191.3 cM) which indicates that it is also a novel QTL region. The length of Chr. 4 in the soybean consensus map is only 136 cM [29,31]. Additionally, no other studies identified QTL that control seed isoflavone contents in the QTL region identified on Chr. 5 (*qDAID01*-(NC-2018), *qGEN03*-(NC-2018), and *qGLY02*-(NC-2018), 152.4–166.4 cM) which indicates the discovery of a novel QTL region. The length of Chr. 5 in the soybean consensus map is only 104 cM [29,31]. Interestingly, within the same QTL region that controls seed genistein and daidzein contents on Chr. 6 (*q-GEN01*-(NC-2018), *qGEN04*-(NC-2018), and *qDAID02*-(NC-2018), other studies identified QTL that control seed genistein, daidzein, glycitein, and total isoflavone contents (see a summary in [30]) which is coherent with our results making it an important genomic region to further investigate for candidate genes. Other studies identified QTL that control seed protein, oil, γ–tocopherol, and amino acids contents, and few other traits [29,31]. Interestingly, within the same QTL region that controls seed genistein and daidzein contents on Chr. 10 (*q-GEN01*-(IL-2020), *qGEN05*-(IL-2020), and *qDAID04*-(IL-2020), 128.8–132.9 cM), another study identified QTL that control seed isoflavone content [41,42] which is consistent with our data making it an important genomic region for gene discovery. In fact, two candidate genes have been previously identified in this region [42,43]. Two QTL regions have been identified on Chr. 12. The first region containing QTL that control seed genistein and daidzein contents (*qGEN02*-(IL-2020), *qGEN06*-(IL-2020), and *qDAID05*-(IL-2020), 51.2–71.9 cM). Interestingly, other studies identified QTL that control seed daidzein, genistein, glycitein, and total isoflavone contents in the same QTL region (see a summary in [30]) which makes it an important genomic region for discovering novel candidate genes. The second region contained QTL that control seed genistein, and glycitein contents (*qGEN02*-(NC-2018), *qGEN05*-(NC-2018), *qGLY01*-(NC-2018), and *qGLY03*-(NC-2018), 169.3–181.3 cM). No other studies identified QTL that control seed isoflavone contents in this second QTL region which indicates that it is also a novel QTL region. The length of Chr. 12 in the soybean consensus map is only

125 cM and the second QTL region identified here falls outside of its current limit [29,31]. No previous studies identified QTL that control seed isoflavone contents in the QTL region identified on Chr. 15 (*qGLY01*-(IL-2020), 212.3–218.8 cM) which indicates that it is a novel QTL region. The length of Chr. 15 in the soybean consensus map is only 85 cM and the QTL region identified here falls outside of its current limit [29,31]. Two QTL regions have been identified on Chr. 19. The first region contains QTL that control seed glycitein content (*qGLY04*-(NC-2018), 34.3–36.3 cM). Interestingly, other studies identified QTL that control seed genistein, daidzein, and isoflavone content within the same QTL region (see a summary in [30]). Previous studies identified also QTL for seed protein content (see a summary in [44]). The second region contained QTL that control (*qDAID03*-(NC-2018), 108.5–110.5 cM). No other studies identified QTL that control seed isoflavone contents in this QTL region, making it a novel QTL region. Two QTL regions have been identified on Chr. 20. Within the first region containing QTL that control seed glycitein content (*qGEN05*-(NC-2018), 0–2 cM), other studies identified QTL that control seed daidzein (*qD20*), genistein (*qG20*), malonyldaidzein (*qMD20*), malonylgenistein (*qMG20*), and total isoflavone content (*qTIF20*) [41,44] which makes it an important region to investigate further for candidate genes. In addition, other studies identified QTL for seed calcium [30,44] and sucrose contents within this QTL region as well (see a summary in [44]). Within the second region containing QTL that control seed daidzein and genistein contents (*qDAID02*-(IL-2020), *qDAID06*-(IL-2020), *qGEN03*-(IL-2020), and *qGEN07*-(IL-2020), 64.3–80.5 cM), other studies identified QTL that control seed genistein content (*qGEN20*, [17]) and seed daidzein and glycitein contents (*qGC | prol_1 and qDZ | prol_2*, [39,41] which makes it another important region to investigate further for candidate genes. In addition, other studies identified QTL that control seed phytate, stearic acid, calcium, alpha-tocopherol, and few amino acids [44].

2.4. Seed Isoflavone Candidate Genes

A total of 130 candidate genes involved in soybean isoflavone biosynthetic pathway have been identified in all 20 soybean Chrs. (Table S1); however, 16 candidate genes have been identified within or close to the seed isoflavone QTL identified in this study on Chrs. 2, 6, 10, 12, 15, 19, and 20 (Figure 4, Table 5).

Among them, the 4′-methoxyisoflavone 2-hydroxylase gene (*Glyma.02G067900*) and the chalcone synthase gene (*Glyma.02G130400*) are located at 3.7 and 11.11 cM, respectively, from qDAID03-(NC-2018) on Chr. 2 (Figure 4, Table 5 and Table S1). *Glyma.06G128200* is a flavonol synthase gene located at 5.52 cM from qGEN01-(NC-2018), qDAID02-(NC-2018), and qGEN04-(NC-2018) on Chr. 6. The flavonol 3-O-methyltransferase genes (*Glyma.06G137100* and *Glyma.06G137300*) as well as the chalcone-flavonone isomerase gene (*Glyma.06G143000*) are located at 6 cM from qGEN01-(NC-2018) on Chr. 6 (Figure 4, Table 5 and Table S1). *Glyma.10G058200* is a phenylalanine ammonia-lyase gene that is located 0.6 cM from qGEN01-(IL-2020), qDAID04-(IL-2020) and qGEN05-(IL-2020) on Chr. 10 (Figure 4, Table 4 and Table S1). *Glyma.12G067000* and *Glyma.12G067100* are located within qDAID01-(IL-2020), qGEN02-(IL-2020) and qDAID05-(IL-2020) on Chr. 12 and near to (<4 cM) qGEN02-(NC-2018), qGLY01-(NC-2018), qGEN05-(NC-2018), qGEN06-(IL-2020) and qGLY03-(NC-2018) on Chr. 12 (Figure 4, Table 5 and Table S1). Both genes are Cytochrome P450 CYP2 subfamily genes; *Glyma.12G067000* was classified as an isoflavone synthase II gene and *Glyma.12G067100* as its duplicate with 95% identical nucleotide positions in the protein coding sequence ([42] Fliegmann et al., 2010). The trans-feruloyl-CoA synthase gene *Glyma.15G001700* is located at 0.56 cM from qGLY01-(IL-2020) on Chr. 15. The Isoflavone 3′-hydroxylase gene *Glyma.15G156300* and the Isoflavone 2′-hydroxylase gene *Glyma.15G156100* are located at about 11 cM from qGLY01-(IL-2020) on Chr. 15 (Figure 4, Table 5 and Table S1). *Glyma.20G027800* is an isoflavone reductase gene that is located within qDAID06-(IL-2020) and at 0.62 cM from qDAID02-(IL-2020), qGEN03-(IL-2020), and qGEN06-(NC-2018) and 2.44 cM from qGEN07-(IL-2020) on Chr. 20 (Figure 4, Table 5 and Table S1). Three genes *Glyma.19G030500*, *Glyma.19G030700*, and *Glyma.19G030800* encoding for an isoflavone 7-O-glucoside-6″-O-malonyltransferase gene family are located at less than 1 cM from qDAID03-(NC-2018) and qGLY04-(NC-2018) on Chr. 19 (Figure 4, Table 5 and Table S1). Interestingly, Wu et al. (2020) identified seven candidate genes including the mitogen-activated protein kinase (MPK) gene (*Glyma.08G309500*) within the seed isoflavone QTL identified on Chr. 8 [29,31]. A summary of seed isoflavone QTL and corresponding candidate genes for over two decades of research (1999–2020) can be found in Kassem [30]. Recently, Yang et al. (2021) [32] identified four candidate genes including GSTT1a (Glyma.05G206900), GSTT1b (Glyma.05G207000), and the transcription factor (TF) GL3 (Glyma.05G208300) on Chr. 5, and GSTL3 (Glyma.13G135600) on Chr. 13 [32].

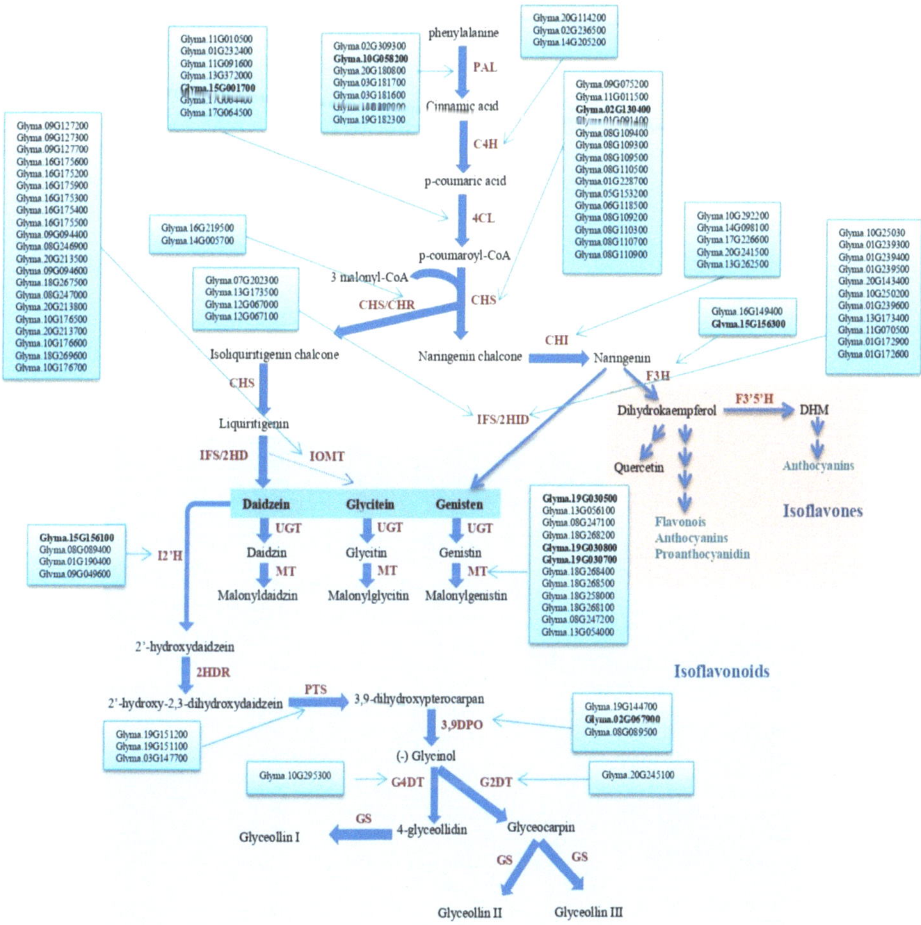

Figure 4. Seed isoflavone metabolic pathway in soybean with identified candidate genes (Vadivel et al., 2010). PAL, phenylalanine ammonia lyase; C4H, cinnamate 4-hydroxylate; 4CL, 4-coumarate-CoA ligase; CHS, chalcone synthase; CHR, chalcone reductase; CHI, chalcone isomerase; IFS, 2-hydroxyisoflavanone synthase; 2HID, 2-hydroxyisoflavanone dehydratase; IOMT, isoflavone O-methyltransferase; UGT, uridine diphosphate glycosyltransferase; MT, malonyltransferase; I2′H, Isoflavone 2′-hydroxylase; 2HDR, 2′-hydroxydaidzein reductase; F3H, flavanone-3-hydroxylase; F3′5′H, flavonoid 3′5′-hydroxylase; DHM, dihydromyricetin; PTS, pterocarpan synthase; 3,9 DPO, 3,9-dihydroxypterocarpan 6a-monooxygenase; G4DT, glycinol 4-dimethylallyltransferase; G2DT, glycinol 2-dimethylallyltransferase; GS, glyceollin synthase.

Table 5. Isoflavone candidate genes located within or close to the isoflavone QTL identified in the FxW82 RIL population two environments over two years (A. Spring Lake, NC (2018) and B. Carbondale, IL (2020)).

(A).

Environment	Trait	QTL	Chr.	Gene	Start	End	Distance (cM)
				2018 CIM QTL with LOD Scores > 2.5			
Spring Lake, NC	Genistein	*qGEN01*	6	Glyma.06G128200	10,543,911	10,545,747	5.52 cM
				Glyma.06G137100	11,225,188	11,228,664	6.21 cM
				Glyma.06G137300	11,237,072	11,239,469	6.22 cM
				Glyma.06G143000	11,642,031	11,644,022	6.62 cM
Spring Lake, NC	Genistein	*qGEN02*	12	Glyma.12G067000	4,909,073	4,911,905	3.47 cM
				Glyma.12G067100	4,919,960	4,922,998	3.48 cM
Spring Lake, NC	Glycitein	*qGLY01*	12	Glyma.12G067000	4,909,073	4,911,905	3.47 cM
				Glyma.12G067100	4,919,960	4,922,998	3.48 cM
				2018 CIM QTL with LOD Scores 2.0 < LOD < 2.5			
Spring Lake, NC	Daidzein	*qDAID01*	5	-	-	-	-
Spring Lake, NC	Daidzein	*qDAID02*	6	Glyma.06G128200	10,543,911	10,545,747	5.52 cM
Spring Lake, NC	Daidzein	*qDAID03*	19	Glyma.19G030500	3,779,017	3,781,453	0.77 cM
				Glyma.19G030700	3,794,404	3,796,426	0.75 cM
				Glyma.19G030800	3,799,941	3,801,335	0.75 cM
Spring Lake, NC	Genistein	*qGEN03*	5	-			
Spring Lake, NC	Genistein	*qGEN04*	6	Glyma.06G128200	10,543,911	10,545,747	5.52 cM
Spring Lake, NC	Genistein	*qGEN05*	12	Glyma.12G067000	4,909,073	4,911,905	3.27 cM
				Glyma.12G067100	4,919,960	4,922,998	3.28 Cm
Spring Lake, NC	Genistein	*qGEN06*	20	Glyma.20G027800	3,179,955	3,183,453	1.47 cM
Spring Lake, NC	Glycitein	*qGLY02*	5	-	-	-	-
Spring Lake, NC	Glycitein	*qGLY03*	12	Glyma.12G067000	4,909,073	4,911,905	3.27 cM
				Glyma.12G067100	4,919,960	4,922,998	3.28 Cm
Spring Lake, NC	Glycitein	*qGLY04*	19	Glyma.19G030500	3,779,017	3,781,453	0.76 cM
				Glyma.19G030700	3,794,404	3,796,426	0.78 cM
				Glyma.19G030800	3,799,941	3,801,335	0.78 cM

(B).

Environment	Trait	QTL	Chr.	Gene	Start	End	Distance (cM)
				2020 CIM QTL with LOD Scores > 2.5			
Carbondale, IL	Daidzein	*qDAID01*	12	Glyma.12G067000	4,909,073	4,911,905	-
				Glyma.12G067100	4,919,960	4,922,998	-
Carbondale, IL	Daidzein	*qDAID02*	20	Glyma.20G027800	3,179,955	3,183,453	0.62 cM
Carbondale, IL	Genistein	*qGEN01*	10	Glyma.10G058200	5,328,963	5,333,501	0.6 cM
Carbondale, IL	Genistein	*qGEN02*	12	Glyma.12G067000	4,909,073	4,911,905	-
				Glyma.12G067100	4,919,960	4,922,998	-
Carbondale, IL	Genistein	*qGEN03*	20	Glyma.20G027800	3,179,955	3,183,453	2.44 cM
				2020 CIM QTL with LOD Scores 2.0 < LOD < 2.5			
Carbondale, IL	Daidzein	*qDAID03*	2	Glyma.02G067900	5,986,285	5,987,684	3.70 cM
				Glyma.02G130400	13,399,253	13,401,493	11.11 cM
Carbondale, IL	Daidzein	*qDAID04*	10	Glyma.10G058200	5,328,963	5,333,501	0.6 cM
Carbondale, IL	Daidzein	*qDAID05*	12	Glyma.12G067000	4,909,073	4,911,905	-
				Glyma.12G067100	4,919,960	4,922,998	-
Carbondale, IL	Daidzein	*qDAID06*	20	Glyma.20G027800	3,179,955	3,183,453	-
Carbondale, IL	Genistein	*qGEN04*	4	-	-	-	-
Carbondale, IL	Genistein	*qGEN05*	10	Glyma.10G058200	5,328,963	5,333,501	0.6 cM
Carbondale, IL	Genistein	*qGEN06*	12	Glyma.12G067000	4,909,073	4,911,905	3.48 cM
				Glyma.12G067100	4,919,960	4,922,998	3.49 cM
Carbondale, IL	Genistein	*qGEN07*	20	Glyma.20G027800	3,179,955	3,183,453	2.44 cM
Carbondale, IL	Glycitein	*qGLY01*	15	Glyma.15G001700	190,985	194,451	0.56 cM
				Glyma.15G156300	13,098,492	13,100,036	11.02 cM
				Glyma.15G156100	13,076,997	13,079,333	11 cM

2.5. Expression Analysis

To gain insight into the role of isoflavone genes in soybean seeds, RNA-Seq analysis was conducted to check the expression levels of the 16 candidate genes that are located within or near the isoflavone QTL identified in FxW82 RIL population. Expression analysis of these genes showed that four genes, Glyma.10G058200, Glyma.06G143000, Glyma.06G137100, and Glyma.06G137300, are highly expressed in seeds of both Forrest and Williams 82 cultivars. Whereas, Glyma.19G030800 is highly expressed in Williams 82 and have a low expression in Forrest cv.; the rest of the 16 genes showed lower expressions; whereas Glyma.02G067900 and Glyma.15G156300 were not expressed neither in Forrest nor in Williams 82 cultivars (Figure 5).

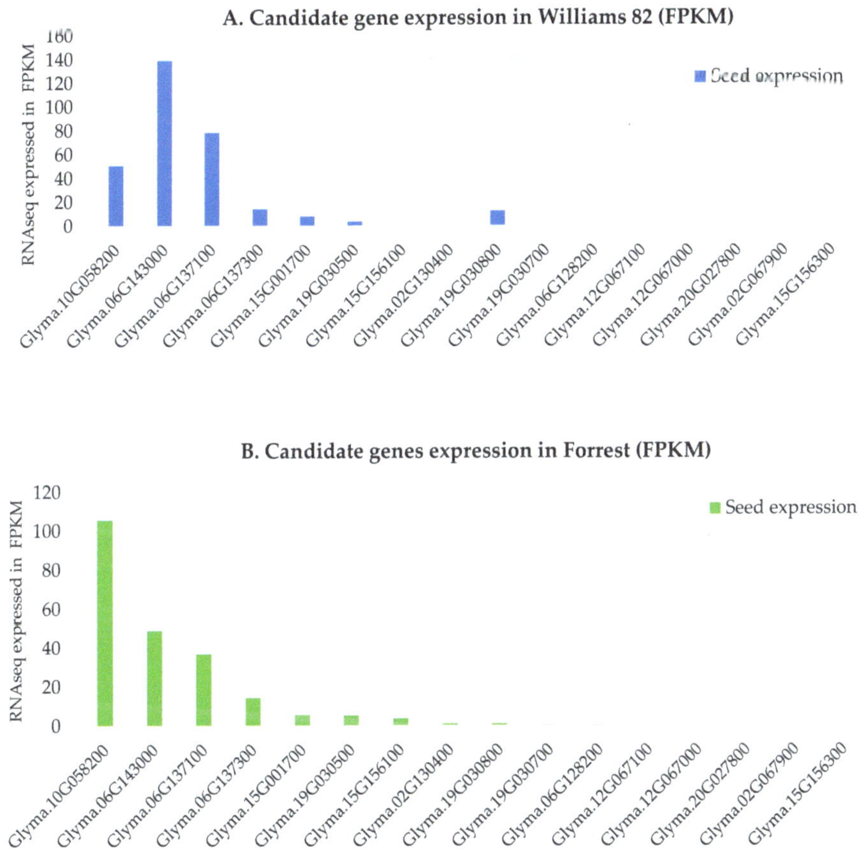

Figure 5. Expression pattern of isoflavone genes in soybean seeds. (**A**) Expression of the 16 isoflavone genes located within isoflavone QTL in Williams 82 (FPKM) were retrieved from publicly available RNA-seq data from Phytozome database [45], in addition to (**B**) the RNAseq data from the cultivar 'Forrest' (FPKM).

Surprisingly, Glyma.10G058200 expression in Forrest cv. is higher than its expression in Williams 82. This could explain the presence of RILs from the FxWI cross that showed higher daidzein, glycitein or genistein content than the parent Williams 82 (parent with the high isoflavones content), these lines inherited most likely the beneficial alleles from both parents Forrest and Williams 82.

2.6. Conclusions

In conclusion, we constructed the FxW82 dense SNP-based genetic linkage map (2075 SNPs and 4029.9 cM covered) and identified 27 QTL that control soybean seed isoflavone contents and 16 candidate genes involved in soybean isoflavone biosynthetic pathways among which four candidate genes are highly expressed in seeds of both Forrest and Williams 82, in addition to Glyma.19G030800 that has a higher expression profile in Williams 82 compared to its expression in Forrest cv. (Figure 5).

A comparison of the Forrest and Williams 82 sequences of these four genes has shown that two of these genes have SNPs between Forrest and Williams 82 sequences, Glyma.10G058200 and Glyma.06G143000. Glyma.10G058200 has ten SNPs, one SNP is upstream 5′ UTR, four SNPs are located at the intron, two SNPs are at the exon 1, one of them caused a missense mutation (A127G) and the other one caused a silent mutation (A32A). The last three are in the 3′ UTR downstream region. For Glyma.06G143000, there is only one SNP located in the 5′UTR upstream region (Figure 6). These SNPs could potentially play a role in the difference of isoflavones content between Forrest and Williams 82 cultivars. Moreover, Glyma.10G058200 and Glyma.06G143000 are highly expressed in the seed tissue of both Forrest and Williams 82 (Figure 5). Glyma.10G058200 is associated with qGEN01-(IL-2020) QTL, qDAID04-(IL-2020) QTL and qGEN05-(IL-2020) QTL. Additionally, Glyma.06G143000 is associated with qGEN01-(NC-2018) QTL. The two genes could be useful for breeding for increased isoflavones content in soybeans.

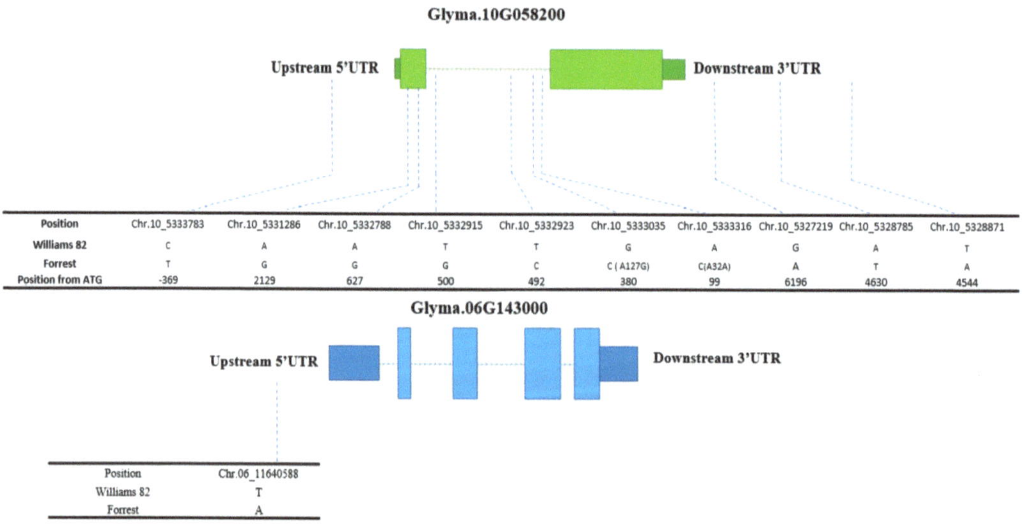

Figure 6. Positions of SNPs between Forrest and Williams 82 cultivars in Glyma.10G058200 and Glyma.06G143000 genes.

3. Materials and Methods

3.1. Plant Material and Growth Conditions

In this study, we used 'Forrest' × 'Williams 82' RIL population (n = 306). The cultivar 'Forrest' was derived from the cross of 'Dyer' and 'Bragg' developed by USDA [46]. The cultivar 'Williams 82' was derived from the cross of 'Williams' and 'Kingwa' [47]. The genomes of soybean cultivars including Forrest and Williams 82 genomes are duplicated polyploid genomes with highly conserved gene-rich regions [48]. Originally, the 'Forrest' × 'Williams 82' RIL population was developed with more than 1000 RILs [49]. The genetic map used in this study was based on 306 RILs and 2075 SNP markers; however, QTL data analysis in Spring Lake, NC-2018 was based on 190 RILs.

The RIL population was evaluated in a farm in Spring Lake, NC (35.17° N, 78.97° W) in 2018 and in a farm in Carbondale, IL (37° N, 89° W) in 2020. Seeds of parents (Forrest and Williams 82) were sown directly in the field in a randomized complete block design (RCB) and 75 cm row spacing between seeds with three replicates. The plants were watered by drip irrigation and kept in the field until maturity. No pesticide, herbicide, or fertilizer were applied. In September 2018, hurricane Florence hit NC and its winds of 90+ mph damaged the fence in the farm in Spring Lake, NC and the deer damaged about 119 RILs; therefore, QTL data analysis for this location involved 187 RILs (n = 187). The plants grown in Carbondale, IL (n = 306) were not damaged.

In Spring Lake, NC (2018) during the growing season (May–Sept.), the temperatures ranged from 7.2 to 35 °C, it was partly (40%) to mostly cloudy (80%), wind speeds ranging from 55 to 90+ mph (hurricane Florence), and humidity comfort level ranged from comfortable to miserable [50]. The soil type in this location is mainly sandy (NC Sandhills). In Carbondale, IL (2020), the temperatures ranged from 7.2 to 29.4 °C, it was mostly clear (25%) to mostly cloudy (80%), wind speeds ranging from 30 to 38 mph, and humidity comfort level ranged from comfortable to miserable (weatherspark.com). The field was treated first using Firestorm (contains Paraquat dichloride) to control annual grass and broad-leaved weeds. As pre-emergent herbicide, Dual II Magnum Herbicide with long-lasting control of most annual grasses and small-seeded broadleaf weeds was used to eliminate early-season weed competition. As post-emergent herbicide, Round Up Pro Concentrate (50.2% Glyphosate) was used/sprayed between the rows to control emerging weed. Weed grown inside the plastic mulch very close to the soybeans were removed manually. The soil type in this location is mainly silty clay loam (Southern IL).

3.2. Isoflavone Quantification

Mature seeds of parents Forrest and Williams 82, and the 190 RILs were analyzed for the concentrations aglycones daidzein, genistein, and glycitein. Approximately 25 g of mature seeds from each plot were ground using a Laboratory Mill 3600 (Perten, Springfield, IL, USA). Concentrations of daidzein, genistein, and glycitein were analyzed using a near-infrared reflectance (NIR) diode array feed analyzer (Perten, Spring Field, IL, USA). The calibration equation has been updated every 6 months to 1 year and developed using the Thermo Galactic Grams PLS IQ software developed by Perten Company (Perten, Springfield, IL, USA). Thermo Galactic Grams PLS IQ from Perten (Perten) was used to develop the calibration equations, which was initially developed by the University of Minnesota. Descriptions of quantifying daidzein, genistein, glycitein and total isoflavone contents was reported by others (Akond et al., 2015 [22]; Bellaloui et al., 2012 [51]; Wang et al., 2019 [38]). The calibration equation development and updating for isoflavones was based on standard laboratory analytical methods (AOAC 2002) using High Performance Liquid Chromatography (HPLC) and use of adequate number of samples, providing sufficiently accurate estimations of isoflavones concentrations. The produced calibration equation was characterized by high correlation, indicating the accuracy of the method. The concentrations were calculated on a seed dry matter basis.

3.3. DNA Isolation, SNP Genotyping, and Genetic Map Construction

Genomic DNA of the RIL population and the parents were extracted using a standard cetyltrimethyl ammonium bromide (CTAB) method with minor modifications as previously described [52]. DNA concentration was quantified with a spectrophotometer (NanoDrop Technologies Inc., Centreville, DE, USA) and then normalized at 50 ng/µL for genotyping. SNP genotyping was performed in the Soybean Genomics and Improvement Laboratory, USDA-ARS, Beltsville, MD, USA, using the Illumina Infinium SoySNP6K BeadChips (Illumina, Inc. San Diego, CA, USA) as previously described [53]. Subsequently, SNP alleles were called using GenomeStudio Genotyping Module 2.0 (Illumina, Inc. San Diego, CA, USA).

JoinMap 4.0 [54] was used to construct the genetic linkage map with a LOD score threshold of 3.0 and a maximum genetic distance of 50 cM to group markers. The linkage groups were assigned to corresponding soybean chromosomes as described in SoyBase [29,31].

3.4. *Isoflavone QTL Detection and Statistical Analysis*

The broad sense (mean based) heritability analysis from two-way ANOVA was conducted using the following equation: $h^2 = \sigma_G{}^2/[\sigma_G{}^2 + (\sigma_{GE}{}^2/e) + (\sigma_e{}^2/re)]$ where $\sigma_G{}^2$ (variance of genetic factor), $\sigma_{GE}{}^2$ (variance of genotype-environment interactions), and $\sigma_e{}^2$ (variance of random effect) were calculated with e (number of environment) and r (number of replicates) normalization [55]. R [56] was employed in the statistical analysis including agronomic traits, histogram of trait distribution, two-way ANOVA, and broad sense heritability using its native packages. The significant level of the assessed traits was showed using R package car (type II Wald chi-square tests) [56].

Both interval mapping (IM) and composite interval mapping (CIM) methods of WinQTL Cartographer 2.5 [39] were used to identify QTL for seed genistein, daidzein, glycitein, and total isoflavone contents in this RIL population. The default parameters of WinQTL Cartographer were chosen (Model 6, 1 cM step size, 10 cM window size, 5 control markers, and 1,000 permutations threshold) [39]. Chromosomes were drawn using MapChart 2.2 [57].

3.5. *Isoflavone Candidate Genes Identification*

The Glyma numbers of the isoflavone genes were obtained by searching the available data at the SoyBase [29,31] and Phytozome database [45]. The name of the isoflavone pathway enzymes (Figure 5) were used as a query in a search of the Glycine max reference genome, version Williams 82. The obtained isoflavone genes were mapped to the identified isoflavone QTL.

3.6. *Expression Analysis*

The expression analysis of the genes that are located within or near the isoflavone QTL was conducted using the publicly available soybean expression database from Phytozome database [45] to infer expression profiles of isoflavone genes in the soybean reference genome Williams 82. Gene expression was estimated in FPKM (Fragments Per Kilobase of transcript per Million mapped reads).

For the Forrest cv., RNA-seq library was prepared by using four plant soybean tissues including seed, leaf, root, flower and pods as shown earlier [58]. From 100 mg of frozen grounded samples, total RNA was extracted using RNeasy QIAGEN Kit (Qiagen, Hilden, Germany). The DNase I (Invitrogen, Carlsbad, CA, USA) was used to treat the total RNA. Using Illumina NovaSeq 6000, RNA-seq libraries preparation and sequencing were performed at Novogene INC. Multiplexing and sequencing of the four libraries were done in two different lanes generating 20 million raw pair end reads per sample (150 bp). Quality of sequenced reads was assessed using fastqc, version 0.11.9. [59]. The low-quality reads and adapters were removed with trimmomatic, version V0.39, the remaining high-quality reads were mapped to the soybean reference genome Wm82.a2.v1 using STAR, version v2.7.9 [60,61]. Uniquely mapped reads were counted using Python package HTseq v0.13.5. [62]. Read count normalization and differential gene expression analysis were conducted using the Deseq2 package v1.30.1 [63] integrated in the OmicsBox platform from BioBam (Valencia, Spain) [58,64].

Supplementary Materials: The following are available online at https://www.mdpi.com/article/10.3390/plants10102029/s1, Figure S1: Positions of major QTL (LOD > 2.5) that control seed genistein (qGEN), daidzein (qDAID), and glycitein (qGLY) contents identified in the FXW82 RIL population. (A) QTL identified in Spring Lake, NC (2018) and (B) QTL identified in Carbondale, IL (2020); Figure S2. The SNP-based genetic linkage map of Forrest by Williams 82 recombinant inbred line (RIL) population ($n = 306$) of soybean. Chrs. were drawn using MapChart 2.2 (Voorrips, 2002 [57]). Positions of QTL that control seed genistein (qGEN), daidzein (qDAID), and glycitein (qGLY) contents are indicated with black bars on Chrs. 2, 4, 5, 6, 10, 12, 15, 19, and 20. QTL names are followed by a number, location, and year in which they are identified. For example, qGEN01-(NC-2018); Table S1: Candidate genes involved in soybean isoflavone biosynthetic pathways and their Phytozome annotation. These 130 candidate genes are identified in all 20 Chrs.

Author Contributions: Conceptualization, M.A.K. and K.M.; methodology, D.K., J.Y., T.V., M.U., Q.S., F.B., T.R., E.W., N.L., N.B. and A.M.; validation, M.A.K., K.M. and H.T.N.; formal analysis, D.K., J.Y. and A.M.; investigation, K.M., D.K. and A.M.; resources, K.M., N.B., A.M., and H.T.N.; genotyping, Q.S., data curation, D.K., J.Y., M.U. and T.V.; writing—original draft preparation, M.A.K. and D.K.; writing—review and editing, D.K., J.Y., N.B., N.L., H.M., T.V., A.M., M.A.K., K.M. and H.T.N.; visualization, J.Y.; supervision, M.A.K., J.Y., K.M., H.M., H.T.N.; project administration, M.A.K., K.M. and H.T.N. All authors have read and agreed to the published version of the manuscript.

Funding: This research was funded by the U.S. Department of Agriculture, Agricultural Research Service Project 6066-21220-014-00D, SIUC, UM, and FSU. All authors have read and agreed to the published version of the manuscript.

Data Availability Statement. Data supporting reported results are available on request from the corresponding author.

Acknowledgments: Technical support provided by S. Mosley is appreciated. Mention of trade names or commercial products in this publication is solely for the purpose of providing specific information and does not imply recommendation or endorsement by the United States Department of Agriculture (USDA). USDA is an equal opportunity provider and employer.

Conflicts of Interest: The authors declare no conflict of interest.

References

1. Cavallini, D.C.U.; Bedani, R.; Bomdespacho, L.Q.; Vendramini, R.C.; Rossi, E.A. Effects of probiotic bacteria, isoflavones and simvastatin on lipid profile and atherosclerosis in cholesterol-fed rabbits: A randomized double-blind study. *Lipids Health Dis.* **2009**, *8*, 1. [CrossRef]
2. Cederroth, C.R.; Zimmermann, C.; Nef, S. Soy, phytoestrogens and their impact on reproductive health. *Mol. Cell. Endocrinol.* **2012**, *355*, 192–200. [CrossRef]
3. Kwon, Y. Effect of soy isoflavones on the growth of human breast tumors: Findings from preclinical studies. *Food Sci. Nutr.* **2014**, *2*, 613–622. [CrossRef]
4. Setchell, K.D.R.; Cassidy, A. Dietary Isoflavones: Biological Effects and Relevance to Human Health. *J. Nutr.* **1999**, *129*, 758S–767S. [CrossRef]
5. Cederroth, C.R.; Nef Soy, S. phytoestrogens and metabolism: A review. *Mol. Cell. Endocrinol.* **2009**, *304*, 30–42. [CrossRef]
6. Jiang, Q.; Payton-Stewart, F.; Elliott, S.; Driver, J.; Rhodes, L.V.; Zhang, Q.; Zheng, S.; Bhatnagar, D.; Boue, S.M.; Collins-Burow, B.M.; et al. Effects of 7-O Substitutions on Estrogenic and Anti-Estrogenic Activities of Daidzein Analogues in MCF-7 Breast Cancer Cells. *J. Med. Chem.* **2010**, *53*, 6153–6163. [CrossRef]
7. Hamilton-Reeves, J.M.; Banerjee, S.; Banerjee, S.K.; Holzbeierlein, J.M.; Thrasher, J.B.; Kambhampati, S.; Keighley, J.; Veldhuizen, P.V. Short-Term Soy Isoflavone Intervention Patients with Localized Prostate Cancer: A Randomized, Double-Blind, Placebo-Controlled Trial. *PLoS ONE* **2013**, *8*, e68331. [CrossRef] [PubMed]
8. Fritz, H.; Seely, D.; Flower, G.; Skidmore, B.; Fernandes, R.; Vadeboncoeur, S.; Kennedy, D.; Cooley, K.; Wong, R.; Sagar, S.; et al. Soy, Red Clover, and Isoflavones and Breast Cancer: A Systematic Review. *PLoS ONE* **2013**, *8*, e81968. [CrossRef]
9. Chen, M.; Rao, Y.; Zheng, Y.; Wei, S.; Li, Y.; Guo, T.; Yin, P. Association between Soy Isoflavone Intake and Breast Cancer Risk for Pre- and Post-Menopausal Women: A Meta-Analysis of Epidemiological Studies. *PLoS ONE* **2014**, *9*, e89288. [CrossRef] [PubMed]
10. Wei, J.; Bhatt, S.; Chang, L.M.; Sampson, H.A.; Masilamani, M. Isoflavones, Genistein and Daidzein, Regulate Mucosal Immune Response by Suppressing Dendritic Cell Function. *PLoS ONE* **2012**, *7*, e47979. [CrossRef] [PubMed]
11. Sakai, T.; Kogiso, M. Soy isoflavones and immunity. *J. Med. Investig.* **2008**, *55*, 167–173. [CrossRef] [PubMed]
12. Subramanian, S.; Stacey, G.; Yu, O. Distinct, crucial roles of flavonoids during legume nodulation. *Trends Plant Sci.* **2007**, *12*, 282–285. [CrossRef]
13. Zhang, Y.; Wang, G.-J.; Song, T.T.; Murphy, P.A.; Hendrich, S. Urinary disposition of the soybean isoflavones daidzein, genistein and glycitein differs among humans with moderate fecal isoflavone degradation activity. *J. Nutr.* **1999**, *129*, 957–962. [CrossRef] [PubMed]
14. Thigpen, J.E.; Setchell, K.D.R.; Ahlmark, K.B.; Locklear, J.; Spahr, T.; Cavines, G.F.; Goelz, M.F.; Haseman, J.K.; Newbold, R.R.; Forsyth, D.B. Phytoestrogen content of purified, open- and closed-formula laboratory animal diets. *Lab. Anim. Sci.* **1999**, *49*, 530–536. [PubMed]
15. Kassem, M.A.; Meksem, K.; Iqbal, M.J.; Njiti, V.N.; Banz, W.J.; Winters, T.A.; Wood, A.J.; Lightfoot, D.A. Definition of Soybean Genomic Regions That Control Seed Phytoestrogen Amounts. *J. Biomed. Biotechnol.* **2004**, *2004*, 52–60. [CrossRef]

16. Kassem, M.A.; Shultz, J.; Meksem, K.; Cho, Y.; Wood, A.; Iqbal, M.J.; Lightfoot, D.A. An updated 'Essex' by 'Forrest' linkage map and first composite interval map of QTL underlying six soybean traits. *Theor. Appl. Genet* **2006**, *113*, 1015–1026. [CrossRef] [PubMed]
17. Gutierrez-Gonzalez, J.J.; Wu, X.; Zhang, J.; Lee, J.-D.; Ellersieck, M.; Shannon, J.G.; Yu, O.; Nguyen, H.T.; Sleper, D.A. Genetic control of soybean seed isoflavone content: Importance of statistical model and epistasis in complex traits. *Theor. Appl. Genet.* **2009**, *119*, 1069–1083. [CrossRef]
18. Liang, H.Z.; Yu, Y.L.; Wang, S.F.; Lian, Y.; Wang, T.G.; Wei, T.L.; Gong, P.T.; Liu, X.Y.; Fang, X.J.; Zhang, M.C. QTL Mapping of Isoflavone, Oil and Protein Contents in Soybean (*Glycine max* L. Merr.). *Agric. Sci. China* **2010**, *9*, 1108–1116. [CrossRef]
19. Gutierrez-Gonzalez, J.J.; Vuong, T.D.; Zhong, R.; Yu, O.; Lee, J.-D.; Shannon, G.; Ellersieck, M.; Nguyen, H.T.; Sleper, D.A. Major locus and other novel additive and epistatic loci involved in modulation of isoflavone concentration in soybean seeds. *Theor. Appl. Genet.* **2011**, *123*, 1375–1385. [CrossRef]
20. Smallwood, C.J.; Nyinyi, C.N.; Kopsell, D.A.; Sams, C.E.; West, D.R.; Chen, P.; Kantartzi, S.K.; Cregan, P.B.; Hyten, D.L.; Pantalone, V.R. Detection and Confirmation of Quantitative Trait Loci for Soybean Seed Isoflavones. *Crop. Sci.* **2014**, *54*, 595–606. [CrossRef]
21. Zhao, G.; Jiang, Z.; Li, D.; Han, Y.; Hu, H.; Wu, L.; Wang, Y.; Gao, Y.; Teng, W.; Li, Y.; et al. Molecular loci associated with seed isoflavone content may underlie resistance to soybean pod borer (*Leguminivora glycinivorella*). *Plant Breed.* **2015**, *134*, 78–84. [CrossRef]
22. Akond, A.; Liu, S.; Kantartzi, S.K.; Meksem, K.; Bellaloui, N.; Lightfoot, D.A.; Yuan, J.; Wang, D.; Anderson, J.; Kassem, M.A. A SNP Genetic Linkage Map Based on the 'Hamilton' by 'Spencer' Recombinant Inbred Line (RIL) Population of Soybean [*Glycine max* (L.) Merr.] Identified QTL for Seed Isoflavone Contents. *Plant Breed.* **2015**, *134*, 580–588. [CrossRef]
23. Watanabe, S.; Yamada, R.; Kanetake, H.; Kaga, A.; Anai, T. Identification and characterization of a major QTL underlying soybean isoflavone malonylglycitin content. *Breed. Sci.* **2019**, *69*, 564–572. [CrossRef] [PubMed]
24. Han, Y.; Teng, W.; Wang, Y.; Zhao, X.; Wu, L.; Li, D.; Li, W. Unconditional and conditional QTL underlying the genetic interrelationships between soybean seed isoflavone, and protein or oil contents. *Plant Breed.* **2015**, *134*, 300–309. [CrossRef]
25. Murphy, S.E.; Lee, E.A.; Woodrow, L.; Seguin, P.; Kumar, J.; Rajcan, I.; Ablett, G.R. Genotype × Environment Interaction and Stability for Isoflavone Content in Soybean. *Crop. Sci.* **2009**, *49*, 1313–1321. [CrossRef]
26. Zhang, H.J.; Li, J.W.; Liu, Y.J.; Jiang, W.Z.; Du, X.L.; Li, L.; Li, X.W.; Su, L.T.; Wang, Q.Y.; Wang, Y. Quantitative trait loci analysis of individual and total isoflavone contents in soybean seeds. *J. Genet.* **2014**, *93*, 331–338. [CrossRef] [PubMed]
27. Li, X.; Kamala, S.; Tian, R.; Du, H.; Li, W.; Kong, Y.; Zhang, C. Identification and validation of quantitative trait loci controlling seed isoflavone content across multiple environments and backgrounds in soybean. *Mol. Breed.* **2018**, *38*, 8. [CrossRef]
28. Wu, D.; Li, D.; Zhao, X.; Zhan, Y.; Teng, W.; Qiu, L.; Zheng, H.; Li, W.; Han, Y. Identification of a candidate gene associated with isoflavone content in soybean seeds using genome-wide association and linkage mapping. *Plant J.* **2020**, *104*, 950–963. [CrossRef]
29. Brown, A.V.; Conners, S.I.; Huang, W.; Wilkey, A.P.; Grant, D.; Weeks, N.T.; Cannon, S.B.; Graham, M.A.; Nelson, R.T. A new decade and new data at SoyBase, the USDA-ARS soybean genetics and genomics database. *Nucleic Acids Res.* **2020**, *49*, D1496–D1501. [CrossRef]
30. Kassem, M.A. Two Decades of QTL Mapping of Isoflavone in Soybean Seed. In *Soybean Seed Composition: Protein, Oil, Fatty Acids, Amino Acids, Sugars, Mineral Nutrients, and Isoflavone*, 1st ed.; Kassem, M.A., Ed.; Springer Nature: Basingstoke, UK, 2021.
31. Grant, D.; Nelson, R.T.; Cannon, S.B.; Shoemaker, R.C. SoyBase, the USDA-ARS soybean genetics and genomics database. *Nucleic Acids Res.* **2010**, *38*, D843–D846. [CrossRef]
32. Yang, C.; Yan, J.; Jiang, S.; Li, X.; Min, H.; Wang, X.; Hao, D. Resequencing 250 Soybean Accessions: New Insights into Genes Associated with Agronomic Traits and Genetic Networks. *bioRxiv* **2021**. [CrossRef]
33. Akond, M.; Liu, S.; Schoener, L.; Anderson, J.A.; Kantartzi, S.K.; Meksem, K.; Song, Q.; Wang, D.; Wen, Z.; Lightfoot, D.A.; et al. A SNP-Based Genetic Linkage Map of Soybean Using the SoyS—NP6K Illumina Infinium BeadChip Genotyping Array. *Plant Genet. Genom. Sci.* **2013**, *1*, 80–89. [CrossRef]
34. Kassem, M.A.; Ramos, L.; Leandro, L.; Mbofung, G.; Hyten, D.L.; Kantartzi, S.K.; Grier, R.L.; Njiti, V.N.; Cianzio, S.; Meksem, K. The 'PI 438489B' by 'Hamilton' SNP-Based Genetic Linkage Map of Soybean [*Glycine max* (L.) Merr.] Identified Quantitative Trait Loci that Underlie Seedling SDS Resistance. *J. Plant Genome Sci.* **2012**, *1*, 18–30. [CrossRef]
35. Azam, M.; Zhang, S.; Abdelghany, A.M.; Shaibu, A.S.; Feng, Y.; Li, Y.; Tian, Y.; Hong, H.; Li, B.; Sun, J. Seed isoflavone profiling of 1168 soybean accessions from major growing ecoregions in China. *Food Res. Int.* **2020**, *130*, 108957. [CrossRef]
36. Hsiao, Y.-H.; Ho, C.-T.; Pan, M.-H. Bioavailability and health benefits of major isoflavone aglycones and their metabolites. *J. Funct. Foods* **2020**, *74*, 104164. [CrossRef]
37. Zhang, J.; Ge, Y.; Han, F.; Li, B.; Yan, S.; Sun, J.; Wang, L. Isoflavone Content of Soybean Cultivars from Maturity Group 0 to VI Grown in Northern and Southern China. *J. Am. Oil Chem. Soc.* **2014**, *91*, 1019–1028. [CrossRef]
38. Wang, X.; Liu, S.; Yin, X.; Bellaloui, N.; McClure, M.A.; Mengistu, A. Soybean seed isoflavones respond differentially to phosphorus applications in low and high phosphorus soils. *Nutr. Cycl. Agroecosyst.* **2019**, *113*, 217–230. [CrossRef]
39. Wang, S.; Basten, C.J.; Zeng, Z.B. *Windows QTL Cartographer 2.5*; Department of Statistics, NCSU: Raleigh, NC, USA, 2012; Available online: http://statgen.ncsu.edu/qtlcart/WQTLCart.htm (accessed on 10 March 2021).
40. Pei, R.; Zhang, J.; Tian, L.; Zhang, S.; Han, F.; Yan, S.; Wang, L.; Li, B.; Sun, J. Identification of novel QTL associated with soy-bean isoflavone content. *Crop. J.* **2018**, *6*, 244–252. [CrossRef]

41. Schmutz, J.; Cannon, S.B.; Schlueter, J.; Ma, J.; Mitros, T.; Nelson, W.; Hyten, D.L.; Song, Q.; Thelen, J.J.; Cheng, J.; et al. Genome sequence of the palaeopolyploid soybean. *Nature* **2010**, *463*, 178–183. [CrossRef] [PubMed]
42. Fliegmann, J.; Furtwängler, K.; Malterer, G.; Cantarello, C.; Schüler, G.; Ebel, J.; Mithöfer, A. Flavone synthase II (CYP93B16) from soybean (*Glycine max* L.). *Phytochemistry* **2010**, *71*, 508–514. [CrossRef] [PubMed]
43. Meng, F.L.; Han, Y.P.; Teng, W.L.; Li, Y.G.; Bin Li, W. QTL underlying the resistance to soybean aphid (Aphis glycines Matsumura) through isoflavone-mediated antibiosis in soybean cultivar 'Zhongdou 27'. *Theor. Appl. Genet.* **2011**, *123*, 1459–1465. [CrossRef] [PubMed]
44. Kassem, M.A. QTL that Control Seed Protein, Oil, and Fatty Acids Contents. In *Soybean Seed Composition: Protein, Oil, Fatty Acids, Amino Acids, Sugars, Mineral Nutrients, and Isoflavone*, 1st ed.; Kassem, M.A., Ed.; Springer Nature: Basingstoke, UK, 2021.
45. Phytozome. Available online: https://phytozome.jgi.doe.gov/pz/portal.html# (accessed on 5 July 2021).
46. Hartwig, E.E.; Epps, J.M. Registration of 'Forrest' soybeans. *Crop. Sci.* **1973**, *13*, 287. [CrossRef]
47. Bernard, R.L.; Cremeens, C.R. Registration of Williams 82 soybean. *Crop. Sci.* **1988**, *28*, 1027–1028. [CrossRef]
48. Shultz, J.L.; Kurunam, D.; Shopinski, K.; Iqbal, M.J.; Kazi, S.; Zobrist, K.; Bashir, R.; Yaegashi, S.; Lavu, N.; Afzal, A.J.; et al. The soybean genome database (SoyGD): A browser for display of duplicated, polyploid, regions and sequence tagged sites on the integrated physical and genetic maps of Glycine max. *Nucleic Acids Res.* **2006**, *34*, D758–D765. [CrossRef]
49. Wu, X.; Vuong, T.D.; Leroy, J.A.; Shannon, J.G.; Sleper, D.A.; Nguyen, H.T. Selection of a core set of RILs from Forrest × Williams 82 to develop a framework map in soybean. *Theor. Appl. Genet.* **2011**, *122*, 1179–1187. [CrossRef]
50. Weather Spark. Available online: https://weatherspark.com (accessed on 5 July 2021).
51. Bellaloui, N.; Mengistu, A.; Fisher, D.K.; Abel, C.A. Soybean Seed Composition Constituents as Affected by Drought andPhomopsisin Phomopsis Susceptible and Resistant Genotypes. *J. Crop. Improv.* **2012**, *26*, 428–453. [CrossRef]
52. Vuong, T.D.; Sleper, D.A.; Shannon, J.G.; Nguyen, H.T. Novel quantitative trait loci for broad-based resistance to soybean cyst nematode (Heterodera glycines Ichinohe) in soybean PI 567516C. *Theor. Appl. Genet.* **2010**, *121*, 1253–1266. [CrossRef]
53. Song, Q.; Yan, L.; Quigley, C.; Fickus, E.; Wei, H.; Chen, L.; Dong, F.; Araya, S.; Liu, J.; Hyten, D.; et al. Soybean BARCSoySNP6K: An assay for soybean genetics and breeding research. *Plant J.* **2020**, *104*, 800–811. [CrossRef] [PubMed]
54. Van Ooijen, J.W. *Joinmap 4.0 Software for the Calculation of Genetic Linkage Maps in Experimental Populations*; Plant Res Intl.: Wageningen, The Netherlands, 2006.
55. Pilet-Nayel, M.L.; Muehlbauer, F.J.; McGee, R.J.; Kraft, J.M.; Baranger, A.; Coyne, C.J. Quantitative trait loci for partial resistance to Aphanomyces root rot in pea. *Theor. Appl. Genet.* **2002**, *106*, 28–39. [CrossRef] [PubMed]
56. R Software. Available online: https://www.r-project.org (accessed on 10 March 2021).
57. Voorrips, R.E. MapChart: Software for the graphical presentation of linkage maps and QTL. *J. Heredity* **2002**, *93*, 77–78. [CrossRef]
58. Lakhssassi, N.; Lopes-Caitar, V.S.; Knizia, D.; Cullen, M.A.; Badad, O.; El Baze, A.; Zhou, Z.; Embaby, M.G.; Meksem, J.; Lakhssassi, A.; et al. TILLING-by-Sequencing+ Reveals the Role of Novel Fatty Acid Desaturases (GmFAD2-2s) in Increasing Soybean Seed Oleic Acid Content. *Cells* **2021**, *10*, 1245. [CrossRef] [PubMed]
59. Bolger, A.M.; Lohse, M.; Usadel, B. Trimmomatic: A flexible trimmer for Illumina sequence data. *Bioinformatics* **2014**, *30*, 2114–2120. [CrossRef] [PubMed]
60. Dobin, A.; Gingeras, T.R. Mapping RNA-seq Reads with STAR. *Curr. Protoc. Bioinform.* **2015**, *51*, 11.14.1–11.14.19. [CrossRef] [PubMed]
61. Trapnell, C.; Pachter, L.; Salzberg, S.L. TopHat: Discovering splice junctions with RNA-Seq. *Bioinformatics* **2009**, *25*, 1105–1111. [CrossRef]
62. Anders, S.; Pyl, P.T.; Huber, W. HTSeq—A Python framework to work with high-throughput sequencing data. *Bioinformatics* **2015**, *31*, 166–169. [CrossRef] [PubMed]
63. Love, M.I.; Huber, W.; Anders, S. Moderated estimation of fold change and dispersion for RNA-seq data with DESeq2. *Genome Biol.* **2014**, *15*, 550. [CrossRef]
64. Lakhssassi, N.; Zhou, Z.; Liu, S.; Piya, S.; Patil, G.B.; Cullen, M.A.; El Baz, A.; Badad, O.; Embaby, M.G.; Meksem, J.; et al. Soybean TILLING-by-Sequencing+ reveals the role of novel GmSACPD members in the unsaturated fatty acid biosynthesis while maintaining healthy nodules. *J. Exp. Bot.* **2020**, *71*, 6969–6987. [CrossRef] [PubMed]

plants

MDPI

Article

Utilization of a Wheat50K SNP Microarray-Derived High-Density Genetic Map for QTL Mapping of Plant Height and Grain Traits in Wheat

Dongyun Lv [1,†], Chuanliang Zhang [1,†], Rui Yv [1], Jianxin Yao [1], Jianhui Wu [1], Xiaopeng Song [2], Juntao Jian [3], Pengbo Song [1], Zeyuan Zhang [1], Dejun Han [1,*] and Daojie Sun [1,*]

[1] College of Agronomy, Northwest A&F University, Xianyang 712100, China; lvdongyun1102@163.com (D.L.); wyanan@nwafu.edu.cn (C.Z.); 18404969211@163.com (R.Y.); bilwqz@163.com (J.Y.); wujh@nwafu.edu.cn (J.W.); wheat5200stem@163.com (P.S.); 18238768351@163.com (Z.Z.)

[2] Zhumadian Academy of Agricultural Sciences, Zhumadian 463000, China; weiduxp@163.com

[3] Nanyang Academy of Agricultural Sciences, Nanyang 473000, China; jjt312024501@163.com

* Correspondence: handj@nwafu.edu.cn (D.H.); chinawheat@163.com (D.S.)

† These authors have equal contribution.

Abstract: Plant height is significantly correlated with grain traits, which is a component of wheat yield. The purpose of this study is to investigate the main quantitative trait loci (QTLs) that control plant height and grain-related traits in multiple environments. In this study, we constructed a high-density genetic linkage map using the Wheat50K SNP Array to map QTLs for these traits in 198 recombinant inbred lines (RILs). The two ends of the chromosome were identified as recombination-rich areas in all chromosomes except chromosome 1B. Both the genetic map and the physical map showed a significant correlation, with a correlation coefficient between 0.63 and 0.99. However, there was almost no recombination between 1RS and 1BS. In terms of plant height, 1RS contributed to the reduction of plant height by 3.43 cm. In terms of grain length, 1RS contributed to the elongation of grain by 0.11 mm. A total of 43 QTLs were identified, including eight QTLs for plant height (PH), 11 QTLs for thousand grain weight (TGW), 15 QTLs for grain length (GL), and nine QTLs for grain width (GW), which explained 1.36–33.08% of the phenotypic variation. Seven were environment-stable QTLs, including two loci (*Qph.nwafu-4B* and *Qph.nwafu-4D*) that determined plant height. The explanation rates of phenotypic variation were 7.39–12.26% and 20.11–27.08%, respectively. One QTL, *Qtgw.nwafu-4B*, which influenced TGW, showed an explanation rate of 3.43–6.85% for phenotypic variation. Two co-segregating KASP markers were developed, and the physical locations corresponding to KASP_AX-109316968 and KASP_AX-109519968 were 25.888344 MB and 25.847691 MB, respectively. *Qph.nwafu-4B*, controlling plant height, and *Qtgw.nwafu-4B*, controlling TGW, had an obvious linkage relationship, with a distance of 7–8 cM. Breeding is based on molecular markers that control plant height and thousand-grain weight by selecting strains with low plant height and large grain weight. Another QTL, *Qgw.nwafu-4D*, which determined grain width, had an explanation rate of 3.43–6.85%. Three loci that affected grain length were *Qgl.nwafu-5A*, *Qgl.nwafu-5D.2*, and *Qgl.nwafu-6B*, illustrating the explanation rates of phenotypic variation as 6.72–9.59%, 5.62–7.75%, and 6.68–10.73%, respectively. Two QTL clusters were identified on chromosomes 4B and 4D.

Keywords: wheat; plant height; grain traits; Wheat50K; genetic map; QTL

Citation: Lv, D.; Zhang, C.; Yv, R.; Yao, J.; Wu, J.; Song, X.; Jian, J.; Song, P.; Zhang, Z.; Han, D.; et al. Utilization of a Wheat50K SNP Microarray-Derived High-Density Genetic Map for QTL Mapping of Plant Height and Grain Traits in Wheat. *Plants* **2021**, *10*, 1167. https://doi.org/10.3390/plants10061167

Academic Editor: Abdelmajid Kassem

Received: 17 April 2021
Accepted: 26 May 2021
Published: 8 June 2021

1. Introduction

Wheat (*Triticum aestivum* L.) is a major food crop globally, providing carbohydrates and protein for 35% of the global population. It is estimated that wheat production will increase by more than 70% in the next 30 years to meet the needs of the growing population [1]. To ensure global food security, genetic improvement of food production will be one of the main goals of wheat breeding programs [2–4].

Both 1000-grain weight (TGW) and the genetic improvement of related traits, which play a vital role in wheat yield, are applicable to increasing wheat yield. TGW is mainly affected by grain morphological parameters, such as grain length and grain width [4–6]. TGW-related genes, including sucrose synthase genes, encode cell wall invertase and cytokinin oxidase/dehydrogenase. The sucrose synthase genes *TaSus1-7A*, *-7B* and *TaSus2-2A*, *-2B* determine TGW and grain size [7,8], *TaGW2-6A*, *-6B* the grain width [9,10], and *TaGS-D1* the grain size [11]. *TaCwi-A1* encodes cell wall invertase [12], *TaCKX6-D1* encodes cytokinin oxidase/dehydrogenase [13], and *TaGASR-A1* is a putative Snakin/GASA protein associated with grain length (GL) (Dong et al., 2014). The inheritance of grain traits is relatively stable, forming a higher heritability than overall yield [14]. The method is suitable for QTL analysis of wheat samples planted and collected from different places and years, and a stable QTL can be retrieved and detected. Over the past 20 years, more than 150 QTLs related to TGW, grain length, and grain width have been identified, which are distributed on 21 chromosomes of wheat [5,15–45]. Some studies have shown that there is a significant positive correlation between plant height and TGW [19,32,33,39,46,47]. The application of *Rht1* (*RHT-B1b*) and *Rht2* (*RHT-D1b*) in the 1960s set off a green revolution in wheat breeding. So far, 25 *Rht* genes have been identified in wheat [48,49]. Amongst these 25 genes, *Rht1* and *Rht2* are dwarfing genes that show insensitivity to gibberellins located on chromosomes 4BS and 4DS, respectively [13]. The wild alleles *Rht-B1a* and *Rht-D1a* also have a significant positive correlation with TGW [32,50]. Another gene, called *Rht8*, is sensitive to gibberellins for reducing plant height and is located on the 2DS chromosome. *Rht8* is another widely applied dwarfing gene that has no obvious negative effect on TGW, but affects panicle length. Thus, *Rht8* is a typical pleiotropism gene [6,51]. The genetic relationship can be investigated by targeting gene loci related to TGW and plant height, obtained by QTL mapping [30,52].

QTL genetic mapping is a crucial means to analyze functional loci [28]. Constructing a saturated genetic map is the key to QTL mapping, and molecular markers are the genetic map carrier. *Triticum aestivum* L. is a typical allohexaploid (AABBDD) composed of three subunits, and it represents the largest crop genome. Moreover, it is also the genome with the highest proportion of repetitive sequences such as transposable elements (84.7%) (IWGSC2018). Multitudes of SNP markers bear abundant polymorphism [53], and mapping results are quite advantageous in terms of accuracy and precision, especially for QTL mapping of quantitative traits [53,54]. By constructing a high-density genetic map to target the SNPs' genetic and physical loction, collinearity analysis is performed, and then the recombination rate in different regions of the chromosome can be judged. After comparing the genetic and physical distances between adjacent markers, the relative changes of recombination rates in each chromosome can be further investigated and analyzed. The range of the mating population required for a recombination event in a specific region can be estimated. Scientific and accurate estimation for breaking the chain of specific target areas can be provided, and accurate judgments for evaluating genetic linkage drag, together with guidance for improving breeding efficiency, can be achieved [54,55].

Until now, couples of common wheat SNP microarrays, including Wheat9K [56], Wheat90K [37,57,58], Wheat820K [59], Wheat660K (http://bioservices.capitalbio.com/index.shtml) [37,57,58], and the Wheat55K SNP array, have been developed based on the 660K SNP array [60–65].The Wheat50K SNP array is a high-efficiency genotyping technology completed by the Institute of Crop Science of the Chinese Academy of Agricultural Sciences and Affymetrix. The technology was developed using high-quality SNP markers selected from Wheat90K SNP arrays, 660K SNP arrays, and 35K SNP arrays. In the 50K SNP array, there are 135 functional markers and 700 SNP markers closely linked to known QTLs [66]. The functional markers covering ten TGW-related genes and two plant height-related genes are shown in Table S1.

In this study, a Kompetitive allele-specific PCR (KASP) marker was used, which is a polymerase chain reaction-based (PCR) technology using fluorescence for single nucleotide polymorphism (SNP) and small insertion and deletion (InDel). KASP markers have the advantage of a low error rate and a relatively low cost compared to other SNP genotyping platforms such as TaqMan systems. According to the method of Ma et al. [63], SNPs located in the main QTL interval were selected to develop KASP markers.

This project aims to determine the chromosome recombination rates in different regions using collinearity analysis of the genetic positions and physical locations of the SNP markers. By mapping the environment-stable QTL region of grain-related traits, whether corresponding loci are located in the recombination-rich or recombination-barren area can be confirmed, and a reasonable judgment for further fine mapping can be fulfilled. By traits and linkage analysis of the relationship between plant height and grain traits, useful insights for the next steps of molecular breeding can also be provided.

2. Results

2.1. Agronomic Traits Analysis

As was shown in Table 1, significant differences when $p = 0.01$ in the four environments appeared in relation to the plant height, TGW, grain length, and grain width of the two-parent materials. In Table 1 and Figure S1, we can see that fluctuations occurred in the same traits in different environments, indicating that these four traits were easily affected by the environment. The agronomic traits failed to accord with a strictly normal distribution ($p < 0.05$). The trait heritability values of plant height, TGW, grain length, and grain width were 0.73, 0.62, 0.61, and 0.72, respectively. As can be seen, those of plant height and grain width were relatively high.

Table 1. Statistical analysis of parent and RIL lines for traits.

Traits	Environment	Xinong1376	Xiaoyan81	Mean ± SD	Minimum	Maximum	*p*-Value	Heritability
Plant height	19NY	65.25	77.75 **	67.08 ± 13.78	32.2	96.8	2.19×10^{-3}	0.73
	20NY	68.24	81.22 **	80.03 ± 14.43	40.2	109.8	5.66×10^{-6}	
	19YL	68.36	78.23 **	65.78 ± 12.78	34.6	90.9	6.14×10^{-4}	
	20YL	72.33	83.25 **	72.24 ± 15.08	38.3	109.2	3.82×10^{-2}	
Thousand Grain Weight	19NY	41.35 **	36.23	40.72 ± 4.37	27.81	52.19	1.12×10^{-1}	0.62
	20NY	42.13 **	39.48	42.62 ± 4.51	26.28	51.76	1.24×10^{-3}	
	19YL	44.32 **	41.75	45.32 ± 4.41	34.22	55.05	2.80×10^{-2}	
	20YL	46.23 **	42.32	45.21 ± 4.40	29.5	54.83	3.68×10^{-1}	
Grain length	19NY	7.12 **	6.87	7.23 ± 0.37	6.27	8.04	3.93×10^{-2}	0.61
	20NY	7.32 **	6.75	7.14 ± 0.35	6.34	8.03	6.79×10^{-2}	
	19YL	7.51 **	7.24	7.44 ± 0.34	6.68	8.23	1.81×10^{-2}	
	20YL	7.36 **	7.14	7.51 ± 0.38	6.67	8.51	1.96×10^{-1}	
Grain width	19NY	3.31	3.21	3.37 ± 0.15	2.88	3.69	1.45×10^{-3}	0.72
	20NY	3.88 **	3.62	3.45 ± 0.18	2.81	3.83	2.18×10^{-3}	
	19YL	3.51 **	3.28	3.60 ± 0.16	3.11	3.9	1.38×10^{-2}	
	20YL	3.66 **	3.42	3.60 ± 0.16	3.16	3.95	3.02×10^{-3}	

Note: ** represents a significant difference between the two parents when $p = 0.01$.

As was shown in Figure S1, there was a significant positive correlation between the same traits and different environments when $p = 0.001$. The correlation between different years in the same place was higher than that in other combinations, indicating that a high degree of environmental similarity was present in the same place but in different years. The correlation between plant height and grain length was negative, but there was a significant positive correlation between TGW and grain width. TGW had a significant positive correlation with the other three traits, and a higher correlation with grain width than that with other traits. The correlation between grain length and grain width was different in different environments.

2.2. Construction of a Genetic Map

2.2.1. Description and Illustration of a Genetic Map

66,832 markers were subject to polymorphism analysis of population genotype by 50K gene microarray. A total of 19,601 SNP markers with differences were screened in the derived RIL populations of Xinong1376 and Xiaoyan81, while the remaining 15,822 markers were filtered by Chi-square test, and redundant markers were eliminated using the bin function of IciMapping. A total of 3136 bin markers, including 15,576 SNP markers, were eventually anchored to the genetic map. In addition, the genotyping, polymorphism marker, data filtering, physical map, genetic map, and bin map are all shown and illustrated in Table S2. Based on the 660K chip labeling, the SNP markers that differed between the two parents were detected and stored in Figure S2. The total length of the linkage map was 4512.79 cM, the average map distance was 1.44 cM, and the maximum gap was 26.86 cM, which covered 21 wheat chromosomes. According to linkage lengths in the homologous groups, their sequence in descending order was the fifth, the seventh, the third, the second, the fourth, the sixth, and the first. The linkage lengths were 813.14 cM, 794.35 cM, 703.96 cM, 631.98 cM, 563.99 cM, 537.27 cM, and 468.12 cM, and the numbers of bin markers were 621 (2947 SNP markers included), 549 (2193 SNP markers included), 524 (2846 SNP markers included), 327 (1865 SNP markers included), 393 (2002 SNP markers included), 372 (1865 SNP markers included), and 272 (2016 SNP markers included), respectively.

The numbers of bin markers located in wheat A, B, and D chromosome groups were 1231, 1197, and 708, respectively. The linkage lengths were 1703.69 cM, 1298.23 cM and 1510.87 cM, and the average map distances were 1.38 cM, 1.08 cM, and 2.13 cM, respectively. Molecular markers in the D genome were no more than those in the other two subgroups. In addition, the longest linkage group corresponding to chromosome 3A was 312.11 cM, and the shortest corresponding to chromosome 1D was 130.85 cM. The maps of each linkage group were shown and illustrated in Table 2 and Figure S2.

2.2.2. Collinearity Analysis of the Genetic Map

In this research, the genetic map and the collinearity map of the reference genome were analyzed as follows: The whole chromosome was included in the genetic map, the genetic map and the physical map were collinear, and the linkage map and the physical map were not linearly related. The recombination exchange on chromosomes was unbalanced, and the collinear diagrams of other chromosomes except for chromosome 1B appeared by and large S-shaped. The genetic positions of chromosomes increased linearly with the increase in physical locations, and the rest of the genetic positions aligned constantly with the increase in physical locations. This indicated that the two ends of the chromosome were recombination-rich areas and that the middle region was a recombination-barren area. A significant correlation of the genetic map and the physical one appeared when p = 0.001, the correlation coefficient ranged from 0.63 to 0.99, and the correlation coefficient of chromosome 1B was 0.63. The distribution presentation of bin markers on the reference genome showed that the number of bin markers on both ends of the chromosome was significantly higher than that of the middle region. The recombination rate of the two sides with a U-shaped distribution was significantly higher than that of the middle region, which confirmed that the ends of the chromosome were recombination-rich areas and the middle was the recombination-barren area. The reason for these findings was the inhibitory effect of centromere recombination.

No markers could be detected in the middle regions (more than 200 MB) of chromosomes 1D, 5A, and 6A. However, the linkage group was not divided into two parts in these regions, which were supposed to be recombination-barren regions. For nine chromosomes (2D, 3D, 5A, 5B, 5D, 6A, 6D, 7A and 7D), each chromosome included two linkage groups. For different linkage groups corresponding to the same chromosome, the grouping regions all appeared at both ends of the chromosome as the recombination-rich area, and the physical distance between the markers was less than 30 MB.

The collinearity map of chromosome 1B from 0 to 480 MB presented as an L-type curve. Although the gradual numerical values of physical location increased, the genetic distances were almost unchanged, and thus homologous recombination hardly occurred in the region. Xinong1376 belonged to 1BL/1RS translocation line, 1RS and 1BS hardly recombined, and the centromere's inhibition of recombination happened in the middle region, making the collinearity map L-shaped.

Table 2. Single-nucleotide polymorphism (SNP) marker statistics about distribution and density on 21 wheat chromosomes derived from crossing between Xinong1376 and Xiaoyan81.

Chromosome	Linkage Group	Length(cM)	Maker Numbers	Bin Number	Insinuation Markers	Maximum Clearance	Average Bin	Bin Density
1A	LG1A	192.66	1064	112	1045	25.68	1.72	0.58
1B	LG1B	144.61	558	118	447	26.86	1.23	0.82
1D	LG1D	130.85	394	42	336	18.01	3.12	0.32
2A	LG2A	215.97	951	140	940	23.46	1.54	0.65
2B	LG2B	244.43	676	173	597	25.44	1.41	0.71
2D	LG2D.1	132.89	161	48	154	25.42	2.77	0.36
	LG2D.2	38.69	77	11	75	10.06	3.52	0.28
3A	LG3A	311.23	1322	285	1301	16.8	1.09	0.92
3B	LG3B	160.61	487	144	458	12.59	1.12	0.9
3D	LG3D.1	17.46	38	8	36	13.71	2.18	0.46
	LG3D.2	214.66	999	87	1026	22.84	2.47	0.41
4A	LG4A	228.42	614	123	592	24.85	1.86	0.54
4B	LG4B	169.56	1185	193	1156	8.57	0.88	1.14
4D	LG4D	166.01	203	77	199	16.52	2.16	0.46
5A	LG5A.1	234.18	969	169	963	16.26	1.39	0.72
	LG5A.2	52.94	139	39	134	9.87	1.36	0.74
5B	LG5B.1	68.44	682	88	675	8.09	0.78	1.29
	LG5B.2	172.4	538	164	529	15.43	1.05	0.95
5D	LG5D.1	223.58	192	119	171	13.69	1.88	0.53
	LG5D.2	61.6	427	42	415	8.03	1.47	0.68
6A	LG6A.1	112.71	154	50	137	20.91	2.25	0.44
	LG6A.2	54.95	161	36	151	17.46	1.53	0.66
6B	LG6B	167.65	852	188	783	7.7	0.89	1.12
6D	LG6D.1	31.08	34	7	34	10.4	4.44	0.23
	LG6D.2	170.88	506	124	497	12.41	1.38	0.73
7A	LG7A.1	75.25	194	76	176	14.91	0.99	1.01
	LG7A.2	225.38	647	201	633	18.9	1.12	0.89
7B	LG7B	170.54	882	129	845	18.19	1.32	0.76
7D	LG7D.1	237.8	453	130	446	15.51	1.83	0.55
	LG7D.2	85.38	17	13	16	24.96	6.57	0.15
1st homologous	3	468.12	2016	272	1828	26.86	1.72	0.58
2nd homologous	4	631.98	1865	372	1766	25.44	1.7	0.59
3rd homologous	4	703.96	2846	524	2821	22.84	1.34	0.74
4th homologous	3	563.99	2002	393	1947	24.85	1.44	0.7
5th homologous	6	813.14	2947	621	2887	15.43	1.31	0.76
6th homologous	4	537.27	1707	405	1602	20.91	1.33	0.75
7th homologous	5	794.35	2193	549	2116	24.96	1.45	0.69
A genome	10	1703.69	6215	1231	6072	25.68	1.38	0.72
B genome	8	1298.23	5860	1197	5490	26.86	1.08	0.92
D genome	12	1510.87	3501	708	3405	25.44	2.13	0.47
TOTAL	30	4512.79	15576	3136	14967	26.44	1.44	0.69

2.2.3. Effects of 1B/1R on Traits Related to Plant Height and TGW

1RS specific marker was used to detect the population, the strains containing 1RS and 1BS were 51 and 147, respectively, and the p value of the chi-square test was 8.95×10^{-12}, which proved to be a severely segregated marker that couldn't be linked to the linkage group. According to the typing of the specific markers, the unpaired data

T test was performed on the traits related to plant height and TGW, and there was no significant difference between 1RS and 1BS. According to the typing of specific markers, a two-factor analysis of variance was performed on the agronomic traits, and the TGW and grain width were not affected by the genotype. According to the results of the variance analysis, Duncan's new multiple range test comparison of plant height and grain length was conducted. In terms of plant height, 1RS contributed to the reduction of plant height by 3.43cm. In terms of grain length, 1RS contributed to the elongation of grain by 0.11mm (shown in Table S3 and Figure S3).

2.3. QTL Mapping Analysis

A total of 43 QTLs for PH, TGW, GL, and GW were identified by QTL mapping analysis (Table 3 and Figure S4). These QTLs with LOD values ranging from 2.51 to 53.34 were distributed on 15 chromosomes and explained 1.36–33.08% of the phenotypic variation (Table 3 and Figure S4). There were 8, 11, 15, and 9 QTLs detected for PH, TGW, GL, and GW, respectively (Table 3 and Figure S4).

Inclusive composite interval mapping (ICIM) for PH identified a total of eight QTLs, which were located on six different chromosomes (Table 3 and Figure S4): 2D(2), 4B, 4D, 5B, 5D, and 6B(2). The QTL on 4B, *Qph.nwafu-4B*, was detected in four environments. *Qph.nwafu-4B* was thus treated as a major QTL, which explained 9.32–13.76% of phenotypic variance with LOD values ranging from 7.93 to 26.85. As was expected, the positive allele of *Qph.nwafu-4B* was contributed by Xiaoyan81 (Table 3 and Figure S4). The QTL on 4D, *Qph.nwafu-4D*, was detected in each of four environments. *Qph.nwafu-4D* was thus treated as a major QTL, which explained 20.11–27.09% of phenotypic variance with LOD values ranging from 16.78 to 42.21. As we expected, the positive allele of *Qph.nwafu-4D* was contributed by Xinong1376 (Table 3 and Figure S4).

One QTL, *Qph.nwafu-2D.1*, for PH was detected in two environments, which explained 3.3–3.73% of phenotypic variance. The remaining QTLs were detected only in a single environment (Table 3 and Figure S4).

ICIM for TGW identified a total of eleven QTLs, which were located on eight different chromosomes (shown in Table 3 and Figure S4): 2A, 2B, 3A, 4B, 4D(2), 5A, 5D(3), and 6A. The QTL on 4B, *Qtgw.nwafu-4B*, was detected in three environments. *Qtgw.nwafu-4B* was thus treated as a stable QTL, which explained 3.43–6.85% of phenotypic variance with LOD values ranging from 2.85 to 4.37. As was expected, the positive allele of *Qtgw.nwafu-4B* was contributed by Xinong1376 (shown in Table 3 and Figure S4). Based on the initial QTL mapping results, we developed two KASP markers, KASP_AX-109316968 and KASP_AX-109333198, and integrated them into the genetic map. When remapping with this integrated KASP marker, it was indicated that *Qtgw.nwafu-4B* was located in a 5 cM interval on chromosome arm 4BS, between the markers of AX-111494900 and AX-94438527, containing the newly developed KASP markers, including KASP_AX-109316968 and KASP_AX-109333198 (Figure S5 and Table S3).Three QTLs, *Qtgw.nwafu-4D.1*, *Qtgw.nwafu-5A*, and *Qtgw.nwafu-5D.1*, for TGW were detected in each of two environments, which explained 2.85–14.79% of phenotypic variance. The remaining QTLs were detected only in a single environment (Table 3).

Table 3. Full genomic QTL mapping results of plant height and grain related traits in the F8 RIL lines between Xinong1376 and Xiaoyan81.

Trait	QTLs Name	Environment	Position	LOD	PVE (%)	Add	Left and Right Marker	Interval	Physical Interval	Reference
PH	*Qph.nwafu-2D.1*	19YL	17	10.7	3.73	−3.77	AX-111561744/AX-179557748	16.5–20.5	23.416254/28.417456	(Zhai et al., 2016)
PH		20YL	17	8.82	3.3	−3.75	AX-111561744/AX-179557748	16.5–20.5	23.416254/28.417456	
PH	*Qph.nwafu-2D.2*	19YL	103	53.34	33.08	11.29	AX-94570302/AX-109998182	102.5–103.5	413.778968/425.474614	
PH	*Qph.nwafu-4B*	19NY	59	7.39	10.23	4.93	AX-179477460/AX-110984065	58.5–59.5	30.805339/32.961929	(Mohler et al., 2016)
PH		20NY	59	9.36	13.76	5.9	AX-179477460/AX-110984065	58.5–59.5	30.805339/32.961929	
PH		19YL	59	23.33	9.32	6.28	AX-179477460/AX-110984065	58.5–59.5	30.805339/32.961929	
PH		20YL	59	26.85	12.26	7.62	AX-179477460/AX-110984065	58.5–59.5	30.805339/32.961929	
PH	*Qph.nwafu-4D*	19NY	62	17.27	27.08	−7.69	AX-86170701/AX-89445201	61.5–62.5	18.781207/19.459614	(Zhang et al., 2013)
PH		20NY	62	16.78	27.09	−7.93	AX-86170701/AX-89445201	61.5–62.5	18.781207/19.459614	
PH		19YL	62	40.17	20.11	−8.84	AX-86170701/AX-89445201	61.5–62.5	18.781207/19.459614	
PH		20YL	62	42.21	23.71	−10.16	AX-86170701/AX-89445201	61.5–62.5	18.781207/19.459614	
PH	*Qph.nwafu-5B*	20YL	55	4.32	1.49	2.52	AX-109908739/AX-86174612	54.5–55.5	422.122099/425.671678	
PH	*Qph.nwafu-5D*	20YL	190	46.92	29.23	−11.18	AX-94390434/AX-110033637	189.5–190.5	466.230408/469.357817	(Quarrie et al., 2005; Hai et al., 2008)
PH	*Qph.nwafu-6B.1*	20YL	139	6.19	2.19	3.07	AX-109987590/AX-86162252	137.5–139.5	687.177084/688.20385	
PH	*Qph.nwafu-6B.2*	19YL	160	4.15	1.36	2.27	AX-110632551/AX-109509377	159.5–160.5	712.125253/711.370298	
TGW	*Qtgw.nwafu-2A*	20YL	186	2.6	5.08	0.95	AX-95103231/AX-94508212	185.5–186.5	733.854404/734.347961	(Cui et al., 2014)
TGW	*Qtgw.nwafu-2B*	20YL	101	4.24	10.14	1.34	AX-108905289/AX-95235626	95.5–106.5	153.585606/568.468886	(Li et al., 2018)
TGW	*Qtgw.nwafu-3A*	19YL	133	2.7	4.14	−0.89	AX-179477407/AX-94457296	132.5–134.5	457.796943/431.074614	
TGW	*Qtgw.nwafu-4B*	20NY	51	4.18	3.43	1.23	AX-111494900/AX-94438527	48.5–53.5	25.847125/26.491497	[67]

Table 3. *Cont.*

Trait	QTLs Name	Environment	Position	LOD	PVE (%)	Add	Left and Right Marker	Interval	Physical Interval	Reference
TGW		19YL	51	4.37	6.85	1.18	AX-111494900/AX-94438527	49.5–53.5	25.847125/26.491497	
TGW		19NY	52	2.85	5.02	1.06	AX-94438527/AX-110383634	48.5–55.5	26.491497/28.71668	
TGW	*Qtgw.nwafu-4D.1*	19NY	60	5.87	9.73	−1.44	AX-89703298/AX-86170701	56.5–60.5	16.926631/18.781207	(Mohler et al., 2016)
TGW		20NY	60	6.2	5.25	−1.48	AX-89703298/AX-86170701	56.5–60.5	16.926631/18.781207	
TGW	*Qtgw.nwafu-4D.2*	19YL	111	3.54	5.55	−1.03	AX-111926032/AX-94818797	107.5–112.5	476.884228/477.371597	
TGW	*Qtgw.nwafu-5A*	19YL	44	7.18	11.94	1.51	AX-95510385/AX-95117188	43.5–45.5	698.508129/702.466804	
TGW		20YL	44	3.51	6.99	1.11	AX-95510385/AX-95117188	43.5–45.5	698.508129/702.466804	
TGW	*Qtgw.nwafu-5D.1*	19NY	37	5.51	9.32	1.4	AX-111543112/AX-110576074	34.5–38.5	38.070293/41.294446	
TGW		20NY	37	14.79	14.24	2.43	AX-111543112/AX-110576074	35.5–38.5	38.070293/41.294446	
TGW	*Qtgw.nwafu-5D.2*	20NY	46	6.46	5.56	−1.51	AX-111019963/AX-110085499	44.5–49.5	42.928674/44.192407	
TGW	*Qtgw.nwafu-5D.3*	19YL	81	3.66	5.83	1.05	AX-110867187/AX-108827297	79.5–81.5	369.202139/370.064947	
TGW	*Qtgw.nwafu-6A*	20YL	29	3.06	6.26	−1.06	AX-109431286/AX-109358667	27.5–30.5	606.979733/608.046298	(Cui et al., 2014)
GL	*Qgl.nwafu-1A*	19YL	150	3.3	3.11	−0.06	AX-95682344/AX-108726119	148.5–150.5	572.350803/572.658176	(Mir et al., 2012)
GL	*Qgl.nwafu-1B.1*	20YL	0	3.62	4.51	0.09	AX-94835306/AX-179476279	0–0.5	59.471177/94.978091	
GL	*Qgl.nwafu-1B.2*	19NY	65	3.51	4.49	0.08	AX-94650293/AX-112288501	64.5–66.5	640.845515/641.632325	
GL		19YL	65	5.23	5.14	0.08	AX-94650293/AX-112288501	64.5–65.5	640.845515/641.632325	
GL	*Qgl.nwafu-3A*	20YL	135	5.71	7.61	−0.1	AX-94426283/AX-110122062	134.5–136.5	511.755031/510.853056	
GL		20NY	137	3.78	6.88	−0.09	AX-179557644/AX-94387510	136.5–137.5	541.482465/540.048345	
GL	*Qgl.nwafu-4A*	20NY	49	4.41	9.11	0.1	AX-111251110/AX-179476673	46.5–53.5	407.389107/129.089816	
GL		19NY	50	4.57	6.07	0.1	AX-111251110/AX-179476673	47.5–52.5	407.389107/129.089816	
GL	*Qgl.nwafu-4B.1*	19YL	51	4.28	4.17	0.07	AX-179476673/AX-110173140	47.5–52.5	129.089816/140.310606	

Table 3. *Cont.*

Trait	QTLs Name	Environment	Position	LOD	PVE (%)	Add	Left and Right Marker	Interval	Physical Interval	Reference
GL	*Qgl.nwafu-4B.2*	19YL	68	5	4.94	−0.08	AX-109507847/AX-109427900	67.5–69.5	114.952789/161.548436	(Wang et al., 2010)
GL		20YL	68	4.03	5.11	−0.08	AX-109507847/AX-109427900	67.5–69.5	114.952789/161.548436	
GL	*Qgl.nwafu-4D*	19YL	16	2.67	2.55	0.05	AX-108892806/AX-109447997	15.5–18.5	6.598631/7.048661	
GL	*Qgl.nwafu-5A*	19NY	44	6.93	9.59	0.12	AX-95510385/AX-95117188	43.5–44.5	698.508129/698.508129	
GL		19YL	44	6.47	6.72	0.09	AX-95510385/AX-95117188	43.5–44.5	698.508129/700.34701	
GL		20YL	44	5.8	7.73	0.1	AX-95510385/AX-95117188	43.5–45.5	698.508129/700.34701	
GL	*Qgl.nwafu-5B*	19YL	6	2.75	2.56	−0.05	AX-112288130/AX-95631525	5.5–6.5	6.654131/8.917454	
GL	*Qgl.nwafu-5D.1*	20NY	37	3.13	5.76	0.08	AX-111543112/AX-110576074	33.5–39.5	38.070293/41.294446	
GL	*Qgl.nwafu-5D.2*	19YL	82	6.13	5.99	0.08	AX-111496494/AX-109707913	81.5–84.5	370.135626/379.028214	
GL		20YL	82	6	7.75	0.1	AX-111496494/AX-109707913	81.5–84.5	370.135626/379.028214	
GL		19NY	89	4.36	5.62	0.09	AX-110558491/AX-111903917	88.5–91.5	385.893875/386.126855	
GL	*Qgl.nwafu-5D.3*	20YL	191	7.45	9.83	0.11	AX-110033637/AX-110830424	190.5–191.5	469.357817/469.523881	
GL	*Qgl.nwafu-5D.4*	19YL	218	6.85	7.84	0.09	AX-110777538/AX-111512534	215.5–221.5	485.909071/491.01105	
GL	*Qgl.nwafu-6B*	19YL	162	8.79	8.95	0.1	AX-110287286/AX-111572797	161.5–162.5	712.125253/712.245125	(Li et al., 2018)
GL		19NY	167	7.88	10.73	0.13	AX-89379712/AX-94499484	166.5–167	704.884934/718.376276	
GL		20NY	167	3.55	6.68	0.09	AX-89379712/AX-94499484	166.5–167	704.884934/718.376276	
GL		20YL	167	7.73	10.33	0.12	AX-89379712/AX-94499484	166.5–167	704.884934/718.376276	
GW	*Qgw.nwafu-2B*	20YL	94	3.86	6.13	0.04	AX-109423066/AX-108990832	93.5–94.5	152.611396/153.128588	
GW	*Qgw.nwafu-2D*	19NY	12	2.55	4.2	0.04	AX-179477408/AX-111367738	11.5–12.5	20.768547/21.405473	(Huang et al., 2006; Guan et al., 2018; Wu et al., 2015)

Table 3. *Cont.*

Trait	QTLs Name	Environment	Position	LOD	PVE (%)	Add	Left and Right Marker	Interval	Physical Interval	Reference
GW	*Qgw.nwafu-3A*	19YL	311	3.18	4.66	0.04	AX-110915909/AX-110475339	308.5–311	746.360221/749.849798	(Lee et al., 2014)
GW	*Qgw.nwafu-4B.1*	19NY	51	4.15	6.95	0.05	AX-111494900/AX-94438527	49.5–54.5	25.847125/26.491497	
GW		20NY	51	3.5	6.85	0.05	AX-111494900/AX-94438527	48.5–54.5	25.847125/26.491497	
GW	*Qgw.nwafu-4B.2*	20YL	68	3.91	6.23	0.04	AX-109507847/AX-109427900	67.5–68.5	114.952789/161.548436	(Wang et al., 2010)
GW	*Qgw.nwafu-4B.3*	19YL	77	9.23	15.24	0.07	AX-179559104/AX-95658798	76.5–77.5	520.214474/523.447693	
GW	*Qgw.nwafu-4D*	20NY	60	3.2	6.32	−0.05	AX-89703298/AX-86170701	59.5–61.5	16.926631/18.781207	
GW		19NY	61	4.15	7.22	−0.05	AX-86170701/AX-110572006	59.5–61.5	18.781207/19.179341	
GW		19YL	63	7.93	12.12	−0.06	AX-86170701/AX-89445201	61.5–64.5	18.781207/19.459614	
GW		20YL	63	6.37	10.37	−0.05	AX-86170701/AX-89445201	61.5–63.5	18.781207/19.459614	
GW	*Qgw.nwafu-5D*	19NY	163	3.5	6.01	0.04	AX-109317498/AX-109855976	159.5–166.5	448.686533/449.292436	
GW	*Qgw.nwafu-6D*	20NY	4	2.51	5.8	0.05	AX-111594857/AX-109406081	0–12.5	12.650045/8.255713	

Note: PH, TGW, GL, and GW represent plant height, thousand-grain weight, grain length, and grain width, respectively. Reference represents that the confidence interval of this study overlaps with that of previous studies.

ICIM for GL identified a total of fifteen QTLs, which were located on ten different chromosomes (Table 3 and Figure S4): 1A, 1B(2), 3A, 4A, 4B(2), 4D, 5A, 5B, 5D(4), and 6B. The QTL on 6B, *Qgl.nwafu-6B*, was detected in four environments. *Qgl.nwafu-6B* was thus treated as a major QTL, which explained 6.68–10.73% of phenotypic variance with LOD values ranging from 3.35 to 8.79. As was expected, the positive allele of *Qgl.nwafu-6B* was contributed by Xinong1376 (Table 3). The QTL on 5A, *Qgl.nwafu-5A*, was detected in in three environments. Qgl.nwafu-5A was thus treated as a stable QTL, which explained 6.72–9.59% of phenotypic variance with LOD values ranging from 5.8 to 6.93. As we expected, the positive allele of *Qgl.nwafu-5A* was contributed by Xinong1376 (Table 3 and Figure S4). The QTL on 5D, *Qgl.nwafu-5D.2*, was detected in three environments. *Qgl.nwafu-5D.2* was thus treated as a stable QTL, which explained 5.62–7.75% of phenotypic variance with LOD values ranging from 4.36 to 6.13. As was expected, the positive allele of *Qgl.nwafu-5D.2* was contributed by Xinong1376 (Table 3 and Figure S4). Four QTLs, *Qgl.nwafu-1B.2*, *Qgl.nwafu-3A*, *Qgl.nwafu-4A*, and *Qgl.nwafu-4B.2*, for GL were detected in two environments, explaining 3.51–6.13% of phenotypic variance. The remaining QTLs were detected only in a single environment (Table 3 and Figure S4).

ICIM for GW identified a total of nine QTLs, which were located on seven different chromosomes (Table 3, Figure S4): 2B, 2D, 3A, 4B(3), 4D, 5D, and 6D. The QTL on 4D, *Qgw.nwafu-4D*, was detected in each of the four environments. *Qgw.nwafu-4D* was thus treated as a major QTL, which explained 6.32–12.12% of phenotypic variance with LOD values ranging from 3.2 to 7.93. As we expected, the positive allele of *Qgw.nwafu-4D* was contributed by Xinong1376 (Table 3). One QTL, *Qgw.nwafu-4B.1*, for GW was detected in two environments, which explained 6.85–6.95% of phenotypic variance. The remaining QTLs were detected only in a single environment (shown in Table 3 and Figure S4).

Two QTL clusters were identified on chromosomes 4B and 4D (Table 3 and Figure S4). For the QTL cluster on chromosome 4B, *Qtgw.nwafu-4B* for TGW was co-localized with *Qgl.nwafu-4B.1* for GL, and *Qph.nwafu-4B* and *Qgl.nwafu-4B.2* for GL were co-localized with *Qgl.nwafu-4B.2* and *Qgl.nwafu-4B.3* for GL in a region ranging from 51 cM to 77 cM. On chromosome 4D, *Qph.nwafu-4D* for PH was clustered with *Qtgw.nwafu-4D.1* for TGW, and *Qgw.nwafu-4D* for GW was clustered with with the alleles from Xiaoyan81 increasing PH, TGW and GW.

3. Discussion

3.1. The Impact of Linkage Map on QTL Mapping

In this research, a linkage map, based on 50K microarray markers, was constructed from 198 F_8 RIL lines derived from the combination of two parents, Xinong1376 and Xiaoyan81. The linkage map had a total length of 4512.79 cM, covering 21 chromosomes of wheat. The reason why no marks could be targeted in the regions of more than 200 MB in the middle of the four chromosomes 1D, 5A, and 6A was that a recombination-barren area near the centromere appeared in the above regions, as was shown in Figure 1. Both parents were derived from the backbone parent Xiaoyan6, and a region with the same haplotype was formed rapidly [68], so that the two parents had no markers with polymorphic differences in the above regions. There was a long, excellent haplotype segment on chromosome 6A [60,69].

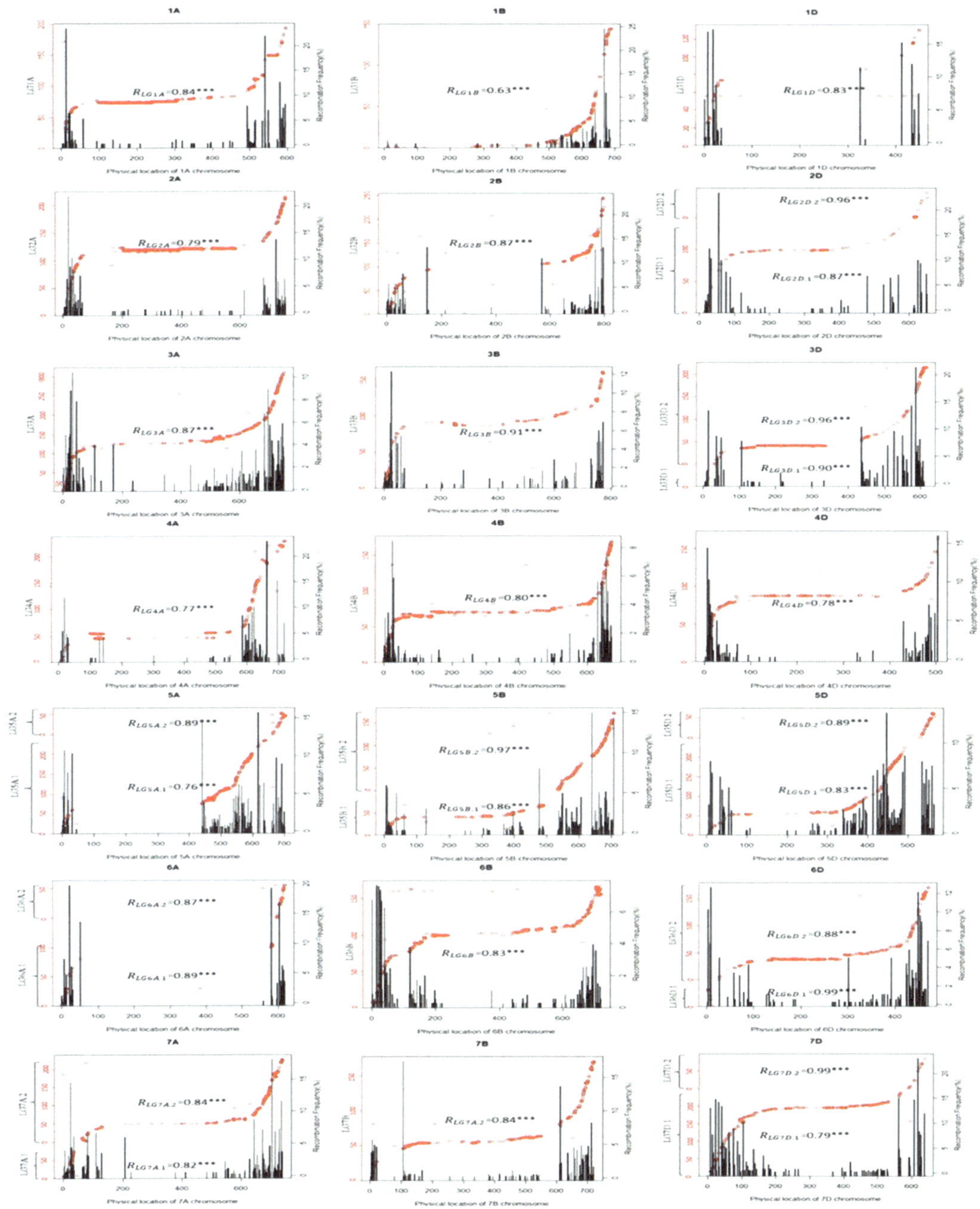

Figure 1. Collinearity analysis of genetic map and reference genome. **NOTE:** The genetic distances of the linkage group are shown as the left Y-axis, the recombination rate of bin markers as the right Y-axis, the physical location of the markers as the x-axis, the collinearity as the red scatter dots, and the recombination rate of bin markers on the reference genome as the black histogram. A, B, D are the three subgroups of common wheat.

In this study, 43 QTLs were located. The genetic distance confidence interval was 0.5–12.5 cM, and the physical distance of the markers on both sides was 0.0201 MB–414.88328 MB. As was shown in Table 2, the genetic distance confidence interval was not proportional to the physical distance, which reflected the imbalance of the recombination exchange on the chromosomes.

By combining Figure S4 and Figure 1, it appeared that there were 5 QTLs located in the recombination-barren region of the reference genome, and more than 20 MB QTLs were distributed in this candidate region. The linkage interval of *Qgl.nwafu-1B.1* was 0–0.5 cM, while the physical interval was 59.47117 MB–94.978091 MB and the interval physical distance was 35.506914 MB. The reason was that Xinong1376 belonged to the 1BL/1RS translocation line, and there was almost no recombination or recombination disorder between 1RS and 1BS [6,38,70,71]. Although the genetic distance of the confidence interval was short, the corresponding physical distance of it was far. As was shown in Figure S4, the linkage region of *Qtgw.nwafu-2B* was 95.5 cM–106.5 cM, and no marks could be targeted in this region. This area belongs to the reorganization cold spot area, and the corresponding physical distance was 153.585606 MB–568.468886 MB. The linkage regions of *Qtgw/gl.nwafu-3A*, *Qgl.nwafu-4A*, and *Qgw.nwafu-4B.2* were 132.5 cM–134.5 cM, 46.5 cM–53.5 cM, and 67.5 cM–69.5 cM, respectively, and the corresponding regions were 457.796943 MB–431.074614 MB, 407.389107 MB–129.089816 MB, and 114.952789 MB–161.548436 MB, respectively. As was shown in Table 2, the above three QTLs all fell in the recombination-barren region of linkage groups with a large physical interval. The confidence interval of *Qgw.nwafu-6D*, which was the largest, was 0 cM–12.5 cM, but the corresponding physical region was 12.650045 MB–8.255713 MB, and the interval was only 4.4 MB. *Qgw.nwafu-6D* was located at the top of the chromosome, and belonged to the recombination-rich region, with a big genetic distance but a short corresponding physical distance.

3.2. Comparison with Previous Research Results

Two loci as environment-stable QTLs, targeted in three or four kinds of environments, were *Qph.nwafu-4B* and *Qph.nwafu-4D*, which control plant height. In the confidence interval, the function markers including *Rht-1* and *Rht-2* were AX-179477460 and AX-86170701, respectively. According to the additive effect, the effect of the *Qph.nwafu-4D* mutant in lowering plant height was stronger than that of the *Qph.nwafu-4B* mutant, which was consistent with the results of Zhai et al. [6] The locus, *Qgl.nwafu-5A*, which controlled the grain length, corresponded to the physical location of 698.508129 MB–700.34701 MB, which was located at the end of the chromosome. Compared with the results of previous studies [23,29–42], Qgl.nwafu-5A was a new QTL. The location of *Qgl.nwafu-5D.2* which controlled the length of the grain corresponded to the physical location of 370.135626 MB-386.126855 MB. Based on previous research [22,24,35,42,43], *Qgl.nwafu-5D.2* was defined as a new QTL as well. The location of *Qgl.nwafu-6B*, which controls grain length, corresponded to the physical location of 704.884934 MB–718.376276 MB. Compared with the results of previous studies [35], the physical location marked by IWB2746 was 701.387367 MB. As was shown in Figure S4, the collinearity between the linkage group and the physical position was relatively disordered at the end of chromosome 6B, and it was not clear whether they were the same QTL.

Qph.nwafu-4B (controlling plant height) and *Qtgw.nwafu-4B* (controlling TGW) had an obvious linkage relationship, with a distance of 7–8 cM. The physical location corresponding to this location of *Qph.nwafu-4B* was 30.805339 MB–32.961929 MB, and the physical position corresponding to the location of *Qtgw.nwafu-4B* was 25.847125 MB–26.491497 MB. Guan's QTL mapping results were marked as *BS00084904_51* and *BS00011338_51* on both sides, and the physical location was 28.954526 MB–66.811785 MB [30]. Cui Fa's QTL mapping results were marked as *Rht-B1* and *Xmag2055* on both sides, and the physical location was 30.860778 MB–20.741542 MB [70]. Quarrie's QTL mapping results were marked as *Rht-B1* and *gwm165.1* on both sides, and the physical location was 30.860778 MB–269.948831 MB [42] (The results of previous studies on chromosome 4B and the specific

QTL information related to TGW are shown and illustrated in Table S4). From the QTL mapping results in this study and the above three research results, it was suggested that the confidence interval had this overlap while the confidence interval of this study was the shortest. Based on heredity Doumai/Shi 41875, Li mapped the plant height and TGW. The physical location on chromosome 4B was 46.621203 MB [35], which was not the same QTL. The confidence intervals of *Qph.nwafu-4D*, *Qtgw.nwafu-4D.1*, and *Qgw.nwafu-4D* had clear overlaps and were stably expressed in multiple environments. The mutant at this locus lowered plant height while also decreasing TGW and grain width. *Rht2* had a significant effect on TGW, as previously shown by Mohler et al. [32]. There was a significant overlap in the confidence interval of *Qph.nwafu-5D* controlling plant height and *Qgl.nwafu-5D.3* controlling grain length, with a typical pleiotropism. This locus's physical position was 466.230408 MB–469.357817 MB, and its additive effect was opposite, so physiological antagonism occurred. The location of *wmc215* targeted by Hai et al. was 472.369175 MB, and that of *gwm212* targeted by Quarrie was 472.630187 MB, which was in line with previous localization results [42,43]. The difference in physical location was 3 MB. Since subgroup D had a large linkage disequilibrium [72], it was impossible to determine whether these loci were the same one. *Qtgw.nwafu-5D.1* controlling TGW and *Qgl.nwafu-5D.1* controlling grain length were located in the region from 38.070293 MB–41.294446 MB, neither of which belonged to the same region of the 5D chromosome, compared with the results of previous studies [35,42,43,73].

3.3. *Qtgw.Nwafu-4B Molecular Marker Development*

Based on the confidence interval of the parental 660K chip marker, two co-segregating KASP markers were developed. Two KASP molecular markers were inserted into the original genetic map, and the genetic map of chromosome 4B maintained a high degree of collinearity. Two KASP molecular markers were inserted into the original genetic map, and the genetic map of chromosome 4B maintained a high degree of collinearity. The primer sequences and typing information of the two molecular markers of KASP_AX-109316968 and KASP_AX-109333198 are shown in Figure S5 and Table S5. *Qph.nwafu-4B* (controlling plant height) and *Qtgw.nwafu-4B* (controlling TGW) had an obvious linkage relationship, with a distance of 7–8 cM. Breeding is based on molecular markers that control plant height and thousand-grain weight to select strains with low plant height and large grain weight.

4. Materials and Methods

4.1. *Plant Materials, Experimental Design, and Investigation of Agronomic Traits*

Xinong 1376 is the female parent and Xiaoyan 81 is the male parent. Based on the single-grain transmission method, 198 RIL lines were generated. There were planted in Yangling, Shaanxi province and Nanyang, Henan province, from October 2018 to June 2019 and from October 2019 to June 2020, respectively. A randomized block design (repeated five times, with two rows of districts, 2 m row length, 70 plants per row, and 0.3 m row spacing) was adopted in each experimental site. The other field managements were subject to the same treatment as the local. During the wax maturity period of wheat, five individual plants were sampled in sequence from the fifth plant of each family. Plant height, TGW, grain length and grain width were also measured. By R/lme4 [73], each environment's agronomic traits were obtained for W-test, and then multiple comparisons of parental traits and calculation of heritability were completed. The heritability of the two traits was calculated by using the formula as follows:

$$H^2 = V_G/(V_G + V_{GY}/y + V_{GE}/e + V_E/nr) \times 100\%$$

where y is the number of years, e is the number of environments, and n is the number of repetitions.

The pedigrees of Xinong1376 and Xiaoyan81 are illustrated in Figure S2.

4.2. Construction and Evaluation of Genetic Maps

The wheat genomic DNA, with tender wheat leaves as the plant material, was extracted by CTAB, and the quality and quantity of DNA were detected and confirmed. Meanwhile, the DNA of each line was hybridized on the wheat 50K SNP array containing 66,832 markers using Burdock Biotechnology (Beijing, China).

The course of constructing the map was conducted as follows: The BIN function of IciMapping 4.1 [70] was utilized to analyze the markers, and the markers with partial separation rate ($p < 0.001$) and missing rate (>15%) were removed. The Kosambi function with LOD ≥ 5 was applied to group the combined marker groups in JoinMap 4.0; Kosambi mapping of MSTmap [74], according to the clustering results, was used in the markers' ordination. The flanking sequences of SNPs were BLAST aligned with the genome of IWGSC RefSeq v1.0 (http://www.wheatgenome.org/News/Latest-news/All-IWGSC-data-related-to-the-reference-sequence-of-bread-wheat-IWGSC-RefSeq-v1.0-publicly-available-at-URGI) to obtain their physical locations. The version of BLAST used was 2.2.31 –outfmt 3–num_alignments 5.

4.3. Identification of 1BL/1RS Translocation

1RS, applied to identify parents and populations as x-sec-p1/x-sec-p2, respectively, was a specific marker [75]. Xinong1376 was identified as a 1BS/1RS translocation line. 1B/1R genotyping and traits data were stored in Table S4 and Figure S3. Analysis of variance and Duncan's new multiple range test comparisons based on genotype and trait were conducted.

4.4. Detection of Quantitative Loci

IciMapping 4.2 based on the biparental population (BIP) module with the inclusive composite interval mapping (ICIM, http://www.isbreeding.net/software/?type=detail&id=28) was used for QTL mapping on data obtained from different environments. QTL mapping of the phenotypic values in the four environments was carried out. The LOD value was determined in 1000 permutation tests with a = 0.05 (Type I Error) as the parameter, and the background was set and controlled by the positive and negative stepwise regression, with the step width set to 1cM. QTLs were named based on the International Rules of Genetic Nomenclature (http://whea.pw.usda.gov/ggpages/wgc/98/Intro.htm). Mapchart2.3 (https://www.wur.nl/en/Research-Results/Research-Institutes/plant-research/Biometris-1/SoftwareService/Download-MapChart.htm) was used for the drawing of the genetic and QTL mapping. The collinearity drawing of genetic and physical maps, and the calculation of correlation coefficient were conducted by package plotrix (https://cran.r-project.org/src/contrib/Archive/plotrix/) and package (https://github.com/braverock/PerformanceAnalytics) of R software.

4.5. Breeding Molecular Marker Development

After obtaining the preliminary QTL mapping results, we anchored the flanking markers to the physical map. In order to develop a competitive allele-specific PCR (KASP) marker that can be used to track stable TGW QTLs, we used the Wheat660K SNP array to further genotype the parents of the Xinong1376/Xiaoyan81 population [63,71]. According to the method of Ma et al. [63], SNPs located in the main QTL interval were selected to develop KASP markers. The developed integrated genetic map of KASP markers was applied to relocate the target QTL.

5. Conclusions

In this research, a genetic map covering the entire wheat genome was constructed, with a total of 3136 bin markers, including 15576 SNP markers, and the total length of the linkage map was 4512.79 cM. Except for chromosome 1B, the ends of chromosomes were identified as recombination-rich areas, while the middle areas were recombination-barren. Both the genetic map and the physical map showed a significant correlation when $p = 0.001$.

The correlation coefficient ranged from 0.63 to 0.99. There was almost no recombination between 1RS and 1BS. Among 43 QTLs indirectly compared by reference genome, only 13 QTLs were consistent with previous mapping results, and 30 QTLs were defined as new QTLs. Seven environment-stable QTLs were detected in this population, including *Qph.nwafu-4B*, *Qtgw.nwafu-4B*, *Qgw.nwafu-4D*, *Qph.nwafu-4D*, *Qgl.nwafu-5A*, *Qgl.nwafu-5D.2*, and *Qgl.nwafu-6B*. *Qtgw.nwafu-4B*, which influenced TGW, showed an explanation rate of 3.43–6.85% for phenotypic variation, with two co-segregating KASP markers developed, and the physical locations corresponding to KASP_AX-109316968 and KASP_AX-109519968 were 25.888344 MB and 25.847691 MB, respectively, for details, see Figure 2. *Qph.nwafu-4B* (controlling plant height) and *Qtgw.nwafu-4B* (controlling TGW) had an obvious linkage relationship, with a distance of 7–8 cM. The physical location corresponding to this location of *Qph.nwafu-4B* was 30.805339 MB–32.961929 MB, and the physical position corresponding to this location of *Qtgw.nwafu-4B* was 25.847125 MB–26.491497 MB. There is a functional marker (AX-179477460) for the control value of plant height in the *Qph.nwafu-4B* confidence interval, and this locus can be determined to be Rht-B1. The physical locations of *Qph.nwafu-4B*, *Qph.nwafu-4D*, and *Qgw.nwafu-4D* were consistent with previous mapping results. For *Qgl.nwafu-6B*, it couldn't be accurately determined whether it was a new QTL or not. Two QTL clusters were identified on chromosomes 4B and 4D (Table 3 and Figure S4).

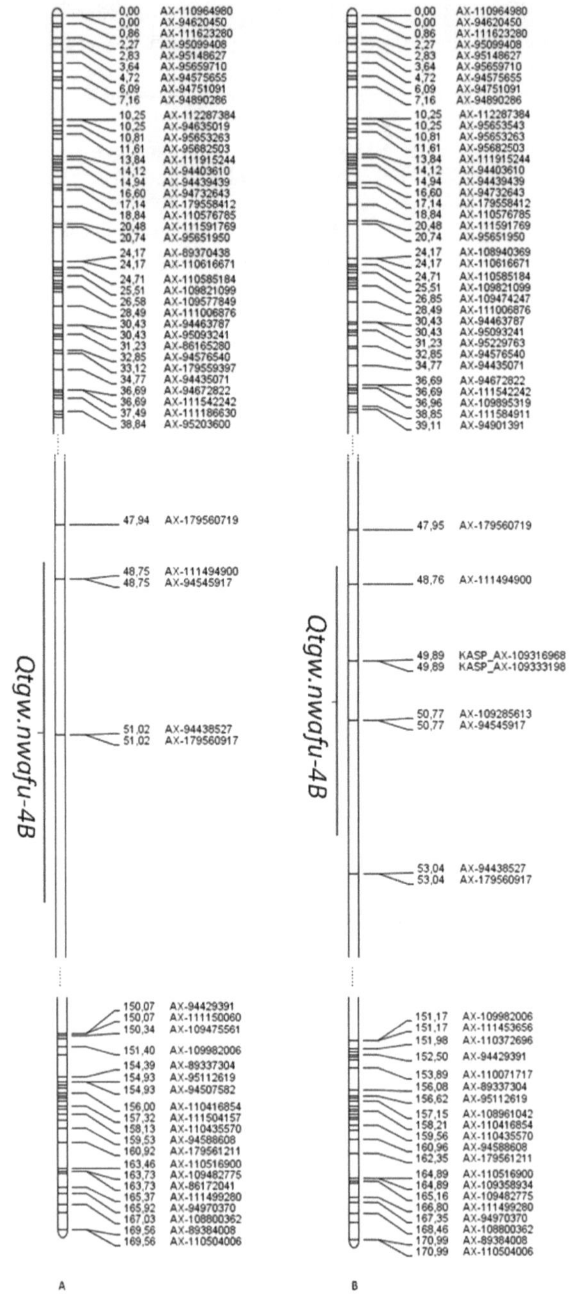

Figure 2. The linkage group corresponding to chromosome 4B. Note: (**A**) represents the original linkage group, and (**B**) the linkage group after the addition of the KASP marker.

Supplementary Materials: The following are available online at https://www.mdpi.com/article/10.3390/plants10061167/s1, Table S1. Statistical analysis of parent and RIL lines for traits, Table S2. Single-nucleotide polymorphism (SNP) marker statistics about distribution and density on 21 wheat chromosomes derived from crossing between Xinong1376 and Xiaoyan81, Table S3. Full genomic

QTL mapping results of plant height and grain related traits in the F8 RIL lines between Xinong1376 and Xiaoyan81, Figure S1. Collinearity analysis of genetic map and reference genome, Figure S2. The linkage group corresponding to chromosome 4B.

Author Contributions: D.S., D.H. and J.W. designed the research. C.Z. and D.L. conducted genotyping of the population. C.Z., D.L., R.Y., J.Y., X.S., J.J., P.S. and Z.Z. conducted phenotyping of the population. C.Z. analyzed all data. D.L. wrote the paper. All authors have read and agreed to the published version of the manuscript.

Funding: This research was funded by grants from the National Key Research and Development Program of China (2016YFD0101802).

Institutional Review Board Statement: Not applicable.

Informed Consent Statement: Not applicable.

Data Availability Statement: Please refer to suggested Data Availability Statements in section "MDPI Research Data Policies" at https://www.mdpi.com/ethics.

Conflicts of Interest: This study did not have any conflict of interest.

References

1. Bailey-Serres, J.; Parker, J.E.; Ainsworth, E.A.; Oldroyd, G.E.D.; Schroeder, J.I. Genetic strategies for improving crop yields. *Nature* **2019**, *575*, 109–118. [CrossRef]
2. Hanif, M.; Gao, F.; Liu, J.; Wen, W.; Zhang, Y.; Rasheed, A.; Xia, X.; He, Z.; Cao, S. TaTGW6-A1, an ortholog of rice TGW6, is associated with grain weight and yield in bread wheat. *Mol. Breed.* **2016**, *36*, 1. [CrossRef]
3. Kumari, S.; Jaiswal, V.; Mishra, V.K.; Paliwal, R.; Balyan, H.S.; Gupta, P.K. QTL mapping for some grain traits in bread wheat (*Triticum aestivum* L.). *Physiol. Mol. Biol. Plants* **2018**, *24*, 909–920. [CrossRef]
4. Xie, Q.; Mayes, S.; Sparkes, D.L. Carpel size, grain filling, and morphology determine individual grain weight in wheat. *J. Exp. Bot.* **2015**, *66*, 6715–6730. [CrossRef]
5. Guan, P.; Di, N.; Mu, Q.; Shen, X.; Wang, Y.; Wang, X.; Yu, K.; Song, W.; Chen, Y.; Xin, M.; et al. Use of near-isogenic lines to precisely map and validate a major QTL for grain weight on chromosome 4AL in bread wheat (*Triticum aestivum* L.). *Theor. Appl. Genet.* **2019**, *132*, 2367–2379. [CrossRef]
6. Zhai, H.; Feng, Z.; Li, J.; Liu, X.; Xiao, S.; Ni, Z.; Sun, Q. QTL Analysis of Spike Morphological Traits and Plant Height in Winter Wheat (*Triticum aestivum* L.) Using a High-Density SNP and SSR-Based Linkage Map. *Front. Plant Sci.* **2016**, *7*, 1617. [CrossRef]
7. Jiang, Q.; Hou, J.; Hao, C.; Wang, L.; Ge, H.; Dong, Y.; Zhang, X. The wheat (T. aestivum) sucrose synthase 2 gene (TaSus2) active in endosperm development is associated with yield traits. *Funct. Integr. Genom.* **2011**, *11*, 49–61. [CrossRef]
8. Hou, J.; Jiang, Q.; Hao, C.; Wang, Y.; Zhang, H.; Zhang, X. Global selection on sucrose synthase haplotypes during a century of wheat breeding. *Plant Physiol.* **2014**, *164*, 1918–1929. [CrossRef]
9. Su, Z.; Hao, C.; Wang, L.; Dong, Y.; Zhang, X. Identification and development of a functional marker of TaGW2 associated with grain weight in bread wheat (*Triticum aestivum* L.). *Theor. Appl. Genet.* **2011**, *122*, 211–223. [CrossRef]
10. Yang, Z.; Bai, Z.; Li, X.; Wang, P.; Wu, Q.; Yang, L.; Li, L.; Li, X. SNP identification and allelic-specific PCR markers development for TaGW2, a gene linked to wheat kernel weight. *Theor. Appl. Genet.* **2012**, *125*, 1057–1068. [CrossRef]
11. Ma, L.; Li, T.; Hao, C.; Wang, Y.; Chen, X.; Zhang, X. TaGS5-3A, a grain size gene selected during wheat improvement for larger kernel and yield. *Plant Biotechnol. J.* **2016**, *14*, 1269–1280. [CrossRef]
12. Ma, D.; Yan, J.; He, Z.; Wu, L.; Xia, X. Characterization of a cell wall invertase gene TaCwi-A1 on common wheat chromosome 2A and development of functional markers. *Mol. Breed.* **2010**, *29*, 43–52. [CrossRef]
13. Zhang, L.; Zhao, Y.L.; Gao, L.F.; Zhao, G.Y.; Zhou, R.H.; Zhang, B.S.; Jia, J.Z. TaCKX6-D1, the ortholog of rice OsCKX2, is associated with grain weight in hexaploid wheat. *New Phytol.* **2012**, *195*, 574–584. [CrossRef] [PubMed]
14. Gupta, P.K.; Rustgi, S.; Kumar, N. Genetic and molecular basis of grain size and grain number and its relevance to grain productivity in higher plants. *Genome* **2006**, *49*, 565–571. [CrossRef]
15. Varshney, R.K.; Prasad, M.; Roy, J.K.; Kumar, N.; Harjit, S.; Dhaliwal, H.S.; Balyan, H.S.; Gupta, P.K. Identification of eight chromosomes and a microsatellite marker on 1AS associated with QTL for grain weight in bread wheat. *Theor. Appl. Genet.* **2000**, *100*, 1290–1294. [CrossRef]
16. Cao, P.; Liang, X.; Zhao, H.; Feng, B.; Xu, E.; Wang, L.; Hu, Y. Identification of the quantitative trait loci controlling spike-related traits in hexaploid wheat (*Triticum aestivum* L.). *Planta* **2019**, *250*, 1967–1981. [CrossRef]
17. Campbell, B.T.; Baenziger, P.S.; Gill, K.S.; Eskridge, K.M.; Budak, H.; Erayman, M.; Dweikat, I.; Yen, Y. Identification of QTLs and Environmental Interactions Associated with Agronomic Traits on Chromosome 3A of Wheat. *Crop Sci.* **2003**, *43*, 1493–1505. [CrossRef]
18. Borner, A.; Schumann, E.; Furste, A.; Coster, H.; Leithold, B.; Roder, S.; Weber, E. Mapping of quantitative trait loci determining agronomic important characters in hexaploid wheat (*Triticum aestivum* L.). *Theor. Appl. Genet.* **2002**, *105*, 921–936. [CrossRef]

19. Huang, X.Q.; Coster, H.; Ganal, M.W.; Roder, M.S. Advanced backcross QTL analysis for the identification of quantitative trait loci alleles from wild relatives of wheat (*Triticum aestivum* L.). *Theor. Appl. Genet.* **2003**, *106*, 1379–1389. [CrossRef] [PubMed]
20. Huang, X.Q.; Cloutier, S.; Lycar, L.; Radovanovic, N.; Humphreys, D.G.; Noll, J.S.; Somers, D.J.; Brown, P.D. Molecular detection of QTLs for agronomic and quality traits in a doubled haploid population derived from two Canadian wheats (*Triticum aestivum* L.). *Theor. Appl. Genet.* **2006**, *113*, 753–766. [CrossRef] [PubMed]
21. Narasimhamoorthy, B.; Gill, B.S.; Fritz, A.K.; Nelson, J.C.; Brown-Guedira, G.L. Advanced backcross QTL analysis of a hard winter wheat × synthetic wheat population. *Theor. Appl. Genet.* **2006**, *112*, 787–796. [CrossRef] [PubMed]
22. Breseghello, F.; Sorrells, M.E. Association mapping of kernel size and milling quality in wheat (*Triticum aestivum* L.) cultivars. *Genetics* **2006**, *172*, 1165–1177. [CrossRef] [PubMed]
23. Breseghello, F.; Sorrells, M.E. QTL analysis of kernel size and shape in two hexaploid wheat mapping populations. *Field Crops Res.* **2007**, *101*, 172–179. [CrossRef]
24. Ramya, P.; Chaubal, A.; Kulkarni, K.; Gupta, L.; Kadoo, N.; Dhaliwal, H.S.; Chhuneja, P.; Lagu, M.; Gupta, V. QTL mapping of 1000-kernel weight, kernel length, and kernel width in bread wheat (*Triticum aestivum* L.). *J. Appl. Genet.* **2010**, *51*, 421–429. [CrossRef] [PubMed]
25. Mir, R.R.; Kumar, N.; Jaiswal, V.; Girdharwal, N.; Prasad, M.; Balyan, H.S.; Gupta, P.K. Genetic dissection of grain weight in bread wheat through quantitative trait locus interval and association mapping. *Mol. Breed.* **2012**, *29*, 963–972. [CrossRef]
26. Rasheed, A.; Xia, X.; Ogbonnaya, F.; Mahmood, T.; Zhang, Z.; Mujeeb-Kazi, A.; He, Z. Genome-wide association for grain morphology in synthetic hexaploid wheats using digital imaging analysis. *BMC Plant Biol.* **2014**, *14*, 128. [CrossRef] [PubMed]
27. Williams, K.; Munkvold, J.; Sorrells, M. Comparison of digital image analysis using elliptic Fourier descriptors and major dimensions to traits seed shape in hexaploid wheat (*Triticum aestivum* L.). *Euphytica* **2012**, *190*, 99–116. [CrossRef]
28. Zhang, Y.; Liu, J.; Xia, X.; He, Z. TaGS-D1, an ortholog of rice OsGS3, is associated with grain weight and grain length in common wheat. *Mol. Breed.* **2014**, *34*, 1097–1107. [CrossRef]
29. Jia, H.; Wan, H.; Yang, S.; Zhang, Z.; Kong, Z.; Xue, S.; Zhang, L.; Ma, Z. Genetic dissection of yield-related traits in a recombinant inbred line population created using a key breeding parent in China's wheat breeding. *Theor. Appl. Genet.* **2013**, *126*, 2123–2139. [CrossRef] [PubMed]
30. Guan, P.; Lu, L.; Jia, L.; Kabir, M.R.; Zhang, J.; Lan, T.; Zhao, Y.; Xin, M.; Hu, Z.; Yao, Y.; et al. Global QTL Analysis Identifies Genomic Regions on Chromosomes 4A and 4B Harboring Stable Loci for Yield-Related Traits Across Different Environments in Wheat (*Triticum aestivum* L.). *Front. Plant Sci.* **2018**, *9*, 529. [CrossRef]
31. Cuthbert, J.L.; Somers, D.J.; Brule-Babel, A.L.; Brown, P.D.; Crow, G.H. Molecular mapping of quantitative trait loci for yield and yield components in spring wheat (*Triticum aestivum* L.). *Theor. Appl. Genet.* **2008**, *117*, 595–608. [CrossRef] [PubMed]
32. Mohler, V.; Albrecht, T.; Castell, A.; Diethelm, M.; Schweizer, G.; Hartl, L. Considering causal genes in the genetic dissection of kernel traits in common wheat. *J. Appl. Genet.* **2016**, *57*, 467–476. [CrossRef]
33. Brinton, J.; Simmonds, J.; Minter, F.; Leverington-Waite, M.; Snape, J.; Uauy, C. Increased pericarp cell length underlies a major quantitative trait locus for grain weight in hexaploid wheat. *New Phytol.* **2017**, *215*, 1026–1038. [CrossRef]
34. Onyemaobi, I.; Ayalew, H.; Liu, H.; Siddique, K.H.M.; Yan, G. Identification and validation of a major chromosome region for high grain number per spike under meiotic stage water stress in wheat (*Triticum aestivum* L.). *PLoS ONE* **2018**, *13*, e0194075. [CrossRef]
35. Li, F.; Wen, W.; He, Z.; Liu, J.; Jin, H.; Cao, S.; Geng, H.; Yan, J.; Zhang, P.; Wan, Y.; et al. Genome-wide linkage mapping of yield-related traits in three Chinese bread wheat populations using high-density SNP markers. *Theor. Appl. Genet.* **2018**, *131*, 1903–1924. [CrossRef]
36. Kumar, A.; Mantovani, E.E.; Seetan, R.; Soltani, A.; Echeverry-Solarte, M.; Jain, S.; Simsek, S.; Doehlert, D.; Alamri, M.S.; Elias, E.M.; et al. Dissection of Genetic Factors underlying Wheat Kernel Shape and Size in an Elite x Nonadapted Cross using a High Density SNP Linkage Map. *Plant. Genome* **2016**, *9*, 1. [CrossRef] [PubMed]
37. Wu, Q.H.; Chen, Y.X.; Zhou, S.H.; Fu, L.; Chen, J.J.; Xiao, Y.; Zhang, D.; Ouyang, S.H.; Zhao, X.J.; Cui, Y.; et al. High-density genetic linkage map construction and QTL mapping of grain shape and size in the wheat population Yanda1817 x Beinong6. *PLoS ONE* **2015**, *10*, e0118144. [CrossRef] [PubMed]
38. Cui, F.; Zhao, C.; Ding, A.; Li, J.; Wang, L.; Li, X.; Bao, Y.; Li, J.; Wang, H. Construction of an integrative linkage map and QTL mapping of grain yield-related traits using three related wheat RIL populations. *Theor. Appl. Genet.* **2014**, *127*, 659–675. [CrossRef]
39. Fan, X.; Cui, F.; Ji, J.; Zhang, W.; Zhao, X.; Liu, J.; Meng, D.; Tong, Y.; Wang, T.; Li, J. Dissection of Pleiotropic QTL Regions Controlling Wheat Spike Characteristics Under Different Nitrogen Treatments Using Traditional and Conditional QTL Mapping. *Front. Plant Sci.* **2019**, *10*, 187. [CrossRef]
40. Liu, G.; Jia, L.; Lu, L.; Qin, D.; Zhang, J.; Guan, P.; Ni, Z.; Yao, Y.; Sun, Q.; Peng, H. Mapping QTLs of yield-related traits using RIL population derived from common wheat and Tibetan semi-wild wheat. *Theor. Appl. Genet.* **2014**, *127*, 2415–2432. [CrossRef]
41. Liu, J.; Wu, B.; Singh, R.P.; Velu, G. QTL mapping for micronutrients concentration and yield component traits in a hexaploid wheat mapping population. *J. Cereal Sci.* **2019**, *88*, 57–64. [CrossRef] [PubMed]
42. Quarrie, S.A.; Steed, A.; Calestani, C.; Semikhodskii, A.; Lebreton, C.; Chinoy, C.; Steele, N.; Pljevljakusic, D.; Waterman, E.; Weyen, J.; et al. A high-density genetic map of hexaploid wheat (*Triticum aestivum* L.) from the cross Chinese Spring x SQ1 and its use to compare QTLs for grain yield across a range of environments. *Theor. Appl. Genet.* **2005**, *110*, 865–880. [CrossRef]

43. Hai, L.; Guo, H.; Wagner, C.; Xiao, S.; Friedt, W. Genomic regions for yield and yield parameters in Chinese winter wheat (*Triticum aestivum* L.) genotypes tested under varying environments correspond to QTL in widely different wheat materials. *Plant Sci.* **2008**, *175*, 226–232. [CrossRef]
44. Wang, J.; Liu, W.; Wang, H.; Li, L.; Wu, J.; Yang, X.; Li, X.; Gao, A. QTL mapping of yield-related traits in the wheat germplasm 3228. *Euphytica* **2010**, *177*, 277–292. [CrossRef]
45. Lee, H.S.; Jung, J.-U.; Kang, C.-S.; Heo, H.-Y.; Park, C.S. Mapping of QTL for yield and its related traits in a doubled haploid population of Korean wheat. *Plant Biotechnol. Rep.* **2014**, *8*, 443–454. [CrossRef]
46. Roder, M.S.; Huang, X.Q.; Borner, A. Fine mapping of the region on wheat chromosome 7D controlling grain weight. *Funct. Integr. Genom.* **2008**, *8*, 79–86. [CrossRef]
47. Tian, X.; Wen, W.; Xie, L.; Fu, L.; Xu, D.; Fu, C.; Wang, D.; Chen, X.; Xia, X.; Chen, Q.; et al. Molecular Mapping of Reduced Plant Height Gene Rht24 in Bread Wheat. *Front. Plant Sci.* **2017**, *8*, 1379. [CrossRef]
48. Mo, Y.; Vanzetti, L.S.; Hale, I.; Spagnolo, E.J.; Guidobaldi, F.; Al-Oboudi, J.; Odle, N.; Pearce, S.; Helguera, M.; Dubcovsky, J. Identification and characterization of Rht25, a locus on chromosome arm 6AS affecting wheat plant height, heading time, and spike development. *Theor. Appl. Genet.* **2018**, *131*, 2021–2035. [CrossRef]
49. Agarwal, P.; Balyan, H.S.; Gupta, P.K. Identification of modifiers of the plant height in wheat using an induced dwarf mutant controlled by RhtB4c allele. *Physiol. Mol. Biol. Plants* **2020**, *26*, 2283–2289. [CrossRef] [PubMed]
50. Zhang, J.; Dell, B.; Biddulph, B.; Drake-Brockman, F.; Walker, E.; Khan, N.; Wong, D.; Hayden, M.; Appels, R. Wild type alleles of Rht-B1 and Rht-D1 as independent determinants of thousand-grain weight and kernel number per spike in wheat. *Mol. Breed.* **2013**, *32*, 771–783. [CrossRef]
51. Chai, L.; Chen, Z.; Bian, R.; Zhai, H.; Cheng, X.; Peng, H.; Yao, Y.; Hu, Z.; Xin, M.; Guo, W.; et al. Dissection of two quantitative trait loci with pleiotropic effects on plant height and spike length linked in coupling phase on the short arm of chromosome 2D of common wheat (*Triticum aestivum* L.). *Theor. Appl. Genet.* **2019**, *132*, 1815–1831. [CrossRef]
52. Chen, Z.; Cheng, X.; Chai, L.; Wang, Z.; Bian, R.; Li, J.; Zhao, A.; Xin, M.; Guo, W.; Hu, Z.; et al. Dissection of genetic factors underlying grain size and fine mapping of QTgw.cau-7D in common wheat (*Triticum aestivum* L.). *Theor. Appl. Genet.* **2020**, *133*, 149–162. [CrossRef]
53. Sun, C.; Dong, Z.; Zhao, L.; Ren, Y.; Zhang, N.; Chen, F. The Wheat 660K SNP array demonstrates great potential for marker-assisted selection in polyploid wheat. *Plant Biotechnol. J.* **2020**, *18*, 1354–1360. [CrossRef] [PubMed]
54. Wang, S.; Chen, J.; Zhang, W.; Hu, Y.; Chang, L.; Fang, L.; Wang, Q.; Lv, F.; Wu, H.; Si, Z.; et al. Sequence-based ultra-dense genetic and physical maps reveal structural variations of allopolyploid cotton genomes. *Genome Biol.* **2015**, *16*, 108. [CrossRef]
55. Schwarzkopf, E.J.; Motamayor, J.C.; Cornejo, O.E. Genetic differentiation and intrinsic genomic features explain variation in recombination hotspots among cocoa tree populations. *BMC Genom.* **2020**, *21*, 332. [CrossRef]
56. Cavanagh, C.R.; Chao, S.; Wang, S.; Huang, B.E.; Stephen, S.; Kiani, S.; Forrest, K.; Saintenac, C.; Brown-Guedira, G.L.; Akhunova, A.; et al. Genome-wide comparative diversity uncovers multiple targets of selection for improvement in hexaploid wheat landraces and cultivars. *Proc. Natl. Acad. Sci. USA* **2013**, *110*, 8057–8062. [CrossRef]
57. Wang, S.; Wong, D.; Forrest, K.; Allen, A.; Chao, S.; Huang, B.E.; Maccaferri, M.; Salvi, S.; Milner, S.G.; Cattivelli, L.; et al. Characterization of polyploid wheat genomic diversity using a high-density 90,000 single nucleotide polymorphism array. *Plant Biotechnol. J.* **2014**, *12*, 787–796. [CrossRef]
58. Sun, C.; Zhang, F.; Yan, X.; Zhang, X.; Dong, Z.; Cui, D.; Chen, F. Genome-wide association study for 13 agronomic traits reveals distribution of superior alleles in bread wheat from the Yellow and Huai Valley of China. *Plant Biotechnol. J.* **2017**, *15*, 953–969. [CrossRef] [PubMed]
59. Winfield, M.O.; Allen, A.M.; Burridge, A.J.; Barker, G.L.; Benbow, H.R.; Wilkinson, P.A.; Coghill, J.; Waterfall, C.; Davassi, A.; Scopes, G.; et al. High-density SNP genotyping array for hexaploid wheat and its secondary and tertiary gene pool. *Plant Biotechnol. J.* **2016**, *14*, 1195–1206. [CrossRef] [PubMed]
60. Liu, S.; Huang, S.; Zeng, Q.; Wang, X.; Yu, R.; Wang, Q.; Singh, R.P.; Bhavani, S.; Kang, Z.; Wu, J.; et al. Refined mapping of stripe rust resistance gene YrP10090 within a desirable haplotype for wheat improvement on chromosome 6A. *Theor. Appl. Genet.* **2021**, *3*, 1–17.
61. Liu, J.; Luo, W.; Qin, N.; Ding, P.; Zhang, H.; Yang, C.; Mu, Y.; Tang, H.; Liu, Y.; Li, W.; et al. A 55 K SNP array-based genetic map and its utilization in QTL mapping for productive tiller number in common wheat. *Theor. Appl. Genet.* **2018**, *131*, 2439–2450. [CrossRef]
62. Ma, J.; Zhang, H.; Li, S.; Zou, Y.; Li, T.; Liu, J.; Ding, P.; Mu, Y.; Tang, H.; Deng, M.; et al. Identification of quantitative trait loci for kernel traits in a wheat cultivar Chuannong16. *BMC Genet.* **2019**, *20*, 77. [CrossRef] [PubMed]
63. Ma, J.; Ding, P.; Liu, J.; Li, T.; Zou, Y.; Habib, A.; Mu, Y.; Tang, H.; Jiang, Q.; Liu, Y.; et al. Identification and validation of a major and stably expressed QTL for spikelet number per spike in bread wheat. *Theor. Appl. Genet.* **2019**, *132*, 3155–3167. [CrossRef]
64. Ren, T.; Hu, Y.; Tang, Y.; Li, C.; Yan, B.; Ren, Z.; Tan, F.; Tang, Z.; Fu, S.; Li, Z. Utilization of a Wheat55K SNP Array for Mapping of Major QTL for Temporal Expression of the Tiller Number. *Front. Plant Sci.* **2018**, *9*, 333. [CrossRef] [PubMed]
65. Huang, S.; Wu, J.; Wang, X.; Mu, J.; Xu, Z.; Zeng, Q.; Liu, S.; Wang, Q.; Kang, Z.; Han, D. Utilization of the Genomewide Wheat 55K SNP Array for Genetic Analysis of Stripe Rust Resistance in Common Wheat Line P9936. *Phytopathology* **2019**, *109*, 819–827. [CrossRef] [PubMed]
66. Rasheed, A.; Xia, X. From markers to genome-based breeding in wheat. *Theor. Appl. Genet.* **2019**, *132*, 767–784. [CrossRef]

67. Cui, F.; Fan, X.; Chen, M.; Zhang, N.; Zhao, C.; Zhang, W.; Han, J.; Ji, J.; Zhao, X.; Yang, L.; et al. QTL detection for wheat kernel size and quality and the responses of these traits to low nitrogen stress. *Theor. Appl. Genet.* **2016**, *129*, 469–484. [CrossRef]
68. Hao, C.; Jiao, C.; Hou, J.; Li, T.; Liu, H.; Wang, Y.; Zheng, J.; Liu, H.; Bi, Z.; Xu, F.; et al. Resequencing of 145 Landmark Cultivars Reveals Asymmetric Sub-genome Selection and Strong Founder Genotype Effects on Wheat Breeding in China. *Mol. Plant* **2020**, *12*, 1733–1751. [CrossRef]
69. Brinton, J.; Ramirez-Gonzalez, R.H.; Simmonds, J.; Wingen, L.; Orford, S.; Griffiths, S.; Wheat Genome, P.; Haberer, G.; Spannagl, M.; Walkowiak, S.; et al. A haplotype-led approach to increase the precision of wheat breeding. *Commun. Biol.* **2020**, *3*, 712. [CrossRef]
70. Meng, L.; Li, H.; Zhang, L.; Wang, J. QTL IciMapping: Integrated software for genetic linkage map construction and quantitative trait locus mapping in biparental populations. *Crop J.* **2015**, *3*, 269–283. [CrossRef]
71. Cui, F.; Zhang, N.; Fan, X.L.; Zhang, W.; Zhao, C.H.; Yang, L.J.; Pan, R.Q.; Chen, M.; Han, J.; Zhao, X.Q.; et al. Utilization of a Wheat660K SNP array-derived high-density genetic map for high-resolution mapping of a major QTL for kernel number. *Sci. Rep.* **2017**, *7*, 3788. [CrossRef] [PubMed]
72. Wu, J.; Yu, R.; Wang, H.; Zhou, C.; Huang, S.; Jiao, H.; Yu, S.; Nie, X.; Wang, Q.; Liu, S.; et al. A large-scale genomic association analysis identifies the candidate causal genes conferring stripe rust resistance under multiple field environments. *Plant Biotechnol. J.* **2021**, *19*, 177–191. [CrossRef] [PubMed]
73. Banta, J.A.; Stevens, M.H.H.; Pigliucci, M. A comprehensive test of the 'limiting resources' framework applied to plant tolerance to apical meristem damage. *Oikos* **2010**, *119*, 359–369. [CrossRef]
74. Wu, Y.; Bhat, P.R.; Close, T.J.; Lonardi, S. Efficient and accurate construction of genetic linkage maps from the minimum spanning tree of a graph. *PLoS Genet.* **2008**, *4*, e1000212. [CrossRef] [PubMed]
75. Yu, L.; He, F.; Chen, G.L.; Cui, F.; Li, X.F. Identification of 1BL·1RS Wheat-Rye Chromosome Translocations via 1RS Specific Molecular Markers and Genomic in situ Hybridization. *Acta Agron. Sin.* **2011**, *37*, 563–569. [CrossRef]

MDPI AG
Grosspeteranlage 5
4052 Basel
Switzerland
Tel.: +41 61 683 77 34

Plants Editorial Office
E-mail: plants@mdpi.com
www.mdpi.com/journal/plants

www.ingramcontent.com/pod-product-compliance
Lightning Source LLC
Chambersburg PA
CBHW042011260926
51850CB00020B/3121
* 9 7 8 3 7 2 5 8 3 3 5 0 4 *